Plant Parasitic Nematodes

Plant Parasitic Nematodes

Mahesh S
Kuruvilla Joseph
Anagha V K

2026
Agri Horti Press
New Delhi-110 084

ISBN: 978-93-49690-49-3 (Hardbound)

Published By

Agri Horti Press

New Delhi

E-mail: agrihortipress@gmail.com

Digitally Printed at : Replika Press Pvt. Ltd.

Preface

Plant parasitic nematodes are among the most significant and widespread pests affecting crops worldwide. Despite their microscopic size, these nematodes cause substantial damage to plants by feeding on roots, disrupting nutrient and water uptake, and creating entry points for secondary pathogens. Their ability to infect a wide range of crops, coupled with their resilience and complex life cycles, makes them a persistent threat to global food security.

This book, titled Plant Parasitic Nematodes, provides a comprehensive exploration of these microscopic invaders, offering insights into their biology, ecology, and the methods used to manage their impact. Aimed at researchers, agricultural practitioners, and students of plant pathology and nematology, this resource is designed to deepen understanding of nematodes and enhance the ability to control them effectively.

The book begins with an introduction to the biology and life cycles of plant parasitic nematodes, detailing their interaction with host plants and the various ways they damage crops. Key nematode species, including Meloidogyne, Heterodera, Globodera, and Pratylenchus, are covered in detail, alongside the symptoms they cause and the diseases they are associated with.

Subsequent chapters explore diagnostic techniques, from traditional methods like root examination to modern molecular tools. Emphasis is placed on integrated pest management (IPM) strategies, combining cultural, chemical, biological, and genetic approaches to control nematode populations. Advances in resistant crop breeding, nematicidal treatments, and the use of natural predators are examined in-depth.

Additionally, the book addresses the impact of plant parasitic nematodes on global agriculture, including their economic consequences and their role in soil health and ecosystem functioning. Real-world case studies are incorporated to demonstrate successful management practices across different crops and regions.

Plant Parasitic Nematodes aims to equip readers with the knowledge and practical tools needed to manage these pests sustainably, minimizing their impact on agricultural productivity and ensuring the health of plants and soils for future generations.

Mahesh S

Kuruvilla Joseph

Anagha V K

Contents

Chapter-1

Ficus microcarpa Bonsai "Tiger bark" Parasitized by the Root-Knot Nematode *Meloidogyne javanica* and the Spiral Nematode *Helicotylenchus dihystera*, a New Plant Host Record for Both Species

1. Introduction

The globalization era opens up new trade routes and increases the volume and complexity of cross-border transactions of goods. The plant sector (plant products, germplasm, grafts and live plants) has been part of the general trend in increased trade. This exchange of species between distant geographical regions of the globe creates new pathways for the introduction of alien plant pests and diseases [1,2].

The introduction of a non-native organism in a new environment produces unpredictable effects. A species may have low impact in its native range, but much greater impact when introduced to new areas, putting native biodiversity and local production systems at risk [1,3,4]. For instance, the introduction of the alien pinewood nematode *Bursaphelenchus xylophilus* in Portugal has caused huge environmental and economic losses in Portuguese pine forests, while in North America, where this nematode is native, it does not cause significant mortality to native conifers [5–7]. These problems are expected to be intensified in the future as climate change is predicted to facilitate the further spread of these species, since many of these new pathogens are of tropical and subtropical origins [8,9]. Recently, the tropical root knot nematode (RKN) *Meloidogyne luci* (Alert List of the European and Mediterranean Plant Protection Organization—EPPO) and *M. enterolobii* (A2 List of Pests EPPO) were detected in Portugal. *Meloidogyne luci* was found to be associated with potato (*Solanum tuberosum*) and tomato (*S. lycopersicum*), the ornamental plant *Cordyline australis*, and the weed *Oxalis corniculata*, whereas *M. enterolobii* was detected in the ornamental plants *Cereus hildmannianus*, *Lampranthus* sp., *Physalis peruviana* and *Callistemon* sp. Taking into account its aggressiveness and distribution, there is a high probability of spread in the Mediterranean region and also in Europe, becoming a potential threat to the agricultural economy [10,11].

The European Commission has proposed an import ban on 35 genera of plants for planting, other than seeds, in vitro material and natural or artificially dwarfed woody plants for planting from countries outside the European Union (EU). *Ficus carica*, common fig, is the only species of the genus *Ficus* included on the list. The ban was put into effect in December 2019 and aims to reduce the probability of the introduction of harmful organisms in the EU [12]. During a survey (3 years) in the Netherlands, around 20% of samples of imported plants for planting and ornamentals from 21 countries showed quarantine nematodes and 11% other important nematodes [13].

Ficus constitutes one of the largest genera of flowering plants (Angiosperms) that, according to the plant list version 1.1 (http://www.theplantlist.org), has 919 accepted species, being primarily found in tropical and subtropical environments throughout the world [14]. During the past few decades, plants from this genus have become quite popular as indoor house plants.

Despite the diversity of *Ficus* species, the research regarding plant-parasitic nematodes (PPN) has been focussed on the edible fig tree *F. carica* native to western Asia and introduced in the Mediterranean region. A number of PPN species have been reported as parasitizing fig trees in many countries, the most prevalent belonging to the genera *Helicotylenchus* (spiral nematodes), *Heterodera* (cyst nematodes), *Meloidogyne* (root-knot nematodes), *Paratylenchus* (pin nematodes), *Pratylenchus* (root lesion nematodes, RLN) and *Xiphinema* (dagger nematodes) [15–24]. Of these, the most common are the RKN species *M. arenaria, M. hapla, M. hispanica, M. incognita* and *M. javanica*, economically important species that directly target plant roots and prevent water and nutrient uptake, resulting in growth or even plant death in extreme cases, and the species *Heterodera fici*, a worldwide parasite of ornamental and cultivated *Ficus* species [24,25].

Concerning *Ficus* bonsai, besides *F. carica*, other species, such as *F. benghalensis, F. macrophylla, F. microcarpa, F. retusa* and *F. rubiginosa*, have been considered suitable for bonsai plants, but records of PPN on them are few and scarce. Some bonsai plants with nematode infections have been intercepted in Europe and other parts of the world [13,26,27].

The species *F. microcarpa*, the Indian laurel tree, sometimes confused with *F. nitida* and *F. retusa*, is widely distributed as an ornamental plant either outdoors or indoors and is known for its pharmacological properties: antioxidant, antibacterial, anticancer, anti-diabetic, anti-diarrhoeal, anti-inflammatory, anti-asthmatic, hepatoprotective and hypolipidemic [28]. This *Ficus* species is a host of many pests, including the Cuban laurel thrips (*Gynaikothrips ficorum*), the Ficus leaf-rolling psyllid (*Trioza brevigenae*), and the Ficus whitefly (*Singhiella simplex*), among others [29]. The PPN found associated with *F. microcarpa* include the genera *Helicotylenchus, Meloidogyne, Pratylenchus, Tylenchorynchus* and *Xiphinema*, and the RKN *M. enterolobii* species have been intercepted in bonsais or plants for planting imported from China or Egypt [27].

The aims of the present study were to find the nematode diversity associated to *F. mircrocarpa* bonsai plant, to characterize/identify the RKN, *Meloidogyne* sp., and the spiral nematode, *Helicotylenchus* sp., parasitizing *F. microcarpa* (Figure 1a) and to enlarge the knowledge on the phytoparasitic nematodes of this *Ficus* species.

Figure 1. ***Meloidogyne javanica*** **parasitizing** ***Ficus microcarpa*****. (a)** ***F. microcarpa*****. (b,c)** ***F. microcarpa*** **infected roots. (d) Polyacrylamide gel stained for esterase activity. J3,** ***M. javanica*** **(*****F. microcarpa*** **isolate); R,** ***M. javanica*** **(reference isolate). (e)** ***Hinf*****I (1),** ***Alu*****I (2) and** ***Dra*****III (3) digestion patterns of the approximately 1800-bp amplification products from** ***M. javanica*****, using C2F3 and MRH106 primers. M, DNA marker (HyperLadder II; Bioline). (f) DNA amplification product using Fjav and Rjav primers. 4, Negative control; 5,** ***M. javanica*****; M, DNA marker (HyperLadder II; Bioline). Scale bar: 1 mm.**

2. Results and Discussion

Meloidogyne females plus egg masses (Figure 1b,c) and *Helicotylenchus* specimens (Figure 2a) were detected in fresh and stained roots. Although *Helicotylenchus* spp. are classified as ectoparasites or semi-endoparasites, they can penetrate the roots and were already found completely embedded in the cortical tissue of the root system of sycamore (*Platanus occidentalis*) [30]. The galls were small and hard, mainly in woody roots (Figure 1b,c), which is common in woody perennial plants. The PPN detected in the soil sample (130 g) of *F. microcarpa* bonsai belonged to four genera: *Helicotylenchus* (94.02%), *Tylenchus* s.l. (4.35%), *Tylenchorynchus* s.l. (1.09%) and *Meloidogyne* J2 (0.54%). The spiral nematode *Helicotylenchus* was the most prevalent PPN, detected in very high numbers, with approximately 3000 nematodes.

For the RKN isolate, the biochemical characterization resulted in three bands of esterase (relative mobility %: 0.38; 0.45; 0.49), which is the characteristic phenotype exhibited by *M. javanica* (J3) isolates (Figure 1d). The mtDNA *COII* and 16S rRNA genes region amplified with the primer set C2F3/MRH106 yielded a single fragment of 1800 bp. When the amplified product was digested with the restriction enzyme *Hinf* I, no digestion occurred. *Alu*I and *Dra*III generated three fragments of approximately 1000, 580, and 240 bp and two fragments of approximately 1000 and 800 bp, which is in accordance with other results for this species [31] (Figure 1e). Additionally, molecular characterization of the RKN species with the species-specific primers Fjav and Rjav produced a fragment size of 600 bp, as expected, thus confirming the presence of *M. javanica* (Figure 1f).

This RKN species is known to parasitize *F. carica* and is one of the most widely distributed species and the second highest in economic importance after *M. incognita* [22,23,32]. *Meloidogyne javanica* was first reported from Portugal on potato in Azores [33]. Since then, it has been found on several economically important crops, including *Humulus lupulus*, *Musa* sp., *Phaseolus vulgaris*, *Prunus persica*, *S. lycopersicum* and *S. tuberosum*, ornamental plants, such as *Cordyline australis* and *Dianthus plumarius*, as well as many other dicots [11,34–40].

Helicotylenchus females were spirally curved, with up to two turns after relation with gentle heat, a key feature of *H. dihystera* (Figure 2b), the tail dorsally convex-conoid to a narrow terminus with a slight projection (Figure 2d) and lateral field with four non-areolated incisures (Figure 2e). Males were not found [41]. Amplification of the D2D3 expansion region of the 28S rDNA gene resulted in a product of ca. 750 bp (Figure 3). Sequences (744 bp) were submitted to the GenBank database with accession numbers MT277384, MT277385 and MT277386. The three sequences of the *Helicotylenchus* sp. from *Ficus microcarpa* (Fm) were compared and nine nucleotide changes at positions 57, 77, 98, 189, 537, 541, 546, 548 and 572 in alignment were identified. The comparison of

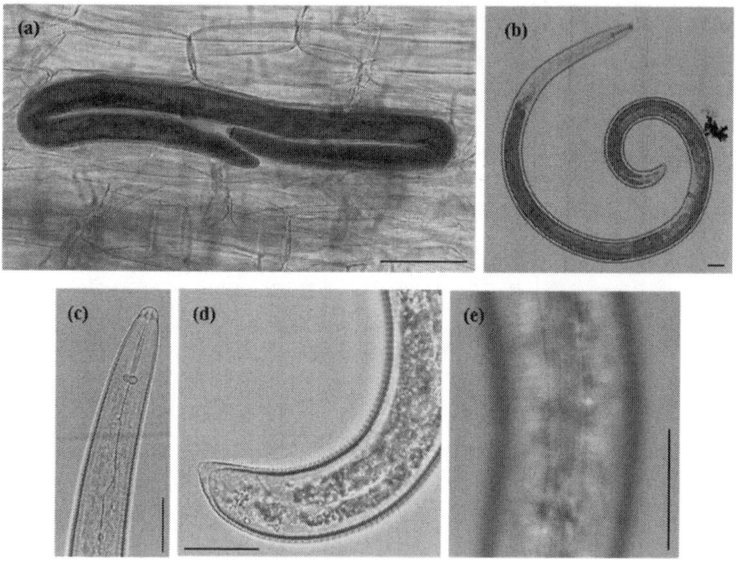

Figure 2. ***Helicotylenchus dihystera*** **(females) light microscope photographs. (a) Infected *Ficus microcarpa* root. (b) Whole specimen. (c) Anterior region in lateral view. (d) Posterior region in lateral view. (e) Lateral field with four lateral lines. Scale bars: 20 μm (a–d), and 50 μm (e).**

this region with the *Helicotylenchus* sequences available in the GenBank database revealed a similarity of 99.99% with *H. dihystera.* Phylogenetic analysis from the alignment of Fm *Helicotylenchus* 28S rDNA sequences with available sequences of similar *Helicotylenchus* spp. [41] revealed that this isolate and all listed *H. dihystera* sequences appeared together in a well-separated clade with 100% bootstrap support (Figure 4), confirming the morphological identification. Considering a common start and end point to *Helicotylenchus* spp. (560 bp), the Fm sequences differed in several positions from at least one *H. dihystera* sequences included in the analysis; however, only three (4, 493 and 519) position changes were completely distinct from *H. dihystera* sequences (Figure S1). Fm *H. dihystera* sequences had divergences ranging from 0.2 to 2.2% when compared with *H. dihystera* sequences and 3.7 to 6.5% to the other *Helicotylenchus* species (Table S1).

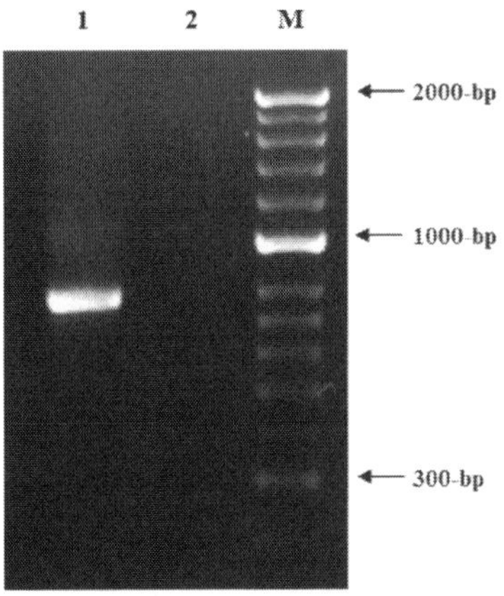

Figure 3. DNA amplification product obtained from *Helicotylenchus dihystera* isolate identified on *Ficus microcarpa* to the D2D3 expansion region of the 28S rDNA gene (1). 2, negative control. M, DNA marker (HyperLadder II; Bioline).

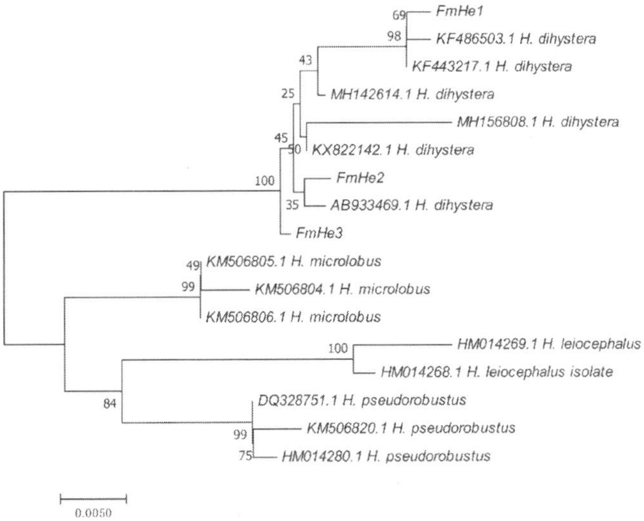

Figure 4. Neighbor-joining tree based on analysis of alignment of D2D3 expansion region of the 28S rDNA gene sequences of the *Helicotylenchus dihystera*

isolate identified on *Ficus microcarpa* (FmHe) with available sequences of close *Helicotylenchus* spp. (*H. leiocephalus, H. microlobus* and *H. pseudorobustus*) [41]. The percentage of replicate trees in which the associated *Helicotylenchus* spp. clustered together in the bootstrap test (1000 replicates) is shown next to the branches. Evolutionary distances were computed using the maximum composite likelihood method and all positions containing gaps and missing data were eliminated.

Although *H. dihystera* is considered as a polyphagous species with a wide distribution, reports on its pathogenicity are very few, and it is rarely recognized as an economically important PPN [30,42–44], even when high population densities are found. This is the first report of *H. dihystera* infecting *F. microcarpa;* however, it has been associated with *F. benjamina, F. carica, F. elastica, F. formosana* and *F. retusa* [21,26,44–47]. In Portugal, *H. dihystera* was reported as being associated with *Begonia* sp., *Colocasia esculenta, Cactus* sp., *Mentha* sp., *Musa* sp., *Pelargonium* sp., Polygonaceae, beans, maize and tomato [45,48–50].

Although no specific quarantine measures are being implemented against *M. javanica* or *H. dihystera*, preventive measures are particularly important to decrease the risk of spread into a region where they do not exist. Once nematodes are established in the soil, their eradication is very difficult. Thus, the use and transportation of clean, healthy, nematode-free planting material is a prerequisite for limiting the spread of nematodes. Plant parts liable to carrying PPN in trade/transport can be bulbs/tubers/corms/rhizomes, growing medium accompanying plants, roots and micropropagated plants. During routine inspections, the detection of nematode infections can be easily overlooked or misdiagnosed, as low to moderate populations of nematodes may cause no visible aboveground symptoms, making it harder to diagnose them [51,52]. Furthermore, above-ground symptoms are non-specific and usually involve stunting, lack of vigour, leaf nutritional deficiencies and temporary wilting in periods of water stress and high temperatures. The examination of roots can reveal the presence of galls that are specific symptoms associated with the occurrence of *Meloidogyne* spp., but the symptoms caused by *Helicotylenchus* spp., when present, can be confused with the damage associated with poor nutrition or injury caused by pathogens that attack the root system (other nematodes, bacteria, fungi and/or virus).

To our knowledge, *M. javanica* and *H. dihystera* are here reported for the first time as parasitizing *F. microcarpa*. Although both PPN species are common and widely distributed, our findings emphasize the importance of inspections by governmental authorities to find out whether imported material is free of PPN. If plants are grown in infested soil and then commercialized, it increases the probability of PPN dissemination to new regions and/or other suitable hosts with potential impact on economically important crops.

3. Materials and Methods

In December 2017, a *F. microcarpa* "Tiger bark" bonsai tree with a phytosanitary certificate was acquired by the first author in a shopping center in Coimbra, Portugal, showing small galls in the protruding roots, which aroused our attention, but without symptoms in the leaves (Figure 1a,b). Consequently, a few roots and a soil sample of 130 g were collected from the pot. Roots were observed directly and stained with acid fuchsin to detect nematode-infected plant tissues [53]. Nematodes were extracted from the soil, according to the Tray Method [54], followed by microscopic examination of nematode diversity and the genera identified and quantified.

3.1 Root Knot Nematode Characterization/Identification

Egg masses from *F. microcarpa* galled roots were propagated on tomato, *Solanum lycopersicum*, cv. Coração-de-Boi, in a growth chamber. After two months, the infected tomato roots were gently rinsed with tap water and 5 young egg-laying females were, individually and randomly, handpicked with their respective egg masses to glass blocks with NaCl 0.9% to obtain pure cultures. Individual young egg-laying females were characterised biochemically by electrophoretic analysis of esterases. Esterase electrophoresis was performed using polyacrylamide gels following the methodology described by Pais and Abrantes [55]. The individual females were transferred to micro-haematocrit tubes containing 5 μL of extraction buffer (20% sucrose and 1% Triton X-100), macerated and stored at −20 °C. Before electrophoresis, the samples were centrifuged at 8905 *g*, at −5 °C for 15 min. Electrophoresis was performed at 6 mA/gel during the first 15 min and then at 20 mA/gel for about 45 min using the Mini-Protean Tetra System (Bio-Rad Laboratories, Hercules, CA, USA). The gels were stained for esterase activity with the substrate α-naphthyl acetate, in the dark at 37 °C. Protein extract from five females of *M. javanica* was included in each gel as a reference. A pure culture (designated as Fm) of RKN was established by inoculating the 5 individual egg masses onto tomato to obtain a sufficient number of second-stage juveniles for molecular characterization. Electrophoretic analysis of esterases was repeated after two months to confirm the biochemical identification and the relative movement of each band calculated taking as reference the buffer front (Relative mobility, Rm%).

Biochemical identification was further confirmed by PCR-RFLP of mtDNA region between COII and 16S rRNA genes with C2F3 and MRH106 primers, and by SCAR-PCR with the species-specific primers Fjav and Rjav, using a pellet of second-stage juveniles (J2) obtained from egg masses of the pure isolate [31,56]. Briefly, for mtDNA region amplification, each PCR contained 1X PCR buffer, 1.8 mM $MgCl_2$, 0.2 mM dNTPs, 0.2 μM of each primer, 2.5 U Taq DNA polymerase (Bioline), and 50 ng DNA. Amplification was conducted using the following conditions: initial denaturation at 94 °C for 4 min, followed by 40 cycles of 94 °C for 30 s, 60 °C for 30 s, and 72 °C for 60 s, and a final extension for 10 min at 72 °C. After amplification, the PCR product was digested separately with 5 U *Hinf* I, *Alu*I and *Dra*III. For SCAR-PCR, the PCR reactions were the same as for the mtDNA region, except the primers (0.3 μM of each primer) and the amplification conditions (35 cycles of denaturation at 94 °C for 30 s, annealing at 52 °C for 30 s and extension at 72 °C for 1 min).

3.2 Spiral Nematode Characterization/Identification

Helicotylenchus specimens from soil and roots were propagated on the same tomato cultivar, in a growth chamber. Two/three months after inoculation, with approximately 3000 specimens, nematodes were extracted from roots/soil, according to the generalist Tray Method [54] and used to *Helicotylenchus* species characterization/identification and isolate maintenance, respectively. The characterization and identification of the

Helicotylenchus species was based on the morphological characters of 10 females (body shape after relaxed with gentle heat and number of incisures) and ribosomal DNA (rDNA) sequencing.

DNA was extracted and purified from 20 spiral nematodes, extracted from tomato roots, using the DNeasy Blood and Tissue kit (QIAGEN, Valencia, CA, USA), according to the manufacturer's instructions, and the D2D3 expansion region of the 28S rDNA gene was amplified using D2A (5›-ACA AGT ACC GTG AGG GAA AGT TG-3›) and D3B (5›-TCG GAA GGA ACC AGC TAC TA-3›) primers [57].

The PCR products were analysed on 1% agarose gel stained with GreenSafe (Nzytech), purified from the gel with the MiniElute Gel Extraction kit (QIAGEN, Valencia, CA, USA), quantified using the NanoDrop 2000C spectrophotometer (Thermo Scientific), cloned and sequenced. Sequences were compared with available close *Helicotylenchus* spp. sequences in GenBank [41]. Sequences were aligned using CLUSTALW multiple alignment in BIOEDIT software [58]. The evolutionary history was inferred using the neighbor-joining (NJ) and maximum likelihood (ML) methods in MEGA 7, as described in Santos et al. [10,59].

References

1. Eschen, R.; Roques, A.; Santini, A. Taxonomic dissimilarity in patterns of interception and establishment of alien arthropods, nematodes and pathogens affecting woody plants in Europe. *Divers. Distrib.* **2015**, *21*, 36–45. [CrossRef]

2. Roques, A.; Blackburn, M.A.T.M.; David, R.; Garnas, J.; Pys, P.; Wingfield, M.J.; Liebhold, A.M.; Duncan, R.P. Temporal and interspecific variation in rates of spread for insect species invading Europe during the last 200 years. *Biol. Invasions* **2016**, *18*, 907–920. [CrossRef]

3. Vovlas, N.; Trisciuzzi, N.; Troccoli, A.; De Luca, F.; Cantalapiedra-Navarrete, C.; Castillo, P. Integrative diagnosis and parasitic habits of *Cryphodera brinkmani* a non-cyst forming heteroderid nematode intercepted on Japanese white pine bonsai trees imported into Italy. *Eur. J. Plant Pathol.* **2013**, *135*, 717–726. [CrossRef]

4. Blackburn, T.M.; Essl, F.; Evans, T.; Hulme, P.E.; Jeschke, J.M.; Kühn, I.; Kumschick, S.; Marková, Z.; Mrugała, A.; Nentwig, W.; et al. A unified classification of alien species based on the magnitude of their environmental impacts. *PLoS Biol.* **2014**, *12*, e1001850. [CrossRef]

5. IFN6. 6º Inventário Florestal Nacional. Instituto da Conservação da Natureza e das Florestas. Lisboa, Portugal. Available online: http://www2.icnf.pt/portal/florestas/ifn/ifn6 (accessed on 2 July 2020).

6. De la Fuente, B.; Beck, P.S.A. Invasive species may disrupt protected area networks: Insights from the pine wood nematode spread in Portugal. *Forests* **2018**, *9*, 282. [CrossRef]

7. EPPO. Bursaphelenchus xylophilus. EPPO Datasheets on Pests Recommended for Regulation; EPPO Global Database: Paris, France. Available online: https://gd.eppo.int (accessed on 2 July 2020).

8. Huang, D.; Haack, R.A.; Zhang, R. Does global warming increase establishment rates of invasive alien species? A centurial time series analysis. *PLoS ONE* **2011**, *6*, e24733. [CrossRef]

9. Fey, S.B.; Herren, C.M. Temperature-mediated biotic interactions influence enemy release of nonnative species in warming environments. *Ecology* **2014**, *95*, 2246–2256. [CrossRef]

10. Santos, D.; Abrantes, I.; Maleita, C. The quarantine root knot nematode *Meloidogyne enterolobii*—A potential threat to Portugal and Europe. *Plant Pathol.* **2019**, *68*, 1607–1615. [CrossRef]

11. Santos, D.; Correia, A.; Abrantes, I.; Maleita, C. New hosts and records in Portugal for the root-knot nematode *Meloidogyne luci*. *J. Nematol.* **2019**, *51*, e2019-03. [CrossRef]

12. European Commission. Commission implementing regulation (EU) 2018/2019. *ANNEX I-List of High Risk Plants, Plant Products and Other Objects*; European Commission: Brussels, Belgium. Available online: http://data.europa.eu/eli/reg_impl/2018/2019/oj (accessed on 20 May 2020).

13. Anthoine, G.; Niere, B.; den Nijs, L.; Prior, T.; Pylypenco, L.; Viaene, N. Nematode interceptions in international trade of plants for planting. *J. Nematol.* **2014**, *46*, 134–135.

14. Harrison, R.D. Figs and the diversity of tropical rainforests. *BioScience* **2005**, *55*, 1053–1064. [CrossRef]

15. McSorley, R. *Plant Parasitic Nematodes Associated with Tropical and Subtropical Fruits*; Technical Bulletin no. 823; Florida Agricultural Experiment Station: Gainesville, FL, USA, 1981; Volume 823, pp. 1–49.

16. McSorley, R. Nematological problems in tropical and subtropical fruit tree crops. *Nematropica* **1992**, *22*, 103–116.

17. Cohn, E.; Duncan, L.W. Nematode parasites of subtropical and tropical fruit trees. In *Plant Parasitic Nematodes in Subtropical and Tropical Agriculture*; Luc, M., Sikora, R.A., Bridge, J., Eds.; CABI Publishing: Wallingford, UK, 1990; pp. 347–362.

18. Campos, V.P. Nematóides na cultura da figueira. *Inf. Agropecu. (Belo Horizonte)* **1997**, *18*, 33–38.

19. Krnjaic, D.; Krnjaic, S.; Bacic, J. Distribution and population density of fig cyst nematode (*Heterodera fici* Kirjanova) in the region of SR Yugoslavia. *Zast. Bilja* **1997**, *48*, 245–251.

20. Li, H.; Xu, J.; Shen, P.; Cheng, H. Distribution and seasonal dynamic changes of nematode parasites in fig main growing areas in Jiangsu Province. *J. Nanjing Agric. Univ.* **1999**, *22*, 38–41.

21. Abrantes, I.M.d.O.; Vieira dos Santos, M.C.; da Conceição, I.L.P.M.; Santos, M.S.N.d.A.; Vovlas, N. Root-knot and other plant-parasitic nematodes associated with fig trees in Portugal. *Nematol. Mediterr.* **2008**, *36*, 131–136.

22. Wohlfarter, M.; Giliomee, J.H.; Venter, E.; Storey, S. A survey of the arthropod pests and plant parasitic nematodes associated with commercial figs, *Ficus carica* (Moraceae), in South Africa. *Afr. Entomol.* **2011**, *19*, 165–172. [CrossRef]

23. Mokbel, A. Nematodes and their associated host plants cultivated in Jazan province, southwest Saudi Arabia. *Egypt. J. Exp. Biol. (Zool)* **2014**, *10*, 35–39.
24. Fanelli, E.; Vovlas, A.; Santoro, S.; Troccoli, A.; Lucarelli, G.; Trisciuzzi, N.; De Luca, F. Integrative diagnosis, biological observations, and histopathology of the fig cyst nematode *Heterodera fici* Kirjanova (1954) associated with *Ficus carica* L. in southern Italy. *ZooKeys* **2019**, *823*, 1–19. [CrossRef]
25. Karssen, G.; Wesemael, W.M.L.; Moens, M. Root-knot nematodes. In *Plant Nematology*; Perry, R.N., Moens, M., Eds.; CABI Publishing: Wallingford, UK, 2013; pp. 73–108.
26. Quénéhervé, P.; Topart, P.; Poliakoff, F. Interception of nematodes on imported bonsai in Martinique. *Nematropica* **1998**, *28*, 101–105.
27. EPPO. *Study on the Risk of Imports of Plants for Planting*; EPPO Technical Document; EPPO Global Database: Paris, France, 2012; Volume 1061.
28. Chan, E.W.C.; Tangah, J.; Inoue, T.; Kainuma, M.; Baba, K.; Oshiro, N.; Kezuka, M.; Kimura, N. Botany, uses, chemistry and pharmacology of *Ficus microcarpa*: A short review. *Syst. Rev. Pharm.* **2017**, *8*, 103–111. [CrossRef]
29. Hodel, D.R. New pests of landscape *Ficus* in California. *Calif. Assoc. Pest Control Adv. (CAPCA)* **2017**, *5*, 58–62.
30. Churchill, R.C., Jr.; Ruehle, J.L. Occurrence, parasitism, and pathogenicity of nematodes associated with sycamore (*Platanus occidentalis* L.). *J. Nematol.* **1971**, *3*, 189–196. [PubMed]
31. Maleita, C.M.; Simões, M.J.; Egas, C.; Curtis, R.H.C.; Abrantes, I.M.d.O. Biometrical, biochemical, and molecular diagnosis of Portuguese *Meloidogyne hispanica* isolates. *Plant Dis.* **2012**, *96*, 865–874. [CrossRef] [PubMed]
32. Lima-Medina, I.; Somavilla, L.; Carneiro, R.M.D.G.; Gomes, C.B. Espécies de *Meloidogyne* em figueira (*Ficus carica*) e em plantas infestantes. *Nematropica* **2013**, *43*, 56–62.
33. Hunt, J. *List of Intercepted Plant Pests*; Bureau of Entomology and Plant Quarantine, United States Department of Agriculture: Washington, DC, USA, 1956.
34. Santos, M.S.N.d.A. Identificação de populações portuguesas de *Meloidogyne* spp. pelas reações induzidas em plantas diferenciadoras-I. In *I Congresso Português de Fitiatria e Fitofarmacologia II*; Instituto Superior de Agronomia: Lisboa, Portugal, 1980; pp. 147–150.
35. Santos, M.S.N.d.A.; Abrantes, I.M.d.O. Root-knot nematodes in Portugal. In Proceedings of the Second Research Planning Conference on Root-Knot Nematodes, *Meloidogyne* spp., Region VII. Athens, Greece, 22–26 March 1982; North Carolina State University Graphics: Raleigh, NC, USA, 1982; pp. 17–23.
36. Reis, L.G.L. Prospeção Nematológica. Relatório das atividades da Estação Agronómica Nacional: Oeiras, Portugal, 1983.
37. Santos, M.S.N.d.A.; Abrantes, I.M.d.O.; Fernandes, M.F.M. Identificação de populações portuguesas de *Meloidogyne* spp. (Nematoda: Meloidogynidae) pelas reações induzidas em plantas diferenciadoras-III. *Ciênc. Biol. Ecol. Syst. (Portugal)* **1987**, *7*, 37–42.

38. da Conceição, I.L.P.M.; da Cunha, M.J.M.; Feio, G.; Correia, M.; Vieira dos Santos, M.C.; Abrantes, I.M.d.O.; Santos, M.S.N.d.A. Root-knot nematodes, *Meloidogyne* spp., on potato in Portugal. *Nematology* **2009**, *11*, 311–313.

39. Esteves, I.; Maleita, C.; Abrantes, I. Root-lesion and root-knot nematodes parasitizing potato. *Eur. J. Plant Pathol.* **2015**, *141*, 397–406. [CrossRef]

40. Rusinque, L.; Inácio, M.L.; Mota, M.; Nóbrega, F. Morphological, biochemical and molecular characterisation of *Meloidogyne javanica*, from North Portugal, in tomato. *Rev. Ciênc. Agrár.* **2018**, *41*, 201–210. [CrossRef]

41. Fortuner, R.; Merny, G.; Roux, C. Morphometrical variability in *Helicotylenchus* Steiner, 1945. 3: Observations on African populations of *Helicotylenchus dihystera* and considerations on related species. *Rev. Nématol.* **1981**, *4*, 235–260.

42. Benson, D.M.; Barker, K.R.; Aycock, R. Effects of density of *Helicotylenchus dihystera* and *Pratylenchus vulnus* on American Boxwood growing in microplots. *J. Nematol.* **1976**, *8*, 322–326. [PubMed]

43. Willers, P.; Grech, N.M. Pathogenicity of the spiral nematode *Helicotylenchus dihystera* to Guava. *Plant Dis.* **1986**, *70*, 352. [CrossRef]

44. Pedersen, J.F.; Rodriguez-Kabana, R. Nematode response to cool season annual graminaceous species and cultivars. *Ann. Nematol.* **1987**, *1*, 116–118.

45. Siddiqi, M.R. *Helicotylenchus dihystera. C.I.H. Descriptions of Plant-Parasitic Nematodes. Set 1–9*; Commonwealth Agricultural Bureaux: Farnham Royal, UK, 1972.

46. CABI/EPPO. Helicotylenchus dihystera. Distribution map. In *Distribution Maps of Plant Diseases*, 1st ed.; CABI Publishing: Wallingford, UK, 2010.

47. Plantwise Knowledge Bank. *Common Spiral Nematode Helicotylenchus Dihystera*; CABI Publishing: Wallingford, UK, 1956. Available online: https://www.plantwise.org/knowledgebank/datasheet/26824 (accessed on 30 March 2020).

48. Macara, A.M. *Contribuição para o estudo de algumas espécies do género Heterodera Schmidt 1871 encontradas em Portugal-Relatório Final do Curso de Engenheiro Agrónomo*; Instituto Superior de Agronomia, Universidade de Lisboa: Lisboa, Portugal, 1962.

49. Sher, S.A. Revision of the Hoplolaiminae (Nematoda) VI Helicotylenchus Steiner, 1945. *Nematologica* **1966**, *12*, 1–56. [CrossRef]

50. Siddiqi, M.R. On the genus *Helicotylenchus* Steiner, 1945 (Nematoda: Tylenchida), with descriptions of nine new species. *Nematologica* **1972**, *18*, 74–91. [CrossRef]

51. Coyne, D.L.; Fourie, H.H.; Moens, M. Current and future management strategies in resource-poor farming. In *Root-Knot Nematodes*; Perry, R.N., Moens, M., Starr, J.L., Eds.; CABI Publishing: Wallingford, UK, 2009; pp. 444–475.

52. Nicol, J.M.; Turner, S.J.; Coyne, D.L.; den Nijs, L.; Hockland, S.; Tahna Maafi, Z. Current nematode threats to world agriculture. In *Genomics and Molecular Genetics of Plant-Nematode Interactions*; Jones, J., Gheysen, G., Fenoll, C., Eds.; Springer: Berlin/Heidelberg, Germany, 2011; pp. 21–44.

53. Byrd, D.W.; Kirkpatrick, T.; Barker, K.R. An improved technique for clearing and staining plant tissues for detection of nematodes. *J. Nematol.* **1983**, *15*, 142–143.

54. Whitehead, A.G.; Hemming, J.R. A comparison of some quantitative methods of extracting small vermiform nematodes from soil. *Ann. Appl. Biol.* **1965**, *55*, 25–38. [CrossRef]

55. Pais, C.S.; Abrantes, I.M.d.O. Esterase and malate dehydrogenase phenotypes in Portuguese populations of *Meloidogyne* species. *J. Nematol.* **1989**, *21*, 342–346.

56. Zijlstra, C.; Donkers-Venne, D.T.H.M.; Fargette, M. Identification of *Meloidogyne incognita, M. javanica* and *M. arenaria* using sequence characterised amplified region (SCAR) based PCR assays. *Nematology* **2000**, *2*, 847–853. [CrossRef]

57. De Leij, O.; Félix, M.A.; Frisse, L.; Nadler, S.; Sternberg, P.; Thomas, W. Molecular and morphological characterisation of two reproductively isolated species with mirror image anatomy (Nematoda: Cephalobidae). *Nematology* **1999**, *1*, 591–612.

58. Hall, T.A. BioEdit: A user friendly biological sequence alignment editor and analyses program for windows 95/98/NT. *Nucleic Acids Symp. Ser.* **1999**, *41*, 95–98.

59. Kumar, S.; Stecher, G.; Tamura, K. MEGA7: Molecular evolutionary genetics analysis version 7. 0 for bigger datasets. *Mol. Biol. Evol.* **2016**, *33*, 1870–1874. [CrossRef] [PubMed]

Chapter-2

First Detection of *Meloidogyne luci* (Nematoda: Meloidogynidae) Parasitizing Potato in the Azores, Portugal

1. Introduction

Potato, *Solanum tuberosum*, is the third most important crop in the world after rice and wheat, with more than 156 countries producing it, and hundreds of millions of people depending on it for survival. According to Food and Agriculture Organization of the United Nations FAO estimates, in 2018, over 368 million metric tons of potatoes were produced worldwide, a substantial increase from 333.6 million metric tons in 2010. China is the biggest producer of potatoes worldwide, with an estimated production of 91 million metric tons, while Europe accounts for 106 million metric tons [1]. In Portugal, this crop has great social and economic importance, since it is grown throughout the country, including the archipelagos of Madeira and the Azores. On average, 430,000 metric tons of potato are produced, with the most representative production areas being Bragança, Chaves, Aveiro, Viseu, Oeste Region, and Montijo [2].

Plant-parasitic nematodes are a hampering factor in potato production and quality. Many species have been reported to be associated with potato, among which are the potato cyst nematodes *Globodera* sp., the root-knot nematodes (RKN) *Meloidogyne* sp., the lesion nematodes *Pratylenchus* sp., the potato-rot nematode *Ditylenchus destructor*, and the false root-knot nematode *Nacobbus aberrans* [3].

RKN are one of the oldest known parasitic nematodes of plants and considered serious pests of economically important crops [4,5]. The genus comprises more than 90 species [6] and many have been reported in Portugal: *Meloidogyne arenaria* (Neal, 1889) Chitwood, 1949; *Meloidogyne chitwoodi* Golden et al., 1980; *Meloidogyne enterolobii* Yang and Eisenback, 1983; *Meloidogyne hapla* Chitwood, 1949; *Meloidogyne hispanica* Hirschmann, 1986; *Meloidogyne incognita* (Kofoid and White, 1919) Chitwood, 1949; *Meloidogyne javanica* (Trub, 1885) Chitwood, 1949); *Meloidogyne luci* Carneiro et al., 2014; *Meloidogyne lusitanica* Abrantes and Santos, 1991; and *Meloidogyne naasi* Franklin, 1965 [7–11].

Meloidogyne luci was first described in 2014 on different plant species in Brazil, Chile, and Iran [12]. Due to its morphological resemblance and similar esterase (EST) phenotype to

M. ethiopica, several populations of *M. ethiopica* in Europe were reclassified and identified as *M. luci* using biochemical and molecular analyses. In Portugal, it was detected in 2013 in a potato field near Coimbra [9] and was recently found parasitizing tomato, *Solanum lycopersicum*, the ornamental plant *Cordyline australis*, and the weed *Oxalis corniculata* [13]. Since *M. luci* is a polyphagous species with many of its host plants being of economic importance, it poses a risk to agricultural production, especially for potato. Furthermore, its detection in Europe shows that it has the potential to enter the region and survive under temperate conditions [14]. For those reasons, in 2017 *M. luci* was added to the European Plant Protection Organization (EPPO) alert list and in 2019 a national survey was implemented aiming to avoid dispersion.

The aim of the present study was morphological, morphometric, biochemical and molecularly characterize the isolate of RKN *M. luci* found in the Azores islands.

2. Results and Discussion

Morphological characterization from the recovered second-stage juveniles (J2), males and females of the isolate of *M. luci* was performed, as were morphometric studies on J2 (Table 1).

Table 1. Morphometric comparison of second-stage juveniles (J2) of Meloidogyne luci from the Azores, Portugal, with the original description (Carneiro et al., 2014). All measurements are in μm and in the format mean ± standard deviation (range).

Character/Ratio	M. luci J2 (Azores) (n = 10)	Carneiro et al., 2014 (n = 30)
Length	404.99 ± 23 (376.3–446)	383 ± 85 (300–470)
Stylet length	14.05 ± 1 (11.4–15.6)	12.5 ± 0.2 (12.0–13.5)
Dorsal oesophageal and	3.04 ± 0.41 (2.2–3.4)	2.9 ± 0.5 (2.3–3.3)
Tail length	46.64± 5.92 (39.7–54.1)	44 ± 4.5 (40.0–48.5)
Hyaline terminus	11.23 ± 1.97 (9.02–14.06)	11.7 ± 3.0 (9–15)
Max. body width	15.93 ± 2.26 (14.6–20.9)	16 ± 1.5 (13–20)
a *	25.76 ± 2.87 (22.2–31.9)	25.6 ± 10.5 (15.0–36.1)
c **	8.77 ± 0.86 (7.5–9.6)	8.7 ± 2.6 (6.2–11.5)

* *length/max. body width; ** length/tail length.*

2.1 Morphological and Morphometric Characterization

The J2 were vermiform, slender, and clearly annulated. The head region was slightly set apart from the body. The stylet was delicate, narrow, and sharply pointed; the knobs were small and oval shaped. The excretory pore was distinct and the hemizonid was anteriorly adjacent to the excretory pore. The tail was conoid with a rounded tip and the hyaline terminus was distinctive (Figure 1a–d).

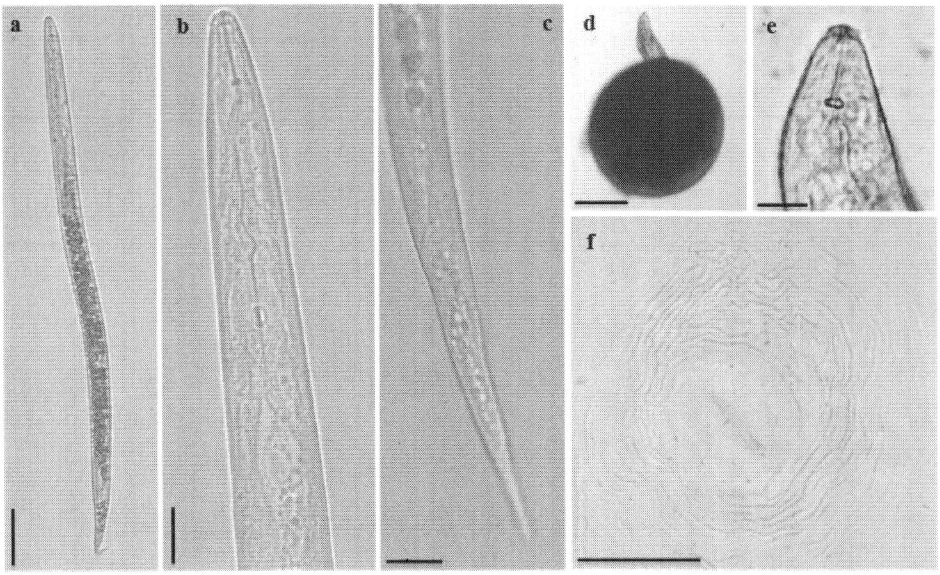

Figure 1. ***Meloidogyne luci*** **light microscope observations. Second-stage juvenile: (a) whole specimen; (b) anterior region, (c) tail region. Female: (d) egg-laying female, whole specimen; (e) anterior end; (f) perineal pattern (bar = 20 μm).**

Females were elongated, ovoid, or pear-shaped, with a prominent neck (Figure 1d). The head was slightly set apart from the body. The stylet was robust, with knobs well developed. The stylet cone was wider near the shaft and the shaft was wider near the junction with knobs (Figure 1e). The perineal pattern was oval to squarish, with the dorsal arch high to low and rounded. The striae were smooth and wavy, widely separated, and continuous. The lateral lines were weakly demarcated and the perivulval region was free from striae (Figure 1f). The patterns were found to be highly variable.

The males were vermiform, bluntly rounded posteriorly, and with an anterior end narrowing. The body cuticle was annulated, with large annuli. The head region was not set off from the body. The stylet was robust, and the cone was larger than the shaft and increased in width near the junction with the shaft. The knobs were rounded and small, merging gradually into the shaft. The tail was short and the spicules were curved.

Morphological features are valuable tools for RKN identification due to their low cost and ease of learning the skills, with accuracy depending on the number of characteristics to be evaluated and the number of specimens. The species identification of *Meloidogyne* based on these characters is nevertheless a challenge because morphological differences between RKN species are in most cases indistinctive and measurements of individual specimens in general overlap.

The morphology and morphometrics were compared to the description of *M. luci* made by [9] and the results were consistent (Table 1). However, due to the intraspecific variability, its identification became difficult; for instance, characteristics such as the morphology of the perineal pattern were highly variable and could be found in more than one species. Therefore, morphological and biometrical diagnostic characteristics need to be supported by other studies, such as biochemical and molecular.

2.2 Biochemical Characterization

The EST phenotype from young egg-laying females exhibited three bands (relative mobility, Rm: 1, 1.10, 1.20), corresponding to the *M. luci* L3 phenotype [11]. *Meloidogyne ethiopica* also presented an EST phenotype of three bands (Rm: 0.93, 1.13, 1.24). The three EST bands observed in *M. javanica* (Rm: 1, 1.17, 1.26) were used as a reference to determine the relative position of *M. luci* and *M. ethiopica* bands (Figure 2).

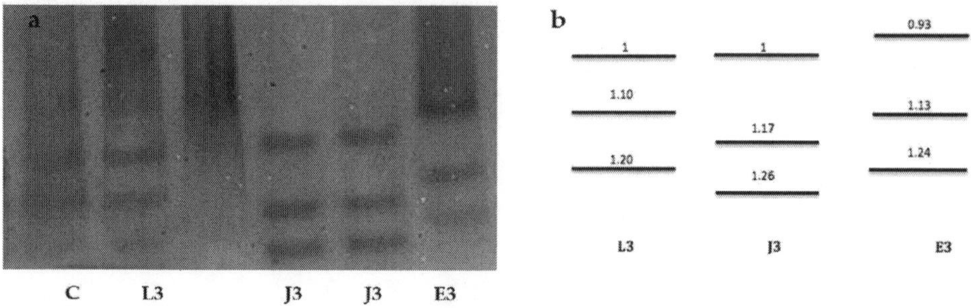

Figure 2. (a) Phenotypes of protein homogenates from one egg-laying female of the *Meloidogyne* species: C—Positive control *M. luci*; 1: L3—*M. luci* esterase (Azores), J3—*M. javanica*, and E3—*M. ethiopica*, and (b) relative mobility L3—*M. luci*, J3—*M. javanica*, and E3—*M. ethiopica*.

Many studies have shown the usefulness of the nonspecific EST phenotype as the quicker, more reliable, and more stable method to identify *Meloidogyne* spp. [15,16]. The EST phenotype found in the isolate from the Azores was similar to the Portuguese isolate found parasitizing potato in Continental Portugal in 2013. The first band was located at the same level of the band of the reference *M. luci* and *M. javanica*. Since *M. ethiopica* was included for comparison, it could be clearly seen that this first band was well above. Therefore, in spite of the similarity between *M. luci* and *M. ethiopica,* the patterns have clear differences and can be consider reliable in the identification of these two species.

2.3 Molecular Characterization

The PCR amplification of mtDNA COII/16S rRNA yielded a single fragment of 1800 bp. The nucleotide sequence obtained in this study was deposited into the GenBank database (NCBI) under the accession number MW160418. A BLAST search of the nu- cleotide sequence showed a similarity of 99.17% with the sequences of *M. luci* available in the database.

The molecular phylogenetic analysis is presented in Figure 3. The phylogram revealed one clade, supported by a bootstrap value of 93%, that included all isolates of *M. luci* from other countries, the isolate from the Azores, and the isolates of *M. ethiopica*. The isolates of *M. javanica, M. incognita*, and *M. arenaria* formed separate major clades with bootstrap values of 81, 97, and 99%, respectively.

According to [17], the region of mtDNA COII/16S rRNA is useful in the identification of the closely related species *M. luci* and *M. ethiopica*. In this study, that region allowed us to identify the isolate from the Azores as *M. luci*. However, due to the closeness between the species, the molecular markers needed to be used in combination with biochemical analysis. In general, the eradication of nematodes is very difficult, and it is even more so when it comes to the RKN species, especially *M. luci*. Its ability to adapt to temperate conditions and a wide variety of hosts make its management a challenge. Additionally, several nematicides have recently been strictly regulated or banned in the EU, due to the adverse impacts on the environment and human health, reducing the alternatives for control. Therefore, to define sustainable management strategies, an accurate diagnosis and knowledge of the species is required, with the combination of biochemical and molecular analysis being the best approach for RKN species identification. Furthermore, not only does the identification have constrains, but the detection does as well, as occurred in this study. Plants may either not present any symptoms, or they can often be misdiagnosed, as symptoms may appear similar to other factors.

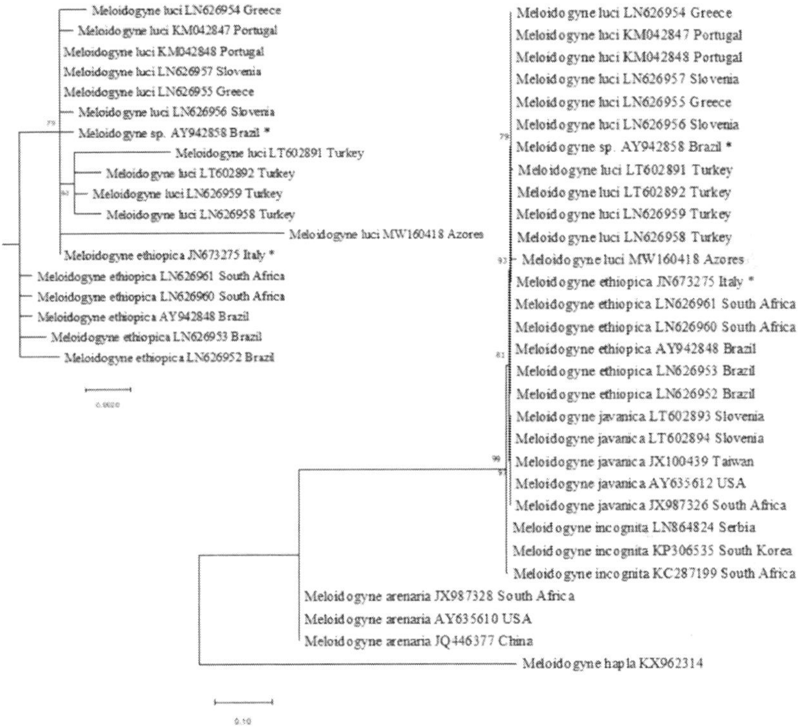

Figure 3. **Phylogenetic relationships of *Meloidogyne luci* isolate collected from the Azores, Portugal, and *M. luci* isolates from other geographical regions, including other species of the *Meloidogyne* group, based on the sequence alignment of the mtDNA region between COII and 16S genes. The dendrogram was inferred by using the maximum likelihood method and the Hasegawa–Kishino–Yano model with 1000 bootstrap replication. Bootstrap values are indicated at the nodes. The analysis involved 30 nucleotide sequences and there was a total of 1596 positions in the final dataset. Evolutionary analyses were conducted in MEGA X. * Recently reclassified as *M. luci*, according to [16].**

Finally, due to the threats and problems presented above, to evaluate the distribution and potential impact of this nematode, a national survey was implemented after 2019 in Continental Portugal and the Azores.

To our knowledge, this is the first report of *M. luci* in the islands of the Azores, Portugal, adding valuable information to the current location of this organism in the EPPO zone.

3. Materials and Methods

3.1 Nematode Isolates

During the 2019 National Survey in the Azores islands, soil samples were collected from the council of Santo António on Pico Island. Each consisted of 5 to 8 cores sampled at roughly equal intervals. Six composite soil samples were placed in polyethylene bags and brought for analysis. A 400 mL subsample was taken from each composite sample and the nematodes were extracted using sieving and decanting together with centrifugal technique according to protocol PM 7/119 (1) [18]. The suspension was observed under a stereomicroscope (Nikon SMZ1500, Tokyo, Japan) and suspect specimens of *Meloidogyne* were observed using a bright-field light microscope (Olympus BX-51, Hamburg, Germany) for confirmation.

For positive detections of *Meloidogyne* it was necessary to perform bioassays in order to obtain material (females, egg masses, and males) for identification. Bioassays were carried out by planting tomato plants cv. Oxheart in the remaining soil from the analyzed sample and maintained in a quarantine greenhouse for two months. Females and egg masses were handpicked from the infected tomato roots.

3.2 Morphological and Morphometric Characterization

Nematodes were placed in a drop of water on a glass slide and gently heat killed for morphological and morphometric characterization using a bright-field light microscope (Olympus BX-51, Hamburg, Germany) and photographed with a digital camera (Leica MC190 HD, Wetzlar, Germany). The measurements were taken using the Leica LAS Live. Perineal patterns of adult females were cut from live specimens in 45% lactic acid and mounted in glycerine.

3.3 Biochemical Characterization

Young egg-laying females were handpicked from infected tomato roots and transferred to micro-hematocrit capillary tubes with 5 μL of extraction buffer (20% sucrose *v*/*v* and 1% Triton X-100 *v*/*v*). The females were macerated with a pestle, frozen, and stored at 20 °C until use. Proteins were separated by polyacrylamide gel electrophoresis (PAGE) on thin-slab 7% separating polyacrylamide gels, in a Mini-Protean II (BioRad Laboratories, Hercules, CA, USA) according to [19]. The gels were stained for EST activity with the substrate α-naphthyl acetate. Protein extracts from young egg-laying females of *M. ethiopica* and *M. luci* were included in each gel for comparison and a protein extract of an isolate of *M. javanica* was used as a reference.

3.4 Molecular Characterization

The mtDNA COII/16S rRNA region was selected for molecular characterization of the *M. luci* isolate from the Azores islands. The total DNA was extracted from the egg masses using the DNeasy Blood & Tissue kit (Qiagen, Hilden, Germany) follow- ing the manufacturer's instructions. The mtDNA COII/16S rRNA region was amplified using the primers C2F3 (5′-GGTCAATGTTCAGAAATTTGTGG-3′) and 1108 5′-TACCT TTGACCAATCACGCT-3′ [20]. PCR reactions were performed in a 50 µL final volume mixture containing 25 µL Supreme NZYTaq II Green Master Mix, 10 µL of isolated DNA, and 0.2 µM of each primer in a Biometra TGradient thermocycler (Biometra, Göttingen, Germany). Thermal cycling conditions were as described by [17]. PCR products were re- solved by electrophoresis at 5 V.cm^{-1} in agarose gel (1.5%) containing 0.5 µg/mL ethidium bromide and 0.5x Tris-borate-EDTA (TBE) running buffer. Amplifications were visualized using the VersaDoc Imaging System (BioRad Laboratories, Hercules, CA, USA). PCR prod- ucts were purified using the DNA clean and concentrator kit (Zymo Research Corp, Irvine, CA, USA), according to the manufacturer's instructions. Amplicons were sequenced in both directions at STABVida Sequencing Laboratory (Lisbon, Portugal) on a DNA analyzer ABI PRISM 3730xl (Applied Biosystems). The newly obtained sequence was manually checked, edited, and assembled. The sequence was compared to those of *M. luci* and other relevant sequences of *Meloidogyne* spp. available in the GenBank database using the BLAST homology search. The multiple alignment of the retrieved sequences was performed using ClustalW multiple alignment in BioEdit (Figure S1).

Phylogenetic analyses were conducted using MEGA X v10.1 [21] and the maximum likelihood (ML) method based on the Hasegawa–Kishino–Yano model. The robustness of the ML tree was inferred using 1000 bootstrap replicates.

References

1. FAO. FAOSTAT. 2020. Available online: http://www.fao.org/faostat/en/#data (accessed on 16 October 2020).
2. Camacho, M.J.; Nóbrega, F.; Lima, A.; Mota, M.; Inácio, M.L. Morphological and molecular identification of the potato cyst nematodes *Globodera rostochiensis* and *G. pallida* in Portuguese potato fields. *Nematology* **2017**, *8*, 883–889. [CrossRef]
3. Lima, F.S.O.; Mattos, S.V.; Silva, S.E.; Carvalho, M.A.S.; Teixeira, R.A.; Silva, J.C.; Correa, V.R. Nematodes Affecting Potato and Sustainable Practices for Their Management. In *Potato—From Incas to All Over the World*, 1st ed.; Yildiz, M., Ed.; IntechOpen: London, UK, 2018; Volume 1, pp. 107–121.
4. Dong, K.; Dean, R.A.; Fortnum, B.A.; Lewis, S.A. Development of PCR primers to identify species of root knot nematodes: *Meloidogyne arenaria*, *M. hapla*, *M. incognta* and *M. Javanica*. *Nematropica* **2001**, *31*, 271–280.
5. Trudgill, D.L.; Blok, V.C. Apomictic, polyphagous root-knot nematodes: Exceptionally successful and damaging biotrophic root pathogens. *Annu. Rev. Phytopathol.* **2001**, *39*, 53–77. [CrossRef] [PubMed]
6. Hunt, D.; Handoo, Z. Taxonomy, identification and principal species. In *Root—Knot Nematodes*; Perry, R.N., Moens, M., Starr, J.L., Eds.; CABI: London, UK, 2009; pp. 55–88.
7. Abrantes, I.D.; dos Santos, M.V.; da Conceição, I.L.; Santos, M.D.; Vovlas, N. Root-knot and other plant-parasitic nematodes associated with fig trees in Portugal. *Nema. Mediterr.* **2008**, *36*, 131–136.

8. Abrantes, I.D.; Santos, M.S.; Correia, M.; da Conceição, I.L.; da Cunha, M.J.; Feio, G.; dos Santos, M.C. Root-knot nematodes, *Meloidogyne* spp., on potato in Portugal. *Nematology* **2009**, *11*, 311–313. [CrossRef]

9. Maleita, C.; Esteves, I.; Cardoso, J.M.S.; Cunha, M.J.; Carneiro, R.M.D.G.; Abrantes, I. *Meloidogyne luci*, a new root-knot nematode parasitizing potato in Portugal. *Plant Pathol.* **2018**, *67*, 366–376. [CrossRef]

10. Santos, D.; Correia, A.; Abrantes, I.; Maleita, C. The quarantine root knot nematode *Meloidogyne enterolobii*—A potential threat to Portugal and Europe. *Plant Pathol.* **2019**, *68*, 1607–1615. [CrossRef]

11. Viera dos Santos, M.; Almeida, M.T.M.; Costa, S.R. First report of *Meloidogyne naasi* parasitizing turfgrass in Portugal. *J. Nematol.* **2020**, *52*, 1–4. [CrossRef]

12. Carneiro, R.M.D.G.; Correa, V.R.; Almeida, M.R.A.; Gomes, A.C.M.M.; Deimi, A.M.; Castagnone-Sereno, P.; Karssen, G. *Meloidogyne luci* n. sp. (Nematoda: Meloidogynidae), a root-knot nematode parasitising different crops in Brazil, Chile and Iran. *Nematology* **2014**, *16*, 289–301. [CrossRef]

13. Santos, D.; Correia, A.; Abrantes, I.; Maleita, C. New Hosts and Records in Portugal for the Root-Knot Nematode *Meloidogyne luci*. *J. Nematol.* **2019**, *51*, 1–4. [CrossRef]

14. EPPO Alert List: Addition of *Meloidogyne luci* together with *Meloidogyne ethiopica*. Available online: https://gd.eppo.int/ reporting/article-6186 (accessed on 23 November 2017).

15. Esbenshade, P.R.; Triantaphyllou, A.C. Identification of major *Meloidogyne* species employing enzyme phenotypes as differentiat- ing characters. In *An Advanced Treatise on Meloidogyne*; Sasser, J.N., Carter, C.C., Eds.; North Carolina State University Graphics: Raleigh, NC, USA, 1985; Volume I, pp. 135–140.

16. Carneiro, R.M.D.G.; Almeida, M.R.A.; Quénéhervé, P. Enzyme phenotypes of *Meloidogyne* spp. isolates. *Nematology* **2000**, *2*, 645–654. [CrossRef]

17. Stare, B.G.; Strajnar, P.; Susicˇ, N.; Urek, G.; Širca, S. Reported populations of *Meloidogyne ethiopica* in Europe identified as *Meloidogyne luci*. *Plan. Dis.* **2017**, *101*, 1627–1632. [CrossRef]

18. *Standard Protocol PM 7/119 (1). Nematode Extraction*; EPPO Bulletin 43:471-95; EPPO: Paris, France, 2013.

19. Pais, C.S.; Abrantes, I.D.; Fernandes, M.F.; Santos, M.D. Técnica de electroforese aplicada ao estudo das enzimas dos nematodes- das-galhas-radiculares, *Meloidogyne* spp. *Ciência Biol. Ecol. Syst.* **1986**, *6*, 19–34.

20. Powers, T.O.; Harris, T.S. A polymerase chain reaction method for identification of five major *Meloidogyne* species. *J. Nematol.*

21. Kumar, S.; Stecher, G.; Li, M.; Knyaz, C.; Tamura, K. MEGA X: Molecular Evolutionary Genetics Analysis across computing platforms. Mol. Biol. Evol. 2018, 35, 1547–1549. [CrossRef]

Chapter-3

Pratylenchus vovlasi sp. Nov. (Nematoda: Pratylenchidae) on Raspberries in North Italy with a Morphometrical and Molecular Characterization

1. Introduction

Raspberries (*Rubus* sp.) have a long history of human consumption and cultivation in Europe, with Russia being the leading producer [1]. Raspberries have been eaten fresh for thousands of years. In the last decades, there has been an increase in the demand for fruit and fruit-based products as consumers seek out healthier dietary options. In particular, berries are considered one of the best dietary sources of bioactive compounds that have important antioxidant properties, with associated health effects such as protective effects against several cancers and cardiovascular disorders [2,3] and thus the berry market is expected to expand during the next years. Raspberries are the most productive in areas with mild winters and long, moderate summers. The production of raspberries is greatly influenced by biotic and abiotic factors. Among the biotic factors, plant-parasitic nematodes have an important role in the reduction of the raspberry yield [4,5]. Several genera of plant- parasitic nematodes have been found to be associated with the raspberry [4–9]. The most frequently observed species in soil and the roots of raspberries belong to the root-lesion nematode genus *Pratylenchus*, mainly *P. crenatus* Loof, 1960, (71.7% of soil samples, 53.4% of root samples), *P. penetrans* (Cobb, 1917) Filipjev and Schuurmans Stekhoven, 1941 (51.6%, 54.2%) and *P. scribneri* Steiner in Sherbakoff and Stanley 1943 (9.2%, 10.2%; [9]). In Italy, raspberries are grown mainly in the northern areas including Trentino Alto Adige, the main production area, Verona province and Piedmont. Other Italian areas growing raspberries are in Romagna in the north and Calabria, Sicily, Campania and Basilicata in the south. Considering the economic significance of root-lesion nematodes on raspberries and the need to accurately distinguish these damaging species for their practical management in the field, we provide here the morphometrical and molecular characterization of a new root-lesion nematode, *Pratylenchus vovlasi* sp. nov. parasitizing raspberries in the Piedmont area. The specific objectives of this paper were: (i) to carry out a comprehensive identification with morphological and morphometric approaches of *P. vovlasi* sp. nov. with a differential diagnosis to closely related species; (ii) to provide a molecular characterization of *P. vovlasi* sp. nov. and estimate its phylogenetic relationships with other representatives of the *Pratylenchus* genus using three molecular markers: two ribosomal markers (the D2-D3 expansion segments of the 28S rRNA and the internal transcribed spacer region (ITS rRNA) and the nuclear region of the partial heat shock protein 90 (*hsp*90) gene.

2. Results

Two species were identified from raspberries in the Piedmont region. Detailed morphometrics and molecular analyses of the two root-lesion nematode populations were consistent with *P. crenatus* and with an undescribed *Pratylenchus* species. As *P. crenatus* is a well-known and well molecular characterized species, we concentrated our efforts on the integrative taxonomic identification of the undescribed species.

2.1 Molecular Characterization

For molecular analyses, a total of five individual specimens for each molecular marker were amplified. The D2-D3 expansion domains of *28S* rRNA, *ITS* rRNA and partial *hsp90* genes of the new species *P. vovlasi* sp. nov. yielded single fragments of ~800 bp, 700 bp and 300 bp, respectively, based on a gel electrophoresis. The D2-D3 for *P. vovlasi* sp. nov. (OA984892–OA984893) showed a very low intraspecific variability with one different nucleotide and no indels (99% similarity). The D2-D3 for *P. vovlasi* sp. nov. differed from the closest related species, *P. pseudopratensis* Seinhorst, 1968 (JX261965) by 18 nucleotides and 0 indels (97% similarity), *P. pratensis* (De Man, 1880) Filipjev, 1936 (KY828298) by 23–26 nucleotides and 0 indels (96–97% similarity) and from *P. vulnus* (JQ003993) by 72–73 nucleotides and four to ten indels (89–90% similarity).

The ITS region for *P. vovlasi* sp. nov. also showed a low intraspecific variability by 19 nucleotides and 6 indels (97% similarity). The ITS1 for *P. vovlasi* sp. nov. (OA984869–OA984870) showed a low similarity with all of the ITS sequences of *Pratylenchus* spp. deposited in NCBI including the most similar species, *P. pratensis* (=*P. lentis*) (AM933158, AM933147, AM933149) and *P. fallax* Seinhorst, 1968, (FJ719921, FJ719917), by 97–98 different nucleotides and 34–39 indels (86% similarity).

The partial *hsp*90 gene amplified products of two individual specimens were cloned and four clones for each individual specimen were sequenced (OA984937–OA984944). The sequence analyses revealed the occurrence of two different fragments for each specimen and both fragments coded for *hsp90* differed in the length of the intron (52 bp vs. 43 bp) and nucleotide variability. The 298 bp fragment showed an 82% similarity (254/310 identities) with the 310 bp fragment at a nucleotide level. At an amino acid level, the two fragments showed 96% identities (81/84) and 98% positives (83/84). The new *hsp90* sequences for *P. vovlasi* sp. nov. showed a high intraspecific variability by 3–60 nucleotides and 0–13 indels (81–99% similarity). The *hsp90* sequences for *P. vovlasi* sp. nov. differed by 13, 16 and 61 nucleotides and 4, 4 and 24 indels from the most closely related species, *P. speijeri* De Luca, Troccoli, Duncan, Subbotin, Waeyenberge, Coyne, Brentu and Inserra, 2012 (HE601547) with a 96% similarity, *P. coffeae* (Zimmermann, 1898) Filipjev and Schuurmans Stekhoven, 1941 (HE601548) with a 95% similarity and *P. hippeastri* Inserra, Troccoli, Gozel, Bernard, Dunn and Duncan, 2007 (HE601549) with an 81% similarity, respectively.

2.2 Restriction Profiles

PCR-RFLP analyses of the ITS region allowed us to determine the species-specific patterns for the Italian population of *P. vovlasi* sp. nov. (Figure 1) that clearly identified this species.

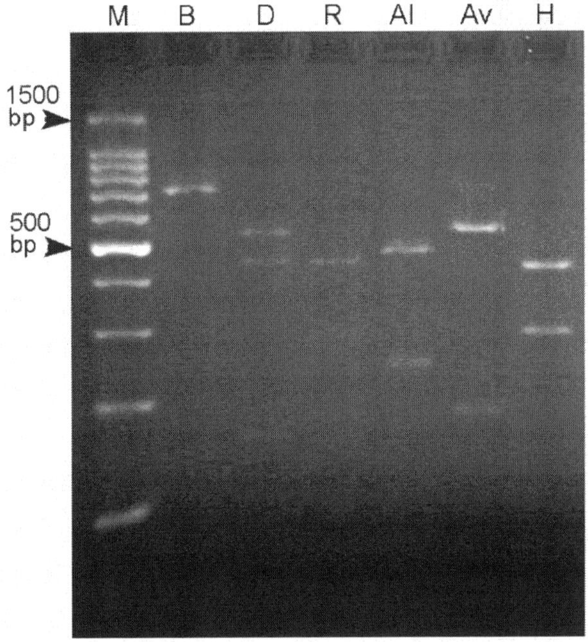

Figure 1. Restriction fragments of an amplified ITS of *Pratylenchus vovlasi* sp. nov. Al: *Alu*I, Av: *Ava*II, B: *Bam*HI, D: *Dde*I, H: *Hinf* I, R: *Rsa*I and M: 100 bp ladder.

2.3 Phylogenetic Relationships

The phylogenetic relationships among *Pratylenchus* species inferred from the analyses of the D2-D3 expansion domains of 28S rRNA, ITS and the partial *hsp90* gene sequences using BI are shown in Figures 2–4, respectively. The phylogenetic trees generated with the ribosomal and nuclear markers included 54, 60 and 31 sequences with 693, 548 and 284 positions in length, respectively (Figures 2–4). The D2-D3 tree of *Pratylenchus* spp. showed a well-supported subclade (PP = 1.00) including *P. vovlasi* sp. nov., *P. pratensis* and *P. pseudopratensis* (Figure 2). All other clades followed the same pattern as previous studies on *Pratylenchus* species.

The 50% majority-rule consensus ITS BI tree showed several clades that were not well-defined (Figure 2) but a well-supported subclade (PP = 1.00) including *P. vovlasi* sp. nov. and *P. pratensis* (Figure 3). This subclade was phylogenetically related to *P. vulnus* Allen and Jensen, 1951 and *P. kumamotoensis* Lal and Khan, 1990 in a moderately supported subclade (PP = 0.95) (Figure 3). Finally, the *hsp90* BI tree confirmed the occurrence of two isoforms within *P. vovlasi* sp. nov. that were clearly separated in two independent well-supported subclades (PP = 0.98 and PP = 1.00, respectively) (Figure 4).

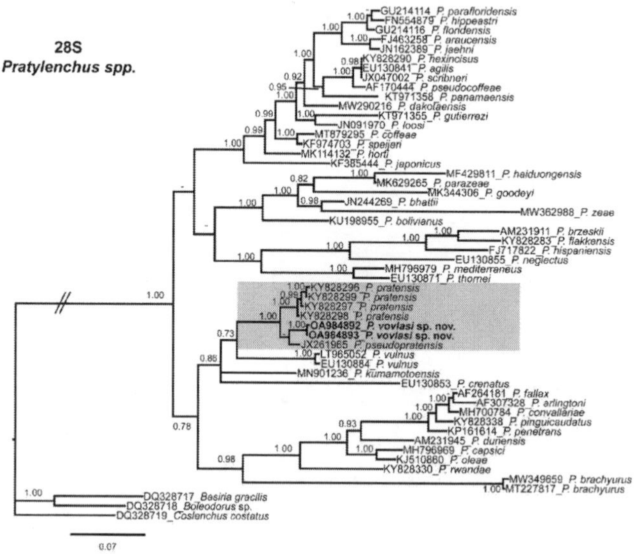

Figure 2. Phylogenetic relationships of *Pratylenchus vovlasi* sp. nov. within the genus *Pratylenchus*. A Bayesian 50% majority-rule consensus tree as inferred from the D2 and D3 expansion domains of the 28S rRNA sequence alignment under the general time-reversible model of the sequence evolution with a correction for invariable sites and a gamma-shaped distribution (GTR + I + G). Posterior probabilities more than 0.70 are given for appropriate clades. Newly obtained sequences in this study are shown in bold. Scale bar = expected changes per site.

2.4 Morphology and Morphometry of *Pratylenchus vovlasi* sp. nov.

***Pratylenchus vovlasi* sp. nov.** (http://zoobank.org/urn:lsid:zoobank.org:act:3B91A2 A8-B809-47BA-A822-CF380EBC0245, accessed on 2 April 2021).

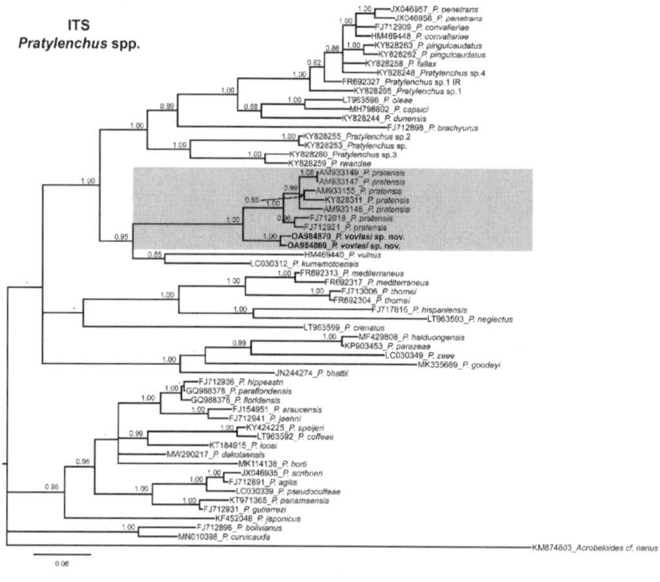

Figure 3. Phylogenetic relationships of *Pratylenchus vovlasi* sp. nov. within the genus *Pratylenchus*. A Bayesian 50% majority-rule consensus tree as inferred from the ITS sequence alignment under a transversional model with a proportion of invariable sites and a rate of variation across sites (TVM+I+ G). Posterior probabilities more than 70% are given for appropriate clades. Newly obtained sequences in this study are in bold letters. Scale bar = expected changes per site.

2.4.1 Description

Female: the body assumes an almost straight to open C posture when heat-killed (Figure 5A). The lip region is slightly offset from body contour and bears three annuli, which narrow in diameter towards the anterior end (Figures 5D and 6D). In the en face SEM view lip region, they appear dumb-bell-shaped (Figure 7A,B,D,I) with an acute pattern (sensu Subbotin et al. [10]) fitting the group II according to the classification scheme of Corbett and Clark [11]. The stylet is relatively small and delicate with conus about 45 ± 2.5 (41–49) % of the entire stylet length (Table 1). The stylet shaft is slender ending with rounded basal knobs, which are slightly anteriorly flattened. The pharyngeal procorpus narrows just anterior to the small, oval metacorpus. The valve of the median bulb is conspicuous. The isthmus is short, encircled by a nerve ring and widening to a pharyngeal lobe with a dorsal nucleus just posterior to the cardia and ventro-sublateral nuclei near the tip of the pharyngeal lobe (Figure 5B,C) and overlapping the intestine ventro-laterally for almost a two-body diameter at the cardia level. The secretory-excretory pore is level with the cardia just posterior to the hemizonid. The body annulation is distinct; the lateral

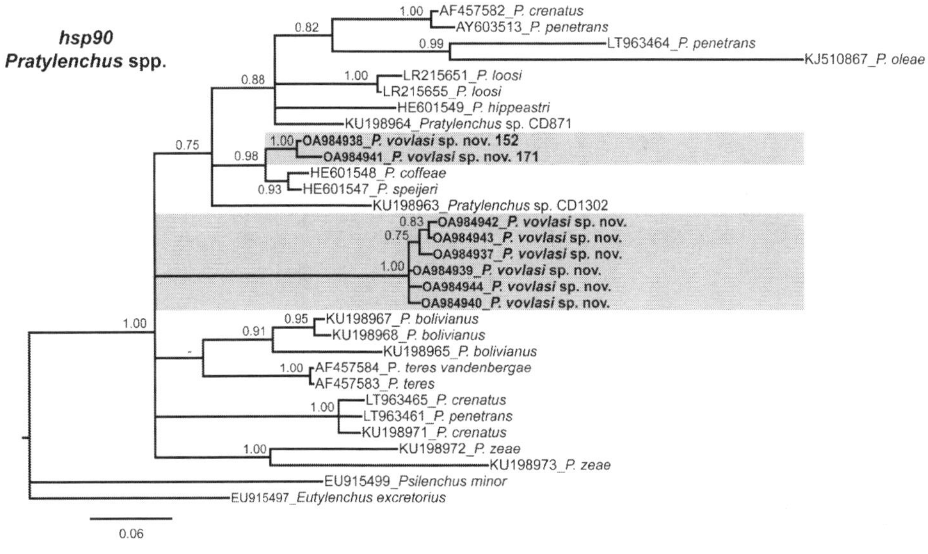

Figure 4. Phylogenetic relationships of *Pratylenchus vovlasi* sp. nov. within the genus *Pratylenchus*. A Bayesian 50% majority-rule consensus tree as inferred from the *hsp90* sequence alignment under the general time-reversible model of the sequence evolution with a correction for invariable sites and a gamma-shaped distribution (GTR + I + G). Posterior probabilities more than 70% are given for appropriate clades. Newly obtained sequences in this study are in bold letters. Scalebar=expectedchanges persite.

field usually with four smooth incisures and a few specimens have an additional line (Figure 6K)

or more rarely a few oblique striae (Figure 6J) in the middle of the central band. The outline of outer bands becomes indented towards the tail end, posterior to phasmid, with the inner lines fusing just posterior to the phasmid. The genital tract is well developed with oocytes arranged in a single row. The spermatheca is large and spherical to oval, usually full of sperm (Figure 5F,E,G); its posterior margin is 50.0 ± 9.6 (36.5–64.5) μm from the vagina (Table 1). The vulva is slightly sunken and the vulval lips are not prominent (Figure 5E,F). The post-uterine sac is about 1.1 vulval body diameter long and is usually undifferentiated. The phasmids are located to the mid-tail, 13.2 ± 2.2 (11–17) μm from the tail tip. The tail is typically subcylindrical, tapering towards the tip with a rounded-truncate terminus in most specimens; a few specimens have broadly rounded, conical tails with coarsely pointed or indented striated termini (Figures 5H,I and 6F,H,I,L).

Male: similar to the female except in the posterior end of the body and in a slightly smaller and usually slender body length. The lip region is usually higher and narrower than in the female (Figure 6B,C). The stylet is slightly smaller than that of female with smaller, more rounded knobs. The pharyngeal bulb is small and round; the isthmus is slender and rather short ending in a long, narrow glandular lobe. The testis is outstretched and filled with round spermatozoa in the vas deferens. The spicules are paired, weakly cephalated and ventrally arcuate. The gubernaculum is simple and slightly curved. The tail is conical and bent on the ventral side with a prominent, crenate bursa.

2.4.2 Type Host and Locality

Pratylenchus vovlasi sp. nov. was found associated with the roots and soil of *Rubus* sp. with a population density of 650 nematodes/250 cm^3 of soil in the locality of Prarostino, Turin province, Piedmont Region, north Italy.

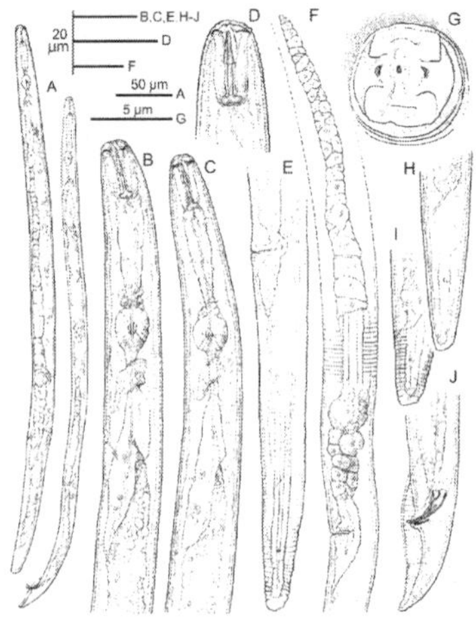

Figure 5. Line drawings of *Pratylenchus vovlasi* sp. nov. (A) entire female and male; (B–C) the female and the male (C) pharyngeal regions; (D) detail of the female lip region; (E) detail of the female posterior region; (F) the female reproductive system with details of the lateral field; (G) en face view; (H,I) the female tail; (J) the male tail.

2.4.3 Type Material

Holotype, female and male paratypes mounted on glass slides were deposited in the nematode collection at the Istituto per la Protezione Sostenibile delle Piante (IPSP), CNR, Bari, Italy (collection numbers IPSP-M-1238-1250). The additional paratypes were distributed to the United States Department of Agriculture Nematode Collection, Beltsville, MD, USA (collection number IPSP-M-1237), WaNeCo Plant Protection Service, Wageningen, The Netherlands, (collection number IPSP-M-1246) and Instituto de Agricultura Sostenible, CSIC, Córdoba, Spain (collection number IPSP-M-1245).

2.4.4 Diagnosis and Relationships

Pratylenchus vovlasi sp. nov. is characterized by a lip region slightly offset with three annuli narrowing towards the anterior end, sub-median sectors fused with the oral disc

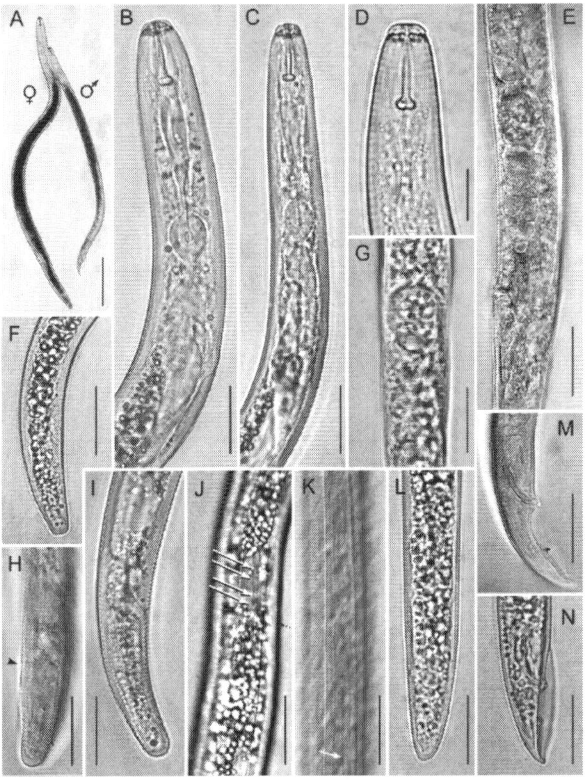

Figure 6. Light photomicrographs of *Pratylenchus vovlasi* sp. nov.: (A) entire female and male; the female (B) and the male (C) anterior regions; (D) the female lip region; (E,G) detail of the vulval region showing the spermatheca; (F,H,I,L) detail of the female tail (arrow in H shows the anus); (J,K), detail of the lateral field incisures showing the oblique striae (arrowed in J) or an additional line in the central band (arrowed in K); (M,N), detail of the male tail. Scale bars: A = 100 μm; B, C, E–M=20 μm; D = 10μm.

and separated by a lateral sector to give a dumb-bell -shaped pattern; the stylet is rather small (15.0 μm long) with rounded knobs and a short pharyngeal overlap. The lateral field has four smooth incisures in most specimens, the spermatheca is round to oval and is usually full of sperm, the vulva is located in a relative anterior position and the tail is subcylindrical with a truncate, smooth terminus and with common males. The matrix code of the new species according to Castillo and Vovlas [12] is A2 B2, C2, D2, E2, F3, G1,2, H1, I2, J1, K1 (Table 2).

Related species sharing with *P. vovlasi* sp. nov. a three lip annuli, a divided face as seen with SEM (Group 2 according to Corbett and Clark [11]), a functional spermatheca and numerous males include *P. bhattii* Siddiqi, Dabur and Bajaj, 1991, *P. kralli* Ryss, 1982, *P. mediterraneus* Corbett, 1983, *P. thornei* Sher and Allen, 1953, *P. pratensis*, *P. pseudopratensis* Seinhorst, 1968, *P. penetrans*, *P. fallax* and *P. convallariae* Seinhorst, 1959.

Table 1. Morphometrics of the holotype and paratypes of Pratylenchus vovlasi sp. nov. All measurements are in μm and in the form: mean ± S.D. (range).

		Paratypes	
Character	**Holotype**	**Females**	**Males**
n	-	16	9
L	463	510 ± 33.4 (459–577)	481 ± 32.1 (411–518)
a	21	23.8 ± 2.2 (20.0–26.5)	28.3 ± 1.8 (26.0–31.9)
b	5.8	6.5 ± 0.5 (5.8–7.3)	6.0 ± 0.3 (5.6–6.4)
b'	4.4	4.7 ± 0.4 (4.3–5.5)	4.5 ± 0.2 (4.1–4.9)
c	22.4	22.3 ± 2.7 (18.2–27.2)	17.9 ± 4.0 (9.6–22.0)
c'	1.7	1.8 ± 0.2 (1.4–2.3)	2.6 ± 0.7 (2.0–4.3)
Lip region height	2.0	2.3 ± 0.3 (2.0–2.8)	2.3 ± 0.2 (2.0–2.7)
Lip region diameter	8.5	8.1 ± 1.0 (6.6–9.0)	7.4 ± 0.6 (6.3–8.0)
Stylet length	14.5	15.0 ± 0.6 (14.3–16.3)	14.0 ± 0.6 (13.5–15.0)
Stylet cone	7.0	6.8 ± 0.6 (6.0–8.0)	6.5 ± 0.4 (6.5–7.5)
Stylet knob width	4.0	4.0 ± 0.6 (3.0–4.5)	2.7 ± 0.4 (2.5–3.5)
DGO from stylet base	2.0	2.3 ± 0.3 (2.0–2.7)	1.7 ± 0.4 (1.5–2.5)
o	13.8	15.2 ± 1.9 (13.6–18.4)	11.7 ± 2.8 (9.9–16.3)
Anterior End to:			
center of metacorpus	48.0	54 ± 2.6 (48–58)	52.0 ± 2.8 (47.5–55.5)
cardia	80.0	79 ± 5.4 (70–85)	80.0 ± 5.1 (73.5–90.0)
end of pharyngeal lobe	105.0	110 ± 9.0 (97–125)	108 ± 6.5 (99.5–120)
vulva	347	397 ± 31.8 (347–462)	-
Pharyngeal overlap	25	32 ± 6.7 (20–43)	28.0 ± 4.3 (22.5–36.5)
Max body diameter	22.0	22.0 ± 1.5 (20–24.5)	17.0 ± 1.0 (15.5–18.5)
Anal body diameter	12.0	13.0 ± 1.1 (11.5–15.0)	11.0 ± 1.0 (9.5–12.5)
Anterior genital tract length	224.0	250 ± 30.8 (203–313)	192 ± 23 (163–232)

Spermatheca to vagina distance	42.0	50.0 ± 9.6 (36.5–64.5)	-
Spermatheca length	17.0	17.0 ± 2.5 (12.5–21.0)	-
Spermatheca width	14.0	15.0 ± 1.9 (11.5–18.0)	-
Vulva to anus distance	96	90.0 ± 10.0 (78.5–110)	-
V or T	75	77.8 ± 2.0 (74–80)	40.1 ± 5.0 (32–50)
G1	48	49.0 ± 6.1 (39–59)	-
PUS	26	21.7 ± 2.4 (18.0–26.0)	-
Tail length	20.7	23.5 ± 3.3 (18.0–30.0)	26.5 ± 4.6 (19.5–31.5)
Number of tail annuli	16	16 ± 2.0 (14–20)	-
Spicule length	-	-	16.5 ± 1.4 (14.5–18.5)
Gubernaculum length	-	-	5.0 ± 0.3 (5.6–6.4)

Abbreviations: a = body length/greatest body diameter; b = body length/distance from the anterior end to the pharyngo-intestinal junction; DGO = distance between the stylet base and the orifice of the dorsal pharyngeal gland; c = body length/tail length; c' = tail length/tail diameter at the anus or cloaca; G1 = anterior genital branch length expressed as a percentage (%) of the body length; L = overall body length; n = number of specimens on which measurements are based; o = distance from the stylet base to the dorsal esophageal gland outlet × 100/total stylet length; T = distance from the cloacal aperture to the anterior end of the testis expressed as a percentage (%) of the body length; V = distance from the body anterior end to the vulva expressed as a percentage (%) of the body length; PUS = Post uterine sac length.

P. vovlasi sp. nov. is most closely related to P. mediterraneus, matching 10 out of 11 characteristics according to the matrix code [12] and differing from it just by a slightly shorter pharyngeal overlap (32 ± 6.7 (20–43) vs. 25–55 µm, code I2 vs. I3 in the tabular key), a face pattern showing transverse incisures in the middle of the sub-medial sectors (not present in P. mediterraneus), lateral sectors fused (vs. separated) with the oral disc, smooth lateral fields, not areolated vs. crenate (outer bands), occasionally areolated and with the middle band variously ornamented (Figure 6J,K and Figure 7K). From P. pratensis, P. fallax, P. penetrans and P. vulnus, it differs in the en face SEM pattern (belonging to Group 2 vs. Group 3, according to Corbett and Clark, [11]) and the tail tip morphology despite a certain degree of overlap (mostly truncate and smooth vs. usually oblique and annulated in P. pratensis, rounded or with a slightly irregular contour in P. fallax, generally rounded in P. penetrans and narrowly rounded to subacute and occasionally irregular in P. vulnus). Furthermore, the new species differs from P. pratensis by a mostly rounded vs. oval to rectangular spermatheca and the number of tail annuli (14–20 vs. 20–28); from P. fallax by a shorter stylet (range: 14.3–16.3 vs. 16–17 µm) and a smaller c' value 1.8 ± 0.2 (1.4–2.3) vs. 2.5 (2.0–3.0); from P. penetrans by a slightly shorter stylet mean length (15 vs. 16), a slightly more anterior position of the vulva (78 (74–80) vs. 78–84%) and in a fewer number of tail annuli on the ventral surface (15.6 ± 2.0 (14–20) vs. 15–27); from P. vulnus by a shorter post-uterine sac (1.1 vs. ca. 2.0 vulval body diameter; code F3 vs. F6 after Castillo and Vovlas, [12] (Table 2) and a more anterior position of the vulva (78 (74–80) vs. 77–82%). From P. pseudopratensis, it differs in the en face SEM view (not dumb-bell-shaped in *P. pseudopratensis*), a shorter stylet (range: 14.3–16.3 vs. 16–17 µm), the shape of the spermatheca (rectangular, sometimes empty in

P. pseudopratensis) and the tail tip (smooth vs. crenated; code H1 vs. H2 after Castillo and Vovlas, [12]).

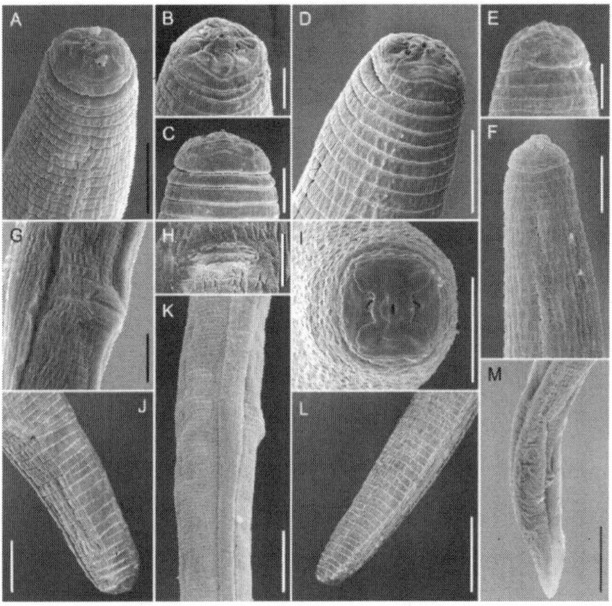

Figure 7. SEM photomicrographs of Pratylenchus vovlasi sp. nov.: (A–D), detail of the female lip region; (E,F), detail of the male anterior region; (I) the female en face view; (G,H) the vulval region in the sub-lateral (G) and the ventral (H) view; (J–L), detail of the female tail; (M) detail of the male tail. Scale bars: (A,D,F–J)=5 μm; (B,C,E) = 2.5 μm; (K–M)= 10 μm.

Other *Pratylenchus* species with three lip annuli, a functional spermatheca and the presence of males and with a matrix code (sensu Castillo and Vovlas, [12]) similar to *P. vovlasi* sp. nov. have been described without examining their lip patterns. These species include *P. convallariae*, *P. kralli* and *P. bhattii*, from which the new species differs by the following characteristics: from *P. convallariae*, by a shorter stylet (15 (14.3–16.3), feature C2 in Table 2, vs. 17 (16–18) μm feature C3), a shorter PUS (1.1, feature F3 vs. more than 1.4–2 vulval body diameter, feature F6 in Table 2) and by a smooth tail tip (H1) vs. coarsely and often irregularly annulated (H2); from *P. kralli*, in the stylet knob shape (mostly rounded vs. anteriorly directed), the post-uterine sac length (1.1 vs. more than 1.5 vulval body diameter), the tail tip is truncated, smooth and more rarely pointed vs. pointed and showing a slight groove; from *P. bhattii* by a slightly longer stylet (15 (14.3–16.3) vs. 13.5 (13–14) μm) and a more posterior vulva (V = 78 vs. 73% mean value) with the vulval lips continuous with the body contour vs. raised in a prominent protuberance. Finally, the new species can be compared with *P. rwandae* Singh, Nyiragatare, Janssen, Couvreur, Decraemer and Bert, 2018, but differs from it by a labial region in the en face view showing clearly separated sub-median sectors with transverse incisures in the middle (vs. slightly separated from the lateral sectors and without transverse incisures), males present (vs. absent), females with mostly round and full spermatheca (vs. oval to rounded and empty), lateral fields with four incisures (vs. six or more at the mid-body) and a tail with a fewer number of ventral annuli (14–20) and usually a truncated tip (vs. 18–28 tail annuli and a highly variable tail tip).

Table 2. Polytomous key of *Pratylenchus vovlasi* sp. nov. and the morphologically most closely related species.

Species	Morphological Characteristics *										
	A_{1-3}	B_{1-2}	C_{1-5}	D_{1-4}	E_{1-4}	F_{1-6}	G_{1-3}	H_{1-4}	I_{1-4}	J_{1-3}	K_{1-2}
	Lip Annuli	Presence of Males	Stylet Length (μm)	Shape of Spermath-eça	Vulva Position (%)	PUS ** (μm)	Female Tail Shape	Female Tail Tip	Pharyngeal Overlap (μm)	Lateral Field	Lateral Field Structures
***Pratylenchus vovlasi* sp. nov.**	A2	B2	C2	D2	E2	F3	G1,2	H1	I2	J1	K1
Pratylenchus bhattii	A2	B2	C2	D2	E1	F2	G2	H1	I2	J1	K1
Pratylenchus mediterraneus	A2	B2	C2	D2	E2	F3	G2	H1	I3	J1	K1
Pratylenchus kralli	A2	B2	C2	D2	E2	F1	G3	H1	I1	J1	K1
Pratylenchus fallax	A2	B2	C3	D2	E2	F3	G3	H2	I2	J1	K1
Pratylenchus convallariae	A2	B2	C3	D2	E2	F6	G2	H2	I3	J1	K1
Pratylenchus penetrans	A2	B2	C3	D2	E3	F4	G2	H1	I3	J1	K1
Pratylenchus pratensis	A2	B2	C2	D4	E2	F3	G3	H2	I1	J1	K1
Pratylenchus pseudopratensis	A2	B2	C3	D4	E2	F3	G3	H2	I3	J1	K1
Pratylenchus thornei	A2	B2	C2	D2	E2	F3	G2	H1	I3	J1	K1
Pratylenchus vulnus	A2	B2	C2	D3	E2	F6	G3	H3	I2	J1	K1

* Morphological characteristics according to Castillo and Vovlas [12]. **Group A:** 1 = two; 2 = three; 3 = four. **Group B:** 1 = absent; 2 = present. **Group C:** 1 = <13; 2 = 13–15.9; 3 = 16–17.9; 4 = 18–20; 5 = >20. **Group D:** 1 = absent or reduced; 2 = rounded to spherical; 3 = oval; 4 = rectangular. **Group E:** 1 = <75; 2 = 75–79.9; 3 = 80–85; 4 = >85. **Group F:** 1 = <16; 2 = 16–19.9; 3 = 20–24.9; 4 = 25–29.9; 5 = 30–35; 6 = >35. **Group G:** 1 = cylindrical; 2 = subcylindrical; 3 = conoid. **Group H:** 1 = smooth; 2 = striated; 3 = pointed; 4 = with ventral projection. **Group I:** 1 = <30; 2 = 30–39.9; 3 = 40–50; 4 = >50. **Group J:** 1 = four; 2 = five; 3 = six to eight. **Group K:** 1 = smooth bands; 2 = partially or completely areolated bands. ** PUS = post-uterine sac length.

2.4.5 Etymology

The species epithet, *P. vovlasi*, is dedicated to Dr. Nikos Vovlas, an eminent Italian nematologist and taxonomist from the Istituto per la Protezione Sostenibile delle Piante (IPSP), Consiglio Nazionale delle Ricerche (CNR), Bari, Italy.

3. Discussion

The identification of the *Pratylenchus* species is difficult because many diagnostic char- acteristics overlap and also due to the increasing number of nominal species. An accurate identification is needed in order to adopt appropriate control strategies. The primary objective of this study was to identify and molecularly characterize root-lesion nematodes parasitizing raspberries cultivated in the Piedmont region. The present study reports on the occurrence of two root-lesion nematodes associated with raspberry fields in the Piedmont region, *P. crenatus*, along with an abundant species herein identified as *P. vovlasi* sp. nov. *Pratylenchus crenatus* was previously reported on raspberries in several European countries [6,9].

Our results demonstrated that the application of rRNA molecular markers integrated with morphological studies could help in the diagnosis and characterization of root-lesion nematode species. Based on the molecular characterization using the D2-D3 expansion domains of the 28S rRNA gene, ITS region and the partial hsp90 gene, the abundant species was clearly identified as *P. vovlasi* sp. nov. This species proved very similar in morphometry and morphology to *P. mediterraneus*, differing only by a shorter pharyngeal overlap, as well as with *P. thornei*, *P. penetrans*, *P. fallax* and *P. convallariae*. By blasting at NCBI the D2-D3 region, it was 96–97% similar to *P. pratensis* and *P. pseudopratensis*, respectively, while by using ITS sequences it was 85–87% similar to *P. pratensis* and *P. fallax*. These results represented an additional confirmation of the extraordinary cryptic diversity of the nematodes of the genus *Pratylenchus*, as reported in previous studies [13–17]. Phylogenetic analyses of the ITS and the D2-D3 sequences confirmed a sister relationship between *P. vovlasi* sp. nov. with *P. pratensis* and *P. pseudopratensis*. Furthermore, in both phylogenetic trees, the new species was closely related to *P. vulnus* and *P. kumamotoensis* (Figures 2 and 3). Major clades for D2-D3 and ITS phylogenetic trees were highly correlated with previous phylogenetic studies carried out by Subbotin et al. [10], Palomares-Rius et al. [14,18] and Araya et al. [17]. We agreed with De Luca et al. [13] who suggested that the ITS-containing region allowed a better discrimination among the closely related species studied because it evolved faster than the D2-D3 expansion segments of 28S rDNA and accumulated more substitution changes.

The current study confirmed the occurrence of different hsp90 isoforms in the *Pratylenchus* species as already reported by Fanelli et al. [19]. In *P. vovlasi* sp. nov. the different isoforms differed from each other in the length of the intron and the nucleotide variability grouping in two well-supported clusters. This finding suggested that the different isoforms of *P. vovlasi* sp. nov. hsp90 arose by gene duplication events relatively recently because the nucleotide variability was low and the gene structure was still conserved [20,21]. Similarly, these results confirmed that the primers for the amplification of the hsp90 gene could also amplify other paralogous genes in Pratylenchus and could be used for species delimitation within this genus [22,23]. Furthermore, the occurrence of different *hsp90* isoforms in *P. vovlasi* sp. nov. confirmed that this gene family could contribute to the adaptation to different hosts and to different environments.

In this study, *P. vovlasi* sp. nov. was isolated from raspberries in north Italy together with *P. crenatus* in the same area. Further research on its pathogenicity and the economic damage on this crop is needed.

4. Materials and Methods

4.1 Nematode Isolate and Morphological Studies

No specific permits, other than that of the farm owner, were required for the indicated fieldwork studies. The soil samples were obtained in raspberry cultivated areas in Piedmont and did not involve any endangered species or those protected in Italy, nor were the sites protected in any way.

The soil samples recovered from the rhizosphere and roots of the raspberries located in the Piedmont region were sent to the IPSP laboratory, Bari, in the autumn of 2010. The samples were collected with a shovel from the upper 50 cm of the soil of raspberries arbitrarily chosen at random. The nematodes were extracted from 500 cm^3 of the soil by centrifugal flotation [24]. The specimens for light microscopy were killed by gentle heat, fixed in a solution of 4% formaldehyde + 1% propionic acid and processed to pure glycerin using Seinhorst's method [25]. The specimens were examined with a Zeiss III compound microscope with a Nomarski differential interference contrast at powers up to 1000. The measurements and drawings were made at the camera lucida on glycerin-infiltrated specimens. All measurements were expressed in micrometers (μm) unless otherwise stated.

For the scanning electron microscope (SEM), fixed specimens were dehydrated in a gradient series of ethanol, critical-point dried, sputter-coated with gold according to the protocol of Abolafia et al. [26] and observed with a Zeiss Merlin Scanning Electron Microscope (5 kV; Zeiss, Oberkochen, Germany).

4.2 DNA Extraction, Polymerase Chain Reaction (PCR) and Sequencing

DNA was extracted from 20 single individual root-lesion nematode specimens. The specimens were handpicked and placed singly on a glass slide in 3 μL of the lysis buffer (10 mM Tris-HCl, pH 8.8, 50 mM KCl, 15 mM $MgCl_2$, 0.1% Triton 100, 0.01% gelatin with 90 μg/mL proteinase K) and then cut into small pieces by using a sterilized syringe needle under a dissecting microscope. The samples were incubated at 65 °C for 1 h and then at 95 °C for 15 min to deactivate the proteinase K [27]. The following sets of primers were used for the amplification of the gene fragments in the present study: (i) D2-D3 expansion segments of the 28S rRNA gene using forward D2A and reverse D3B primers; (ii) ITS1- 5.8-ITS2 rRNA using forward TW81 and reverse AB28 primers [28,29]; (iii) the *hsp*90 gene using forward U831 and reverse L1110 primers [30]. New sequences were submitted to the GenBank database under the accession numbers indicated on the phylogenetic trees.

4.3 PCR-RFLP

Ten μL of each ITS-PCR product from three individual nematodes were digested with five units of the following restriction enzymes: *Alu*I, *Ava*II, *Bam*HI, *Dde*I, *Hin*fI and *Rsa*I. The digested products were separated onto a 2.5% agarose gel by electrophoresis, stained with gel red dye, visualized on a UV transilluminator and photographed with a digital system.

4.4 Phylogenetic Analysis

Sequenced genetic markers in the present study (after discarding primer sequences and ambiguously aligned regions) and several *Pratylenchus* spp. sequences obtained from GenBank were used for the phylogenetic reconstruction. The outgroup taxa for each dataset were selected based on previous published studies [17,18,23]. Multiple sequence alignments of the newly obtained and published sequences were made using the FFT-NS-2 algorithm of MAFFT v. 7.450 [31]. The sequence alignments were visualized using BioEdit [32] and edited by Gblocks ver. 0.91b [33] in the Castresana Laboratory server (http: //molevol.cmima.csic.es/castresana/Gblocks_server.html, accessed on 2 April 2021) using options for a less stringent selection (minimum number of sequences for a conserved or a flanking position: 50% of the number of sequences + 1; maximum number of contiguous no conserved positions: 8; minimum length of a block: 5; allowed gap positions: with half).

The phylogenetic analyses of the sequence datasets were based on Bayesian inference (BI) using MRBAYES 3.2.7a [34]. The best-fit model of DNA evolution was calculated with the Akaike information criterion (AIC) of JMODELTEST v. 2.1.7 [35]. The best-fit model, the base frequency, the proportion of invariable sites and the gamma distribution shape parameters and substitution rates in the AIC were then used in the phylogenetic analyses.

The BI analyses were performed under a general time-reversible model with a proportion of invariable sites and a rate of variation across sites (GTR+I+ G) model for D2-D3 and the partial hsp90 gene and under a transversional model with a proportion of invariable sites and a rate of variation across sites (TVM+I+ G) model for the ITS region. These BI analyses were run separately per dataset with four chains for 4 106 generations. The Markov chains were sampled at intervals of 100 generations. Two runs were conducted for each analysis. After discarding the burn-in samples of 30% and evaluating the convergence, the remaining samples were retained for more in-depth analyses. The topologies were used to generate a 50% majority-rule consensus tree. Posterior probabilities (PP) were given on appropriate clades. The trees from all analyses were visualized using FigTree software version v.1.42 [36].

References

1. FAOSTAT. 2009. Available online: http://www.fao.org/faostat/en/#data/QC (accessed on 2 April 2021).

2. Skrovankova, S.; Sumczynski, D.; Mlcek, J.; Jurikova, T.; Sochor, J. Bioactive compounds and antioxidant activity in different types of berries. *Int. J. Mol. Sci.* **2015**, *16*, 24673–24706. [CrossRef]

3. Tefera, T.; Tysnes, K.R.; Utaaker, K.S.; Robertson, L.J. Parasite contamination of berries: Risk, occurrence, and approaches for mitigation. *Food Waterborne Parasitol.* **2018**, *10*, 23–38. [CrossRef]

4. Gigot, J.; Walters, T.; Zasada, I.A. Impact and occurrence of *Phytophthora rubi* and *Pratylenchus penetrans* in commercial red raspberry (*Rubus ideaus*) fields in Northwestern Washington. *Int. J. Fruit Sci.* **2013**, *13*, 357–372. [CrossRef]

5. Poiras, L.; Cernet, A.; Buvol, A.; Poiras, N.; Iurcu-Straistraru, E. Preliminary analysis of plant parasitic nematodes associated with strawberry and raspberry crops in the Republic of Moldova. *Oltenia. Studii s, i comunicări. S, tiin, tele Naturii* **2014**, *30*, 98–104.

6. Tsolova, E.; Koleva, L. Ecological characteristics of plant parasitic nematodes in conventional and organic production of raspberries. *J. Mt. Agric. Balk.* **2015**, *18*, 727–739.

7. Zasada, I.A.; Weiland, J.E.; Han, Z.; Walters, T.W.; Moore, P. Impact of *Pratylenchus penetrans* on establishment of red raspberry. *Plant Dis.* **2015**, *99*, 939–946. [CrossRef] [PubMed]

8. Kroese, D.R.; Weiland, J.E.; Zasada, I.A. Distribution and longevity of *Pratylenchus penetrans* in the red raspberry production system. *J. Nematol.* **2016**, *48*, 241–247. [CrossRef]

9. Mohamedova, M.; Samaliev, H. Phytonematodes associated with red raspberry (*Rubus idaeus* L.) in Bulgaria. *J. Entomol. Zool. Stud.* **2018**, *6*, 123–127.

10. Subbotin, S.A.; Ragsdale, E.J.; Mullens, T.; Roberts, P.A.; Mundo-Ocampo, M.; Baldwin, J.G. A phylogenetic framework for root lesion nematodes of the genus *Pratylenchus* (Nematoda): Evidence from 18S and D2-D3 expansion segments of 28S ribosomal RNA genes and morphological characters. *Mol. Phylogenet. Evol.* **2008**, *48*, 491–505. [CrossRef]

11. Corbett, D.C.M.; Clark, S.A. Surface feature in the taxonomy of *Pratylenchus* species. *Rev. Nématol.* **1983**, *6*, 85–98.

12. Castillo, P.; Vovlas, N. *Pratylenchus (Nematoda: Pratylenchidae): Diagnosis, Biology, Pathogenicity and Management*; Brill Academic Publishers: Leiden, The Netherlands, 2007; p. 555.

13. De Luca, F.; Troccoli, A.; Duncan, L.W.; Subbotin, S.A.; Waeyenberge, L.; Moens, M.; Inserra, R.N. Characterisation of a population of *Pratylenchus hippeastri* from bromeliads and description of two related new species, *P. floridensis* n. sp. and *P. parafloridensis* n. sp. from grasses in Florida. *Nematology* **2010**, *12*, 847–868. [CrossRef]

14. Palomares-Rius, J.E.; Castillo, P.; Liébanas, G.; Vovlas, N.; Landa, B.B.; Navas-Cortes, J.A.; Subbotin, S.A. Description of *Pratylenchus hispaniensis* n. sp. from Spain and considerations on the phylogenetic relationship among selected genera in the family Pratylenchidae. *Nematology* **2010**, *12*, 429–451.

15. De Luca, F.; Troccoli, A.; Duncan, L.W.; Subbotin, S.A.; Waeyenberge, L.; Coyne, D.L.; Brentu, F.C.; Inserra, R.N. *Pratylenchus speijeri* n. sp. (Nematoda: Pratylenchidae), a new root-lesion nematode pest of plantain in West Africa. *Nematology* **2012**, *14*, 987–1004. [CrossRef]

16. Wang, H.; Zhuo, K.; Ye, W.; Liao, J. Morphological and molecular characterisation of *Pratylenchus parazeae* n. sp. (Nematoda: Pratylenchidae) parasitizing sugarcane in China. *Eur. J. Plant Pathol.* **2015**, *143*, 173–191. [CrossRef]

17. Araya, T.Z.; Padilla, W.P.; Archidona-Yuste, A.; Cantalapiedra-Navarrete, C.; Liébanas, G.; Palomares-Rius, J.E.; Castillo, P. Root-lesion nematodes of the genus *Pratylenchus* (Nematoda: Pratylenchidae) from Costa Rica with molecular identification of *P. gutierrezi* and *P. panamaensis* topotypes. *Eur. J. Plant Pathol.* **2016**, *145*, 973–998. [CrossRef]

18. Palomares-Rius, J.E.; Guesmi, I.; Horrigue-Raouani, N.; Cantalapiedra-Navarrete, C.; Liébanas, G.; Castillo, P. Morphological and molecular characterisation of *Pratylenchus oleae* n. sp. (Nematoda: Pratylenchidae) parasitizing wild and cultivated olives in Spain and Tunisia. *Eur. J. Plant Pathol.* **2014**, *140*, 53–67. [CrossRef]

19. Fanelli, E.; Troccoli, A.; Capriglia, F.; Lucarelli, G.; Vovlas, N.; Greco, N.; De Luca, F. Sequence variation in ribosomal DNA and in the nuclear hsp90 gene of *Pratylenchus penetrans* (Nematoda: Pratylenchidae) populations and phylogenetic analysis. *Eur. J. Plant Pathol.* **2018**, *152*, 355–365. [CrossRef]

20. Chen, B.; Piel, W.H.; Gui, L.; Bruford, E.; Monteiro, A. The HSP90 family of genes in the human genome: Insights into their divergence and evolution. *Genomics* **2005**, *86*, 627–637. [CrossRef] [PubMed]

21. Fanelli, E.; Troccoli, A.; Tarasco, E.; De Luca, F. Molecular characterization and functional analysis of the *Hb-hsp*90-1 gene in relation to temperature changes in *Heterorhabditis bacteriophora. Front. Physiol.* **2021**, *12*, 12. [CrossRef] [PubMed]

22. Skantar, A.M.; Carta, L.K. Molecular characterization and phylogenetic evaluation of the *Hsp90* gene from selected nematodes. *J. Nematol.* **2004**, *36*, 466–480. [PubMed]

23. Troccoli, A.; Subbotin, S.A.; Chitambar, J.J.; Janssen, T.; Waeyenberge, L.; Stanley, J.D.; Duncan, L.W.; Agudelo, P.; Múnera Uribe, G.; Franco, J.; et al. Characterisation of amphimictic and parthenogenetic populations of *Pratylenchus bolivianus* Corbett, 1983 (Nematoda: Pratylenchidae) and their phylogenetic relationships with closely related species. *Nematology* **2016**, *18*, 651–678. [CrossRef]

24. Coolen, W.A. Methods for extraction of *Meloidogyne* spp. and other nematodes from roots and soil. In *Root-Knot Nematodes (Meloidogyne Species). Systematics, Biology and Control*; Lamberti, F., Taylor, C.E., Eds.; Academic Press: New York, NY, USA, 1979; pp. 317–329.

25. Seinhorst, J.W. Killing nematodes for taxonomic study with hot f.a. 4:1. *Nematologica* **1966**, *12*, 175. [CrossRef]

26. Abolafia, J.; Liébanas, G.; Peña-Santiago, R. Nematodes of the order Rhabditida from Andalucia Oriental, Spain. The subgenus *Pseudacrobeles* Steiner, 1938, with description of a new species. *J. Nematode Morphol. Syst.* **2002**, *4*, 137–154.

27. De Luca, F.; Fanelli, E.; Di Vito, M.; Reyes, A.; De Giorgi, C. Comparison of the sequences of the D3 expansion of the 26S ribosomal genes reveals different degrees of heterogeneity in different populations and species of *Pratylenchus* from the Mediterranean region. *Eur. J. Plant Pathol.* **2004**, *111*, 949–957. [CrossRef]

28. Joyce, S.A.; Burnell, A.M.; Powers, T.O. Characterization of *Heterorhabditis* isolates by PCR amplification of segments of mtDNA and rDNA genes. *J. Nematol.* **1994**, *26*, 260–270. [PubMed]

29. Joyce, S.A.; Reid, A.; Driver, F.; Curran, J. Application of polymerase chain reaction (PCR) methods to the identification of entomopathogenic nematodes. In *Genetics of Entomopathogenic Nematodes-Bacterium Complexes, Proceedings of Symposium and Workshop*; Burnell, A.M., Ehlers, R.U., Masson, J.P., Eds.; S. Patrick's College, Maynooth, Co.: Kildrare, Ireland; EC DG XII: Luxembourg, 1994; pp. 178–187.

30. Skantar, A.M.; Carta, L.K. Multiple Displacement Amplification (MDA) of total genomic DNA from *Meloidogyne* spp. and comparison to crude DNA extracts in PCR of ITS1, 28 S D2-D3 rDNA and Hsp90. *Nematology* **2005**, *7*, 285–293. [CrossRef]

31. Katoh, K.; Rozewicki, J.; Yamada, K.D. MAFFT online service: Multiple sequence alignment, interactive sequence choice and visualization. *Brief. Bioinform.* **2019**, *20*, 1160–1166. [CrossRef]

32. Hall, T.A. BioEdit: A user-friendly biological sequence alignment editor and analysis program for Windows 95/98/NT. *Nucleic Acids Symp. Ser.* **1999**, *41*, 95–98.

33. Castresana, J. Selection of conserved blocks from multiple alignments for their use in phylogenetic analysis. *Mol. Biol. Evol.* **2000**, *17*, 540–552. [CrossRef]

34. Ronquist, F.; Huelsenbeck, J.P. MrBayes 3: Bayesian phylogenetic inference under mixed models. *Bioinformatics* **2003**, *19*, 1572–1574. [CrossRef]

35. Darriba, D.; Taboada, G.L.; Doallo, R.; Posada, D. jModelTest 2: More models, new heuristics and parallel computing. Nat. Methods 2012, 9, 772. [CrossRef] [PubMed]

Chapter-4

The Genus Pratylenchus (Nematoda: Pratylenchidae) in Israel: From Taxonomy to Control Practices

1. Introduction

Root-lesion nematodes of the genus *Pratylenchus* are migratory endoparasites belonging to the family Pratylenchidae (Nematoda, Tylenchina), with around 100 species recognized today [1–3]. *Pratylenchus* species can cause yield losses of up to 85% of expected production [4], and even higher losses when nematodes interact synergistically with certain soilborne plant pathogens [5]. Hence, *Pratylenchus* species are highly relevant to agriculture.

Israel is a small semiarid country located in western Asia, only 22,000 km^2 in size. Despite the fact that the geography of the country is not naturally conducive to agriculture, advanced irrigation, cultivation mastery, use of elite varieties, and the introduction of state-of-the-art agricultural technologies contribute, in practice, to intensive and efficient farming. On the other hand, this success in agricultural productivity along with a diversity of climatic conditions have led to the proliferation of devastating plant-parasitic nematodes. Among them, *Pratylenchus* species are widely distributed in vegetable and crop fields in Israel and are associated with a major reduction in quality and yield. The genus *Pratylenchus* was first reported from Israel in 1957 [6]. Since then, several studies related to this nematode have been published [7–14]. However, these studies are largely scattered. Some of them are published in less accessible local journals, such as master's or PhD theses, or in scientific reports written in Hebrew. In this review, we collected all available information on *Pratylenchus* in Israel, spanning the last few decades, from local Hebrew journals to international peer-reviewed ones, revealing that *Pratylenchus* species are major pests in many crops throughout the country. We provide a comprehensive summary of the occurrence, taxonomy, distribution, diagnosis, and pathogenicity of *Pratylenchus* in Israel, along with an overview of the status and perspectives for *Pratylenchus* research in this country.

2. Overview of Israeli Agriculture

Agriculture is an important sector for the Israeli economy, representing around 2.5% of Israel's GDP and about 3.5% of its exports. Agricultural production is especially significant in certain areas, such as the Arava, Jordan Rift Valley, and northern Negev, where it provides almost the sole means of livelihood for the population. Although some of these regions are characterized by semiarid land with varied climatic [15,16], topographical, and soil conditions, determination and farming ingenuity have produced maximum yields and crop quality [17]. Among the most common agricultural sectors, vegetable growing has become a specialized skill of Israeli farmers, on the basis of selecting suitable hybrid varieties, fertilizers, irrigation methods, greenhouse covers designed for specific crops, innovative growing tools, and plant protection management. Moreover, vegetable growing exploits the sunshine and high temperatures, providing high-quality vegetables during the competitors' off seasons in other countries. As a result, vegetables account for about 17% of Israel's total crop output value. About two-thirds of Israel's field crops are grown on non-irrigated land. These rain-fed crops include wheat for grains, silage and hay, legumes for seeds, and sunflower for oil. The remaining field crops are summer crops, including cotton, chickpeas, green peas, beans, corn, groundnuts, and watermelon for seeds, most of which are irrigated. Fruit trees mainly include deciduous fruit orchards that are among the main crops in northern Israel, including grapevine, fig, almond, apple, pear, stone fruit, pomegranate, and persimmon, as well as subtropical varieties (citrus, avocado, mango, olive, litchi) and small fruit (various berries).

3. Occurrence of *Pratylenchus* Species in Israel

The most comprehensive survey of Israel's soil nematodes was performed by Cohn et al. [7], wherein 320 soil samples were taken from natural agro-ecosystems, providing a backbone for soil nematode diversity and distribution in Israel (Table 1, Cohn et al. [7]). This survey suggested that in cultivated crops grown throughout Israel, *Pratylenchus* species were among the three most prevalent plant-parasitic nematodes infecting vegetables (49% of the samples, see below), cereal and pasture grasses (68%), pasture legumes (48%), and deciduous fruit trees (47%). Less commonly, *Pratylenchus* species were found in natural vegetation fruit trees (35%) and forest trees (30%), and in cultivated crops of subtropical and tropical fruit trees (20%), grapevines (29%) and lawns (27%) (Table 1). Geographically, *Pratylenchus* was most prevalent in the Negev, located in southern region of the country (54%), while its abundance in the rest of the geographical locations rangedbetween33and49%[7].

Table 1. Percentageoccurrenceofplant-parasiticnematodegenerainther hizosphere ofplantgroups (generaoccurringin20%ormoreofsamples)coveringnaturalvegetation andcultivatedcropsinIsrael[7].

Natural Vegetation					
Fruit Trees (n = 34)		**Forest Trees (n = 20)**		**Herbaceous Plants (n =**	**12)**
Tylenchorhynchus	65	*Tylenchus*	70	*Helicotylenchus*	58
Xiphinema	62	*Xiphinema*	65	*Tylenchus*	58
Helicotylenchus	53	*Helicotylenchus*	60	*Meloidogyne*	50
Tylenchus	50	*Tylenchorhynchus*	40	*Tylenchorhynchus*	42
Pratylenchus	35	*Pratylenchus*	30	*Xiphinema*	33
Meloidogyne	29	*Rotylenchulus*	20		
Rotylenchulus	21				
Cultivated crops					

Deciduous Fruit Trees (n = 38)		**Subtropical and Tropical Fruit Trees (n = 20)**		**Grapevines (n = 17)**	
Xiphinema	76	Xiphinema	60	Helicotylenchus	65
Tylenchorhynchus	58	Tylenchorhynchus	55	Xiphinema	59
Pratylenchus	47	Helicotylenchus	45	Tylenchorhynchus	53
Helicotylenchus	38	Tylenchus	25	Longidorus	47
		Pratylenchus	20	Meloidogyne	47
		Criconemoides	20	Pratylenchus	29
Vegetable crops (n = 41)		**Cereal crops and pasture grasses (n = 50)**		**Pasture Legumes (n = 27)**	
Tylenchorhynchus	59	Tylenchorhynchus	84	Tylenchorhynchus	85
Pratylenchus	49	Pratylenchus	68	Tylenchus	52
Tylenchus	39	Tylenchus	58	Pratylenchus	48
Helicotylenchus	27	Helicotylenchus	26	Helicotylenchus	30
Longidorus	22	Xiphinema	24	Ditylenchus	20
		Lawns (n = 11)			
Helicotylenchus	82	Criconemoides	36	Pratylenchus	27
Tylenchorhynchus	64	Xiphinema	36		
Tylenchus	55	Trichodorus	27		

4. Taxonomy and Diversity of Pratylenchus Species in Israel

The first species of *Pratylenchus* were identified by Minz in 1957 [6]: *Pratylenchus brachyurus* (Godfrey, 1929), *Pratylenchus neglectus* (=*P. minyus*) (Rensch, 1924), *Pratylenchus penetrans* (Cobb, 1917), and *Pratylenchus scribneri* (Steiner in Sherbakoff & Stanley, 1943). Later, Cohn et al. [7] added three more species: *Pratylenchus pratensis* (de Man, 1880), *Pratylenchus thornei* (Sher and Allen, 1953), and *Pratylenchus vulnus* (Allen & Jensen, 1951), and Corbett (1983) described a new species, *Pratylenchus mediterraneus*. Most recently, Qing et al. [18] added another new species, *Pratylenchus capsici*.

To date, nine species of *Pratylenchus* have been reported throughout the country (Figure 1). Most of these species have only been identified by morphological characteristics, but three of them have been recently confirmed by molecular data. Their distribution and associated plant hosts are detailed below.

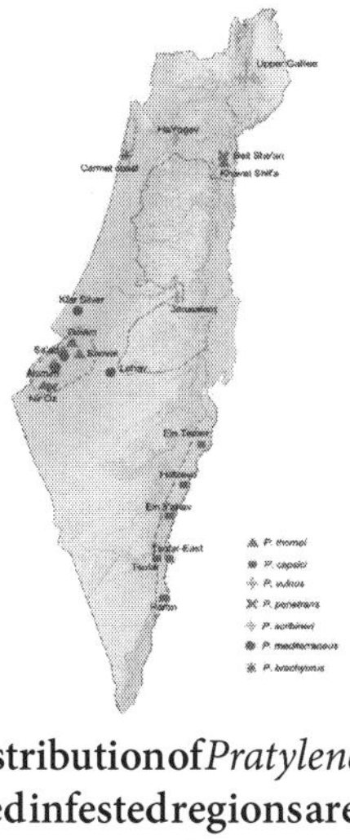

Figure 1. **Map of the known distribution of *Pratylenchus* species recorded in Israel's farming regions. Only recorded infested regions are indicated for each *Pratylenchus* species.**

4.3 Pratylenchus mediterraneus Corbett, 1983

Orion et al. [19] and Krikun and Orion [9] observed an unusual population of *P. thornei* parasitizing potatoes in the northern Negev. After a detailed morphological and morphometric study, Corbett and Clark [20] designated this population as a new species, *P. mediterraneus*. Although the validity of

P. mediterraneus designation was questioned [21], it is generally accepted as a valid species [1,2,22,23], being further supported by a variety of molecular evidence, such as restriction fragment length polymorphism (RFLP) analysis of ribosomal (r)DNA fragments [24–26], sequences of rDNA D3 expansions [27], and sequences of 18S and 28S rDNA [18]. Morphologically, *P. mediterraneus* is closely related to *P. thornei* in labial region shape en face pattern, and only differs in having a shorter stylet, sexual reproduction, and males being common [28]. Therefore, the identities of several *P. thornei* populations reported from various Middle Eastern countries [29,30] are suspected to be *P. mediterraneus*. The matrix code for *P. mediterraneus* is A2, B2, C2, D2, E2, F3, G2, H1, I3, J1, K1 ([23]; Supplementary Material Table S1; Supplementary Material File S2).

Pratylenchus mediterraneus was originally found in the northern Negev region of Israel [14,19]. Later, this species was recorded on chickpea in Turkey [31,32]; chickpea and lentil in Syria [33,34]; legumes in Algeria, Tunisia, and Morocco [33,35]; and chrysanthemum in Korea [36]. In Israel, *P. mediterraneus* primarily parasitizes legumes and cereals, which are the prevalent crops in the northern Negev, but carrot and potato can also be hosts [37]. Hosts reported by the Plant Protection and Inspection Services (PPIS) of the Israeli Ministry of Agriculture and Rural Development currently include alfalfa, barley, beans, broad beans, cabbage, carrot, chickpea, clover, coriander, lovage, sweet potato, vetch, and wheat [38].

4.2 Pratylenchus thornei Sher and Allen, 1953

In Israel, *P. thornei* has been reported on potato [13], cereals such as wheat and barley [8,10,19], carrots [37], legumes such as *Vicia sativa*, alfalfa and trifolium [39], watermelon [19], and cabbage [38], all in the northern Negev. However, most of these are likely to be *P. mediterraneus* [13,14,37]. Given the similarity between these two species, and the fact that *P. thornei* can occur in a wide range of soil types and is commonly found in mixed populations [39,40], the existence of *P. thornei* as part of a mixed population alongside *P. mediterraneus* in these studies is suspected. Notably, 28S rDNA-based molecular identification in recent samplings (2018 and 2019) has suggested the wide distribution of *P. thornei* in barley fields in Gevim, Alumim, and Nir-Oz located in the northern Negev (Table 2, Figure 1), and wheat fields in the Khavat Shif'a region and Avuka (Bet Shean Valley) located in the northofIsrael[41].

Table 2. Recentidentificationofplant-parasiticnematodesinseveralcultivatedcropsin Israel,recorded during2018–2019,accordingtointernaltranscribedspacer(ITS).

Cultivated Crops					
Grapevines					
Helicotylenchus	Tomer				
pseudorobustus					

Xiphinema index	Tomer				
Aphelenchoides sp.	Tomer				
Vegetable Crops		**Cereal Crops and Pasture Grasses**		**Pasture Legumes**	
Pratylenchus mediterraneus	Kfar Silver	Pratylenchus thornei	Gevim	Pratylenchus	Shif'a Gevim,
Pratylenchus nanus	Shoval	Merlinius nanus	Sde Eliyahu	Merlinius nanus	Shif'a
Pratylenchus thornei	Shoval, Alumim	Heterodera avenae	Nirim	Tylenchorhynchus *clarus*	Nir David
Neodolichorhynchus sulcatus	Arava	Heterodera sp.	Nirim	Rotylenchus macrosoma	Shif'a
Pratylenchus capsici	Arava	Geocenamus brevidens	Nirim	Tylenchorhynchus zeae	Shif'a
Meloidogyne javanica	Mivtachim				

This confirms the presence of *P. thornei* but fails to support the co-existence of *P. mediterraneus* and *P. thornei* within the same field populations. Diagnostic parameters described a labial region with three annuli, not offset from the body, an outer margin of sclerotized labial framework extending conspicuously around two annuli into the body, and one annulus into the labial region; lateral fields with four lines—the outer ones straight or weakly crenate; medium-length stylet (17–19 μm), a spermatheca that is difficult to see and does not contain spermatozoa; and males being very rare. The matrix code for *P. thornei* is A2, B2, C3, D1, E2, F2, G3, H1, I3, J1, K1. According to Castillo and Vovlas [23], it can be distinguished from the closely related species *P. penetrans* and *P. mediterraneus* by labial region shape, stylet length, the low proportion of males, and spermatheca and tail shapes.

4.3 Pratylenchus neglectus (Rensch, 1924) Filipjev and S. Stekhoven, 1941

In Israel, *P. neglectus* was first recorded by Minz [6] in association with fig tree roots. It is also known as the California meadow nematode, and has been reported by the Israeli Society of Plant Pathology (ISPP) on cotton crops and fig trees [38]. *Pratylenchus neglectus* is characterized by a labial region with two annuli, the second annulus wider than the first, anteriorly indented stylet knobs, a post-vulval uterine sac that is less than or equal to the body diameter, a variably shaped tail that is usually conoid with a little curvature of the ventral surface, and a tail terminus without annulation that is usually rounded but may be obliquely truncate or slightly digitate [23]. The matrix code for *P. neglectus* is A1, B2, C3, D1, E2, F1, G3, H1, I1, J1, K1.

4.4 Pratylenchus vulnus Allen and Jensen, 1951

Pratylenchus vulnus was first recorded by Cohn et al. [7] It is reported to be the most frequently encountered nematode associated with several pome and stone fruit trees, e.g., cherry, pear, plum, olive, apricot, nectarine, mango, persimmon, almond, citrus, fig, peach, and avocado, as well as some ornamentals including roses [38]. It is frequently found in rose nurseries, as well as in loquat, stone fruit, and apple trees in the north of Israel, very often in dense populations [42].

Pratylenchus vulnus is characterized by a labial region that is almost continuous with the body contour, with three or four annuli, a pharynx overlapping the intestine ventrally in a long lobe, an oblong spermatheca, a post-vulval uterine sac that is around two vulval body diameters long with a rudimentary ovary, and a tapering tail with a narrowly rounded subacute smooth tip; males are common. The matrix code for *P. vulnus* is A2, B2, C2, D3, E2, F6, G3, H3, I2, J1, K1.

4.5 Pratylenchus pratensis (de Man, 1880) Filipjev, 1936

Pratylenchus pratensis was first recorded by Cohn et al. [7], being described as *Anguillulina pratensis*. *Pratylenchus pratensis* has been found on Chinese cabbage, turnip, cauliflower, kohlrabi, white cabbage, radish, and cabbage by the ISPP [38]. This nematode species is characterized by a finely annulated cuticle, a labial region with three annuli, an oval to rectangular spermatheca, a post-vulval uterine sac length similar to the body diameter, and a tail with 20–28 annuli that are annulated to the terminus [23]. The matrix code for *P. pratensis* is A2, B2, C2, D4, E2, F3, G3, H2, I1, J1, K1. This species can be differentiated from closely related species by stylet length, the position of the vulva, shape of the spermatheca, shape of the tail, tail annuli, tail tip, and the presence of males.

4.6 Pratylenchus capsici Qing, Bert, Gamliel, Bucki, Duvrinin, Alon, Braun-Miyara, 2019

Pratylenchus capsici is an endemic Israeli species that has been recently identified from the roots of pepper [18], currently its only known host, with substantial damage observed. With the type population recovered from Tsofar farm, this species is widely spread across the pepper-growing region in the Arava Rift Valley of Israel. *Pratylenchus capsici* has been shown to be a cryptic species of *Pratylenchus oleae*, as they are almost indistinguishable morphologically. In fact, in the tabular key for *Pratylenchus* species identification proposed by Castillo and Vovlas [23], 10 out of 11 traits were identical for the two species. However, *P. capsici* differs from *P. oleae* in several molecular markers, as well as by several minor morphological differences, including the presence of males in the former, a functional spermatheca (vs. nonfunctional and empty in the latter), a larger body (559–642 for *P. capsici* vs. 412–511 μm for *P. oleae*), and a shorter stylet (14–15 vs. 15–17 μm, respectively) [18]. The matrix code for *P. capsici* is A2, B2, C2, D2, E1–3, F4–5, G2–3, H2, I1–2, J1, K2.

4.7 Pratylenchus penetrans (Cobb, 1917) Filipjev and Stekhoven, 1941

Pratylenchus penetrans was first recorded by Minz [6] in soil from a banana plantation. The ISPP has reported this species on lily, olive, nectarine, buttercup, apple, ruscus, strawberry, and peach [38]. It is also associated with grasses, cereals, and potatoes [42]. It was associated with the decline in pepper plants in the last decade in the Arava in a study carried out from 2004–2007, aimed at elucidating the causal agent of pepper collapse in that region [43]. Later on, *P. penetrans* continued to be identified in other studies as well [44]. Notably, during our intensive sampling of the Arava Rift Valley, *P. capsici* was the only root-lesion nematode associated with pepper. Given that *P. capsici* is morphologically similarto *P. penetrans* and species identification in these studies relied solely on morphology, here we consider that *P. penetrans* reported from the Arava might be *P. capsici*. Further morphology and molecular analyses are needed to confirm the distribution and host range of the former species. *Pratylenchus penetrans* is characterized by a labial region that is slightly offset, low, and flat in front with rounded outer margins, with three annuli; a pharynx overlapping the intestine ventrally; a lobe of around 1.5 body diameters in length; a short, undifferentiated post-vulval uterine sac, and a tail that is generally rounded with a smooth tip. The matrix code for *P. penetrans* is A2, B2, C3, D2, E3, F4, G2, H1, I3, J1, K1. It can be distinguished from closely related species by body and stylet length, number of lip annuli, labial framework, position of the vulva, and shape of the spermatheca and tail terminus [23].

4.8 Pratylenchus scribneri Steiner in Sherbakoff and Stanley, 1943

Pratylenchus scribneri was first recorded by Minz [6] in soils of banana, fig, plum, and quince trees. It has also been found on strawberry by the ISPP [38]. According to ISPP nematologists, *Pratylenchus* occurrence in banana plantations throughout Israel is very sparse[42].

This species is characterized by a labial region with two annuli that is slightly offset from the body, a stout stylet with rounded knobs, a pharyngeal overlap of medium length, an oblong spermatheca, and a slightly tapering tail with a smooth terminus. The matrix code for *P. scribneri* is A1, B2, C2, D3, E2,F4,G3,H1,I2,J1-3,K1.

4.9 Pratylenchus brachyurus (Godfrey, 1929) Filipjev and Stekhoven, 1941

Pratylenchus brachyurus was first recorded by Minz [6] and was found associated with other nematodes in soil from Cavendish banana. This species has also been reported on citrus[38].

This species is characterized by a labial region with two annuli, the anterior one showing an angular contour; a stylet with stout, rounded basal knobs; a vulva that is 82–89% of the body length; a post-vulval uterine sac that is less than one body diameter long; an inconspicuous nonfunctional spermatheca; and a tail that is broadly conoid, smooth, and broadly rounded, and truncate or spatulate at the tip. Males are rare. The matrix code for *P. brachyurus* is A1,B2,C4,D1,E4,F3,G3,H1,I4,J2–3,K1.

5. Biology and Pathogenicity of *Pratylenchus* Species

Pratylenchus species are polyphagous, migratory root endoparasites, developing and reproducing in the soil or roots. Their life cycle is simple and direct. The female lays its eggs singly or in small groups in the host root or in the soil near the root surface. Although little information is available about the true length of the *Pratylenchus* life cycle, on the basis of laboratory observations, research has estimated it to last from 45 to 65 days [45]. Symptoms caused by *Pratylenchus* are variable and depend on the host; they can include stunted and inefficient plant growth with reduced numbers of tillers and yellowed leaves.

Pathogenicity studies indicate that *Pratylenchus* species are very well adapted to parasitism, as extremely high populations in the soil do not kill their host plants. Nevertheless, damage thresholds range from 0.05 to 30 nematodes/cm^3 of soil. Apart from direct damage to the roots, *Pratylenchus* species may also predispose plants to other pathogens (e.g.,

Verticillium and *Fusarium*), thereby increasing the damage extent [46,47]. Consequently, elimination of the nematode or reduction of its population causes a marked reduction in the incidence of fungi and an increase in crop yield. In Israel, the synergistic relationship between *P. thornei* and the fungus *Verticillium dahliae* caused a significant increase in the populations of both pathogens and in their damage to potato crops in the northern Negev [48].

Among the nine species recorded in Israel, *P. mediterraneus*, *P. thornei*, and *P. capsici* have been relatively more studied, and their biology and pathogenicity are discussed below.

Pratylenchus mediterraneus parasitism occurs mainly in the winter, but the nematode can survive for 7–8 months in a state of anhydrobiosis during the hot and dry season [8,49]. It is reactivated by the subsequent winter rains. In a field observation conducted by Orion et al. [10] from 1974 to 1983, the highest population of *P. mediterraneus* (as *P. thornei* in their paper) was recorded in the drought of 1978 and partial drought of 1982, and the lowest population in the unusually wet years of 1980 and 1983. Moreover, nematode populations with auxiliary irrigation treatments were extremely low. These data suggest that low moisture level—the natural condition in the northern Negev region—is a major ecological factor required for *P. mediterraneus* to build up its population, supporting the notion that *P. mediterraneus* is native to the semiarid zones of the Middle East [8,19] or, more specifically, the eastern Mediterranean region [50]. During the long hot season (April–November), the nematode population level remains stable due to anhydrobiosis [8]. In this state, the nematode can withstand conditions of 0% relative humidity, and desiccated nematodes can withstand temperatures of up to 40°C. This characteristic enables their survival and facilitates their field or regional transmission in the northern Negev, where soil temperatures typically reach 40 °C in the hot season. This species is also likely to require the higher temperatures found in the Mediterranean region for its development, but this needs to be further studied.

In contrast to *P. mediterraneus*, the optimal temperature for *P. thornei* reproduction seems to be lower, ranging between 20 and 25° C [51,52], suggesting that the northern Negev may not be a suitable area for its survival. However, our molecular- and morphological-based analyses suggested that *P. thornei* is present not only in the mild northern Israel (Mesilot, Avuka, Shif'a), but also in the hot and dry region of the northern Negev [41]. In comparison, *P. mediterraneus* was only recovered from the northern Negev, suggesting that *P. thornei* may be able to adapt to a wider range of environmental conditions than *P. mediterraneus*, with the latter being more specialized for the hot and dry northern Negev.

The pathogenic effect of *P. mediterraneus* is limited to the early plant stages, resulting in reduced foliage and root growth of cereals and legumes, and thus influencing final plant density at harvest [12,14]. *Pratylenchus mediterraneus* was shown to be most concentrated in the root-tip region of hosts *Vicia sativa* and *Trifolium alexandrinum*. A histopathological study using scanning electron microscopy (SEM) showed nematodes penetrating the root epidermis and the cortical parenchyma through a clean-cut hole, probably a result of enzymatic activity and mechanical force [53]. When passing through parenchyma cells, *P. mediterraneus* can consume the cell contents, and these cells are thus void of cytoplasmic structures compared to the prominent nuclei and cytoplasmic structures in adjacent intact cells [12]. Typical symptoms caused by *P. mediterraneus* on common vetch were lesions produced along roots. These lesions lacked root hairs, with necrotic epidermal cells consisting of many holes, leading to severely deformed roots. Similar to *P. penetrans* [54], Orion and co-workers [12,37] speculated that *P. mediterraneus* can infect root tips as ectoparasites as well. Further SEM analysis showed the collapse of the parenchyma cells in the root lesion as the result of nematode feeding activity. The observed destruction was limited to the root cortex with an intact central cylinder, while nematode egg deposition was observed in cavities of the root cortex. These findings were similar to observations of *P. vulnus* in sour orange [55], *P. penetrans* in broad beans [56], and *Pratylenchus zeae* in maize [57].

Pratylenchus penetrans and *P. crenatus* Loof, 1960 have been reported worldwide as the major causal agent of carrots and Kuroda-type carrots [58–61]. In an investigation of carrot nematodes in Shoval, located in the northern Negev, we failed to detect these species. Instead, the field was infested with *P. thornei*, resulting in significant quality loss due to forking of carrot taproots [41]. However, whether *P. thornei* is the causal agent of these symptoms still needs to be confirmed, as continuous sampling from carrots demonstrated that the forking symptoms were not necessarily related to nematode occurrence [41].

Pratylenchus capsici is an endemic Israeli species that is widely distributed in the Arava Rift Valley, causing significant yield reduction of pepper (Figure 2).

Figure 2. Symptoms caused by *Pratylenchus capsici*. (A) Pepper plant decline in the Arava Rift Valley characterized by stunted growth and wilting. (B) Heavily infected roots, with pronounced lesions along primary and secondary roots. (C) Photograph of developed root lesion taken under a dissecting microscope.

The emergence of this species was surprising, as this remote region is isolated from the country's other agricultural areas. Moreover, until 1995, the entire region was free of reported nematodes, mainly due to intensive soil fumigation with methyl bromide [62]. Since the phase-out of this fumigant, certain species of *Meloidogyne* and *Pratylenchus* have become established in the soil, causing substantial damage to vegetable crops. Further

Figure 3. Weed distribution and function as a reservoir for *Pratylenchus capsici* during and in between growing seasons. (A) Weeds emerging early after pepper seedling planting, and (B) throughout the pepper-growing season. (C) Lesions caused by *P. capsici* on *Chenopodium album* growing alongside the pepper plants.

biogeographical analysis suggested that a *P. capsici* population in weeds (*Chenopodium album* and *Sonchus oleraceus*) was an important source for *P. capsici* dispersal, either as the original nematode source or in maintaining the population between growing seasons (Figure 3).

Similar findings were observed for *P. penetrans* [63], *P. brachyurus* [64,65], *Pratylenchus coffeae* [66], *P. zeae* [67], *P. scribneri* and *P. vulnus* [68], and *P. thornei* and *P. neglectus* [69].

Pratylenchus capsici has been shown to survive through the seasons with no host from April to July. During this period, nematode activation is prevented by the high temperature and low moisture in the soil. Extensive nematode extraction from roots and soils yielded a high number of nematodes in the former and low numbers in the latter, supporting its exclusive endoparasitic life strategy. Therefore, these observations raise the question of whether *P. capsici* is ever anhydrobiotic, and if so, whether it goes through anhydrobiosis in the roots or in the soil. Similarly, *P. capsici*'s capacity to migrate to lower soil levels during the off seasons is not known. Further study is needed to clarify this question.

6. Phylogeny and Evolution of *Pratylenchus* Species Occurring in Israel

To date, nine species of *Pratylenchus* have been reported from Israel, with molecular data available for only three of them (*P. thornei*, *P. mediterraneus*, and *P. capsici*) (Figure 4). The

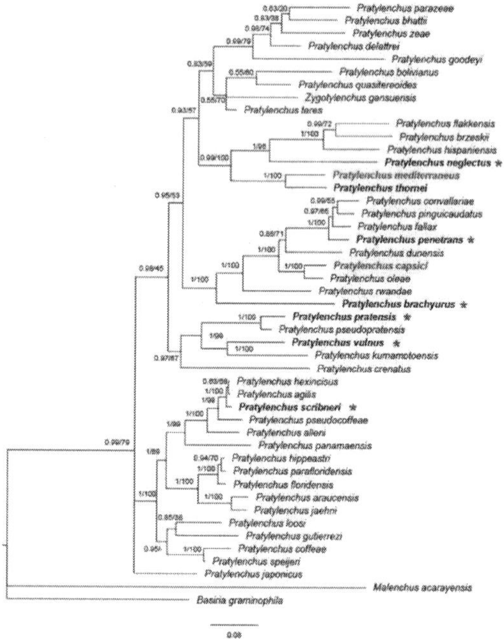

Figure 4. Bayesian 50% majority rule consensus tree inferred on concatenated sequences of 28S; asterisks indicate species that were only identified by morphology. The dataset was aligned by MAFFT v. 7.205 [70] using the G-INS-i algorithm. The phylogeny was reconstructed by maximum likelihood (ML) and Bayesian inference (BI) using RAxML v.8.1.11 [71] and MrBayes 3.2.3. [72]. Branch support is indicated in the following order: posterior probability (PP) value from BI analysis/bootstrap (BS) value from ML analysis. Red marked species indicate local Israeli isolates.

concatenated phylogeny based on 18S and 28S rDNA and internal transcribed spacer (ITS) suggests that *P. thornei* and *P. mediterraneus* form a well-supported (posterior probability (PP) = 1, bootstrap (BS) = 100) monophyletic group, concurring with previous studies [18,24].

Orion [50] suspected that *P. mediterraneus* is a native or at least old inhabitant of the semiarid region of the Eastern Mediterranean. Given the similarities in morphology and morphometric features, the overlapping geographical area (Mediterranean region), the same hosts (mostly cereal and legumes), and the anhydrobiotic survival properties, *P. thornei* and *P. mediterraneus* could be derived from recent speciation events, with insufficient time to attain complete morphological differentiation.

Similarly, *P. capsici* is sister to *P. oleae* in concatenated phylogeny (Figure 4, PP = 1, BS = 100), as well as in a previous study [18]. *Pratylenchus oleae* was found in the Mediterranean region, parasitizing both wild and cultivated olive trees in Spain and Tunisia, with the presence of the nematode in wild olive not showing any clear symptoms in the aboveground plant or roots [3]. Interestingly, *P. capsici* was found in both pepper and weeds, markedly damaging the pepper but causing very mild symptoms on the weeds. Later, population genetic analysis revealed that *P. capsici* is likely to have been native to wild grass and transmitted to pepper by a recent expansion [18]. The adjacent distribution, similar morphology and presumably similar transmission background give rise to the idea that the two closely related species, *P. capsici* and *P. oleae*, may be native to the Mediterranean region.

7. Control and Management Practices

Plant growth and yield losses in any nematode–plant interactions depend primarily on soil nematode densities at planting. In the last few decades, intensive studies in Israel have been dedicated to the development of systems-based approaches to reducing soilborne pathogen densities at planting in different climatic regions [73–77]. These studies have shown that soil fumigants with nematicidal properties can reduce nematode infestation level but fail to eradicate the soil nematode, whereas a combination of fumigants with solarization can enhance the killing of soilborne pathogens [73,78,79], emphasizing the importance of using an appropriate combined application of pesticidesandsolarization.

7.1 The Use of Soil Fumigants

Three commercial soil fumigants are registered and commonly used in Israel: (i) 1,3-dichloropropene (1,3-D), a liquid fumigant (boiling point 104–112 °C) that is considered to be highly effective against nematodes and has been adopted as an alternative to methyl bromide [80]; 1,3,-D is registered for use in the control of all plant-parasitic nematodes and bacterial plant diseases, insects, and weeds. In practice, nematodes are the main target of 1,3-D use on most crops; 1,3,-D is labeled as a pre-planting soil treatment, and its effectiveness is dependent on environmental factors such as length of the growing season, moisture, temperature and soil type. (ii) Metam sodium (sodium N-methyldithiocarbamate, metam-Na) is widely used to control soilborne plant pathogens, mainly fungi and weeds, while its efficacy in the control of plant-parasitic nematodes is limited [81,82]; because metam sodium undergoes rapid decomposition in moist soils to the active compound methyl isothiocyanate [83], soil fumigation of vegetable crops with metam sodium or metam potassium results in inconsistent control, particularly against root-knot nematodes, while intensive experience indicates its efficiency toward migratory plant-parasitic nematodes but no effect on root-knot nematodes. (iii) Dimethyl disulfide (DMDS), which was registered in the last decade and is effective at controlling both sedentary and migratory nematodes, as well as weeds and soilborne fungal pathogens. Unfortunately, the performance of these three fumigants is inferior to that of methyl bromide. In Israel, the prevalent treatment for nematode management in vegetables is targeted to reducing nematode population density primarily through soil fumigation with 1,3-D or DMDS. However, these fumigants do not provide adequate protection of crop health throughout the entire growing season. Therefore, an integrated approach is needed to achieve successful management of lesion nematodes.

7.2 Common Control Methods in Used to Manage Plant Parasitic Nematodes

Currently recommended soil disinfestation approaches against soilborne plant-parasitic nematodes in conventional farming—mainly *Pratylenchus* species and *Meloidogyne* species root-knot nematodes—include the following steps [84]: (i) destruction of the plant roots at the end of the crop season before plant removal (Figure 5A); (ii) plant and root removal followed by tillage, although this latter recommendation is not always followed (Figure 5B); (iii) soil disinfection approaches using effective soil fumigants combined with soil solarization for a minimal period of 4 weeks during the summer (Figure 5C). At this time, nets above protected houses are removed to increase soil solarization efficiency, and shade nets are then reinstalled at seedling planting time (Figure 5D).

Figure 5. Integrated nematode management. Protocol used in practice to control migratory or sedentary plant-parasitic nematodes. (A) Destruction of previous crop's roots before removal to reduce primary inoculum. (B) Root removal, tilling, and soil preparation for fumigation and solarization requirements. Soil-disinfection approaches using different soil fumigants in combination with soil solarization for at least 4 weeks during the summer. (D) Planting of seedlings and reinstallation of shade net.

A combination of solarization with organic material (biosolarization) can reduce nematode densities but not achieve full eradication [85]. Similarly, Oka and Pivonia [86] explored the possibility of using ammonia for controlling soilborne diseases under variable environmental conditions in the Arava region of Israel. Given that soil pH may be the most important factor affecting the nematicidal activity of ammonia, where alkaline soils support better activity [87], as well as the fact that neutral to weakly alkaline sandy soils are common in Israel, the use of ammonia for nematode control is promising [86]. As expected, the use of NH_4OH (at 500 and 1000 kg N/ha) increased tomato yield and reduced the galling index (at 1000 kg N/ha). However, despite its positive control effect, a high percentage of ammonia may be deleterious to the environment. This needs to be further evaluated under different soil conditions, nematicidal activities, and application methods. Another approach to exploitingammoniafornematodecontrolistheapplication of ammonia generators such as chicken manure, soy bean meal, and other organic materials [88]. Further studies by Oka et al. [89] demonstrated that application of ammonium sulfate, chicken litter and chitin, or neem (*Azadirachta indica*) extract alone failed to reduce the root galling index of tomato plants, but application of the amendments in combination with the neem extract reduced root galling significantly. Soil analysis indicated that the neem extract inhibits the nitrification of the ammonium released from the amendments and extends the persistence of the ammonium concentrations in the soil. In addition, biosolarization using chicken compostresultedineffectivecontrolofroot-knotnematodes inalettucecrop[88].

Field crops that are not under intensive production pose a challenge for nematode management. Orion et al. [10] found that leaving the soil fallow for 2 years reduced the *P. mediterraneus* population by 90% and increased wheat grain yields by 40–90%. By monitoring a 30-year rotation trial over several seasons of wheat-cropping systems, researchers found that the use of legumes (vetch, lentil) can increase *P. thornei* populations, whereas sunflower or safflower followed by a fallow period provided the best reduction of *P. thornei* [90]. Alternatively, soil treatment with metam sodium controlled *P. mediterraneus* by 90% and increased yield by 50–70% [91]. The biannual fallowing system was the most desirable environmentally, but it occupied 50% of the land, which in practice is problematic because cultivated land is quite limited in Israel. Since metam sodium treatment is less feasible in dryland agriculture, several alternative control methods were evaluated. Those studies suggested that nitrogen fertilizer does not affect *P. mediterraneus* populations in either dry farming or as a supplement in irrigation treatments [10]. Use of the nematicide formulation of furathiocarb, a systemic soil insecticide, as a seed dressing could reduce *P. mediterraneus* population level and increase yield, while the best nematode killing was achieved by soil application [11,14].

7.3 Resistance to Root-Lesion Nematodes

The wide host range of *Pratylenchus* species, and the restrictions, cost, and inefficiency of chemical nematicides have raised the importance of developing resistant cultivars as a control measure [92,93]. Unfortunately, only a few studies have considered the effects of resistance on *Pratylenchus* biology. Talavera and Van Stone [94] demonstrated that *P. thornei* is able to penetrate resistant cultivars. Farsi [95] observed equal root penetration by *P. neglectus* in both resistant and susceptible wheat lines. Other studies in various plant hosts have shown that, in other *Pratylenchus* species, resistance is associated with reduced motility and reproduction [96]. While the major studies of resistance have focused on wheat varieties [5], vegetable crops have been less investigated. The use of resistant cultivars is advantageous in integrated control programs because an accurate assessment of nematode infestations and infections is critical for the evaluation of plant resistance and tolerance to *Pratylenchus* species.

8. Challenges and Perspectives for *Pratylenchus* Research in Israel

In the last decade, several studies have been implemented toward the development of an integrated nematode management system that includes available and efficient means. Like elsewhere, most soil fumigants and nematicides belonging to containing organophosphates and carbamates have been withdrawn from the market or have strict use restrictions, mainly for environmental and safety reasons [97]. In general, there appears to be little prospect for the management of nematodes in many susceptible crops without repeated application of nematicides, which is economically justified in only a few cases. Alternatively, a number of products and formulations of fumigant–nematicides are available for use [98]. However, the effectiveness of traditional fumigants and nematicides with broad biocidal activity is declining, and the development of new classes of nematicides with novel activity and specific pest targets is perhaps an idealistic pipe dream. Recent research carried out in Israel has shown that the incorporation of nematicidal fluensulfone into the soil can reduce the populations of several migratory nematodes under laboratory conditions [44]. An additional new nematicidal compound, fluopyram, has been evaluated in vitro against root-knot nematodes [99], but its effect on migratory nematodes hasnotyetbeenconfirmedinthefield.

8.1 Taxonomy and Diagnosis of Pratylenchus Species

Given the wide distribution and severe damage caused by *Pratylenchus*, its taxonomy and diagnosis are crucial for *Pratylenchus* research and agricultural production in Israel. Despite its importance, the morphological diagnosis is greatly hampered by phenotypic plasticity, interspecific similarities, and a lack of molecular taxonomy specialists. Today, routine plant-parasitic nematode identification is conducted by the PPIS of the Israeli Ministry of Agriculture and Rural Development using only diagnostic morphological characteristics. The information provided to farmers, agronomists, nurseries, and inspectorsconsistsmostlyofidentificationatthegenuslevelandthedensityofthenematode population found in the soil or root samples. Similarly, identification of *Pratylenchus* species is limited in most instances to the genus level, while species identification relies on the host from which they were recovered. Thus, molecular barcoding is a powerful, efficient, and reliable tool to simplify and standardize nematode identification, but such a method is not yet fully established for routine identification of *Pratylenchus* species, especially for basic research stations and production departments. Further effort is needed to expand *Pratylenchus* diagnostic techniques and improve farms' awareness ofthem.

8.2 Control/Management of Pratylenchus Species

Extensive research is being performed on alternative chemical and nonchemical methods for controlling nematode diseases. However, these methods are generally less effective than soil fumigation in reducing soil nematode densities, and many have not proven consistent enough when used in intensive crop farming. Long-term field trials comparing the nematicide efficacies of several soil disinfestation methods would provide valuable information for nematode management. New nematicides are continually being introduced to the market although their efficiency against *Pratylenchus* species is not always known, and if it is, their label should refer to specific hosts, soils, and environmental conditions. Thus, the participation of professional nematologists is crucial in laboratory and field experiments evaluating nematicides. Symptoms caused by *Pratylenchus* species are frequently overlooked and a lack of nematological knowledge might lead to erroneous interpretations. Moreover, the migratory endoparasitic lifestyle, which might support the association of additional plant pathogens, should be studied for each plant–*Pratylenchus* interaction. In such cases, control strategies need to target both the nematode and the associated pathogen. A study of the etiology underlying nematode survival between seasons under extreme conditions is required to address important questions regarding the occurrence of anhydrobiosis, migration ability to lower soil levels, and factors required for these nematodes' recovery. Exploration of these aspects is expected to contribute to the development of efficient integrated control management of *Pratylenchus*.

9. Conclusions

Delimitation of the various *Pratylenchus* species is considered to be very complicated, especially because of the small number of diagnostic features available at the species level and the intraspecific variability of some of these characteristics [23]. Nevertheless, due to the difficulty in separating species, the number of new proposed species of *Pratylenchus* has increased almost linearly, with a slope of 1.1 species per year between 1940 and 2006 [23]. Although morphology continues to be the basis for identification of *Pratylenchus* species, new technologies based on biochemical and molecular analyses are becoming increasingly important for nematode systematics and practical diagnoses [27,100–102]. New species are continuously being described through extensive morphological and molecular studies of the 28S D2-D3 expansion domains and ITS. The highest biodiversity of the genus is found in Asia, where 40 species have been reported, followed by Europe with 32, North America with 27, Central and South America with 22, Africa with 16, Oceania with 12, and Antarctica with a single species. The most widely distributed and common species are *P. neglectus*, *P. penetrans*, *P. thornei*, and *P. vulnus*, which have been reported on every continent with the exception of Antarctica. Thirty-seven species (54% of the 68 nominal species) in the genus have only been reported from a single continent, while the remaining 31 species (46%) have been reported from two or more continents. Nevertheless, despite the global distribution of the genus, some 32 of the described species have thus far only been recorded from their type locality. Along these lines, it will be interesting to determine whether, similar to *P. mediterraneus*, which was first found in Israel and later in other Middle Eastern countries, the occurrence of *P. capsici* will be identified in neighboring countries as well.

References

1. Geraert, E. *The Pratylenchidae of the World: Identification of the Family Pratylenchidae (Nematoda: Tylenchida)*; Academia Press: Ghent, Belgium, 2013.
2. Hodda, M.; Collins, S.J.; Vanstone, V.A.; Hartley, D.; Wanjura, W.; Kehoe, M. Pratylenchus quasitereoides n. sp. from cereals in Western Australia. *Zootaxa* **2014**, *3866*, 277–288. [CrossRef]

3. Palomares-Rius, J.E.; Guesmi, I.; Horrigue-Raouani, N.; Cantalapiedra-Navarrete, C.; Liébanas, G.; Castillo, P. Morphological and molecular characterisation of *Pratylenchus oleae* n. sp.(Nematoda: Pratylenchidae) parasitizingwildandcultivatedolivesinSpain andTunisia. *Eur. J. Plant Pathol.* **2014**, *140*, 53–67. [CrossRef]

4. Nicol, J.; Turner, S.; Coyne, D.; Den Nijs, L.; Hockland, S.; Maafi, Z.T. Current nematode threats to world agriculture. In *Genomics and Molecular Genetics of Plant-Nematode Interactions*; Springer: Dordrecht, TheNetherlands,2011;pp.21–43.

5. Jones, M.; Fosu-Nyarko, J. Molecular biology of root lesion nematodes (*Pratylenchus* spp.) and their interaction withhostplants. *Ann. Appl. Biol.* **2014**, *164*, 163–181. [CrossRef]

6. Minz, G. Free-living plant-parasitic and possible plant-parasitic nematodes in Israel. *Plant Dis. Rep.* **1957**, *41*, 92–94.

7. Cohn, E.; Sher, S.; Bell, A.; Minz, G. *Soil Nematodes Occurring in Israel*; The Volcani Center: Rishon Letzion, Israel,1973.

8. Glazer,I.;Orion,D.StudiesonanhydrobiosisofPratylenchusthornei. *J. Nematol.* **1983**, *15*, 333.

9. Krikun, J.; Orion, D. Verticillium wilt of potato: Importance and control. *Phytoparasitica* **1979**, *7*, 107. [CrossRef]

10. Orion, D.; Amir, J.; Krikun, J. Field observations on *Pratylenchus thornei* and its effects on wheat under arid conditions. *Rev. Nématol.* **1984**, *7*, 341–345.

11. Orion, D.; Glazer, I. Nematicide seed dressing for *Pratylenchus mediterraneus* control in wheat. *Phytoparasitica* **1987**, *15*, 225–228. [CrossRef]

12. Orion, D.; Lapid, D. Scanning Electron-Microscope Study on the Interaction of Pratylenchus-Mediterraneus andVicia-SativaRoots. *Nematologica* **1993**, *39*, 322–327. [CrossRef]

13. Orion, D.; Nachmias, A.; Lapid, D.; Orenstein, J. Observations on the parasitic behavior of *Pratylenchus mediterraneus* onexcisedpotatoroots. *Nematropica* **1995**, *25*, 71–74.

14. Orion, D.; Shlevin, E. Nematicide seed dressing for cyst and lesion nematode control in wheat. *J. Nematol.* **1989**, *21*, 629. [PubMed]

15. Goldreich,Y.;Karni,O.ClimateandprecipitationregimeintheAravaValley,Israel. *Isr. J. Earth Sci.* **2001**, *50*, 53–59. [CrossRef]

16. Yair, A. Runoff generation in a sandy area—The Nizzana sands, Western Negev, Israel. *Earth Surf. Process. Landf.* **1990**, *15*, 597–609. [CrossRef]

17. Pasternak, D.; De Malach, Y.; Borovic, I.; Shram, M.; Aviram, C. Irrigation with brackish water under desert conditions IV. Salt tolerance studies with lettuce (*Lactuca sativa* L.). *Agric. Water Manag.* **1986**, *11*, 303–311. [CrossRef]

18. Qing, X.; Bert, W.; Gamliel, A.; Bucki, P.; Duvrinin, S.; Alon, T.; Braun Miyara, S. Phylogeography and Molecular Species Delimitation of Pratylenchus capsici n. sp., a New Root-Lesion Nematode in Israel on Pepper(*Capsicum annuum*). *Phytopathology* **2019**, *109*, 847–858. [CrossRef]

19. Orion, D.; Krikun, J.; Sullami, M. The distribution, pathogenicity and ecology ofPratylenchus Thornei in the northernnegev. *Phytoparasitica* **1979**,*7*,3–9. [CrossRef]
20. Cobertt, D.; Clark, A. Surface features in the taxonomy of Pratylenchus species. *Rev. Néinatol* **1983**,*6*,85–98.
21. Ryss, A.Y. *Root Parasitic Nematodes of the Family Pratylenchidae (Tylenchida) of the World Fauna*; Nauka: Leningrad, Russia,1988.
22. Café Filho, A.C.; Huang, C. Description of *Pratylenchus pseudofallax* n. sp. with a key to species of the genus *Pratylenchus* Filipjev, 1936 (Nematoda: Pratylenchidae). *Rev. Nématol.* **1989**,*12*,7–15.
23. Castillo, P.; Vovlas, N. Pratylenchus (Nematoda: Pratylenchidae): Diagnosis, biology, pathogenicity and management. In *Nematology Monographs and Perspectives*; Brill: Leiden,TheNetherlands,2007;Volume6.
24. De Luca, F.; Reyes, A.; Troccoli, A.; Castillo, P. Molecular variability and phylogenetic relationships among different species and populations of *Pratylenchus* (Nematoda: Pratylenchidae) as inferred from the analysis oftheITSrDNA.*Eur. J. Plant Pathol.* **2011**,*130*,415–426. [CrossRef]
25. Troccoli, A.; De Luca, F.; Handoo, Z.; Di Vito, M. Morphological and molecular characterization of *Pratylenchus lentis* n. sp.(Nematoda: Pratylenchidae) from Sicily. *J. Nematol.* **2008**,*40*,190.
26. Waeyenberge,L.;Ryss,A.;Moens,M.;Pinochet,J.;Vrain,T.Molecularcharacterisationof18 *Pratylenchus* speciesusingrDNArestrictionfragmentlengthpolymorphism. *Nematology* **2000**,*2*,135–142. [CrossRef]
27. de Luca, F.; Fanelli, E.; Di vito, M.; Reyes, A.; De giorg, C. Comparison of the sequences of the D3 expansion of the 26S ribosomal genes reveals different degrees of heterogeneity in different populations and species of PratylenchusfromtheMediterraneanregion. *Eur. J. Plant Pathol.* **2004**,*110*,949–957. [CrossRef]
28. Corbett, D. Three new species of *Pratylenchus* with a redescription of *P. andinus* Lordello, Zamith & Boock, 1961(Nematoda: Pratylenchidae). *Nematologica* **1983**,*29*,390–403.
29. Greco, N.; Di Vito, M.; Saxena, M.; Reddy, M. Investigation on the root lesion nematode, *Pratylenchus thornei*, in Syria. *Nematol. Mediterr.* **1988**,*16*,101–105.
30. Saxena, M.C.; Singh, K. *The Chickpea [Cicer Arietinum]*; Commonwealth Agricultural Bureaux International: Wallingford,UK,1987.
31. Di Vito, M.; Greco, N.; Ores, G.; Saxena, M.; Singh, K.; Kusmenoglu, I. Plant parasitic nematodes of legumes in Turkey. *Nematol. Mediterr.* **1994**,*22*,245–251.
32. KepenekcI˙, I. Plant parasitic nematode species of Tylenchida (Nematoda) associated with sesame (*Sesamum indicum* L.) growing in the Mediterranean region of Turkey. *Turk. J. Agric. For.* **2002**, *26*, 323–330.
33. Greco, N.; Di Vito, M. Nematodes of food legumes in the Mediterranean Basin 1. *Eppo Bull.* **1994**, *24*, 393–398. [CrossRef]
34. Greco, N.; Di Vito, M.; Saxena, M. Plant parasitic nematodes of cool season food legumes in Syria. *Nematol. Mediterr.* **1992**,*20*,37–46.

35. Troccoli, A.; Di Vito, M. Root lesion and stem nematodes associated with faba bean in North Africa. *Nematol. Mediterr.* **2002**,*30*,79–81.

36. Choi, D.-R.; Lee, J.-K.; Parte, B.-Y.; Han, H.-R.; Choi, Y.-E. A new and one unrecorded species of *Pratylenchus* from Korea (Nematoda: Pratylenchidae). *J. Asia-Pac. Entomol.* **2006**,*9*,5–9. [CrossRef]

37. Orion, D.; Shlevin, E.; Yaniv, A. Controlling the migratory nematode *Pratylenchus mediterraneus* improves carrot yield quality. *Hassadeh* **1988**,*69*,72–74.

38. Development MoAaR. *Database of Plant Pests in Israel: Ministry of Agriculture and Rural Development*; Development MoAaR: Rishon Letzion, Israel, 2020.

39. Nicol, J. *The Distribution, Pathogenicity and Population Dynamics of Pratylenchus Thornei on Wheat in South Australia*; Department of Crop Protection, University of Adelaide: Adelaide, Australia, 1996.

40. Nicol, J.; Rivoal, R.; Taylor, S.; Zaharieva, M. (Eds.) Global importance of cyst (*Heterodera* spp) and lesion nematode (Pratylenchus spp.) on cereals: Distribution, yield loss, use of host resistance and integration of molecular tools. In Proceedings of the Fourth International Congress of Nematology, Tenerife, Spain, 8–13June2002.

41. Braun Miyara, S.; Nematology Unit, Plant Protection Institute, ARO, The Volcani Center, Rishon Letzion, Israel. Personal communication, 2020.

42. Kozodoi, E.; Plant Protection and Inspection Services, Rishon Letzion, Israel. Personal Communication, 2020.

43. Pivonia, S.; Central-and Northern-Arava Research and Development, Arava Sapir, Israel. Personal Communication, 2020.

44. Oka, Y. Nematicidal activity of fluensulfone against some migratory nematodes under laboratory conditions. *Pest Manag. Sci.* **2014**,*70*,1850–1858. [CrossRef]

45. Agrios, G.N. *Introduction to Plant Pathology*; Academic Press: New York, NY, USA, 1988; pp. 3–39.

46. Rotenberg, D.; MacGuidwin, A.; Saeed, I.; Rouse, D. Interaction of spatially separated *Pratylenchus penetrans* and *Verticillium dahliae* on potato measured by impaired photosynthesis. *Plant Pathol.* **2004**, *53*, 294–302. [CrossRef]

47. Rowe, R.C.; Powelson, M.L. Potato early dying: Management challenges in a changing production environment. *Plant Dis.* **2002**,*86*,1184–1193. [CrossRef]

48. Siti, E. The interrelationships between *Pratylenchus tornei* and *Verticilium daliae*, and their effect on potatoes. In *Jerusalem HUo*; Faculty of Agriculture—Hebrew University of Jerusalem: Rehovot, Israel, 1978.

49. Talavera, M.; Valor, H.; Tobar, A. Post-anhydrobiotic viability of *Pratylenchus thornei* and *Merlinius brevidens*. *Phytoparasitica* **1998**,*26*,293. [CrossRef]

50. Orion, D. Nematodes of agricultural importance in Israel. *Nematology* **2000**,*2*,735–736. [CrossRef]

51. Castillo, P.; Trapero-Casas, J.; Jiménez-Díaz, R. Effect of time, temperature, and inoculum density on reproduction of *Pratylenchus thornei* in carrot disk cultures. *J. Nematol.* **1995**,*27*,120.

52. Thompson, J.; Clewett, T.; O'Reilly, M. Temperature response of root-lesion nematode (Pratylenchus thornei) reproduction on wheat cultivars has implications for resistance screening and wheat production. *Ann. Appl. Biol.* **2015**, *167*, 1–10. [CrossRef]

53. Kurppa, S.; Vrain, T.C. Penetration and feeding behavior of *Pratylenchus penetrans* in strawberry roots. *Rev. Nématol.* **1985**, *8*, 273–276.

54. Zunke, U. Observations on the invasion and endoparasitic behavior of the root lesion nematode *Pratylenchus penetrans*. *J. Nematol.* **1990**, *22*, 309. [PubMed]

55. Inserra, R.; Vovlas, N. Effects of *Pratylenchus vulnus* on the growth of sour orange. *J. Nematol.* **1977**, *9*, 154.

56. Vovlas, N.; Troccoli, A. Histopathology of broad bean roots infected by the lesion nematode *Pratylenchus penetrans*. *Nematol. Mediterr.* **1990**, *18*, 239–242.

57. De Waele, D.; Jordaan, E. Plant-parasitic nematodes on field crops in South Africa. 1. Maize. *Rev. Nématol.* **1988**, *11*, 65–74.

58. Coosemans, J. The influence of Pratylenchus penetrans on growth of *Impatiens balsamina* L. *Daucus carota* L., *Linum usitatissimum* L. and *Crysanthemum indicum* L. *Symp. Int. Phytopharm Gent Belg.* **1975**, *27*, 465–471.

59. Hay, F.; Pethybridge, J. Nematodes associated with carrot production in Tasmania, Australia, and the effect of *Pratylenchus crenatus* on yield and quality of Kuroda-type carrot. *Plant Dis.* **2005**, *89*, 1175–1180. [CrossRef]

60. Potter, J.; Olthof, T.H. Nematode pests of vegetable crops. In *Plant Parasitic Nematodes in Temperate Agriculture*; CABI: Wallinford, UK, 1993; pp. 171–207.

61. Vrain, T.; Belair, G. Symptoms induced by the lesion nematode, *Pratylenchus penetrans* on carrot taproots in organic soil. *Phytoprotection* **1981**, *62*, 79–81.

62. Gamliel, A.; ARO-Volcani Center, Rishon LeTsiyon, Israel. Personal communication, 2020.

63. Townshend, J.; Davidson, T. Some weed hosts of Pratylenchus penetrans in premier strawberry plantations. *Can. J. Bot.* **1960**, *38*, 267–273. [CrossRef]

64. Hogger, C.; Bird, G. Weed and indicator hosts of plant-parasitis nematodes in georgia cotton and soybean fields. *Plant Dis. Report.* **1976**, *60*, 223–226.

65. Koen, H. Notes on the host range, ecology and population dynamics of Pratylenchus brachyurus. *Nematologica* **1967**, *13*, 118–124. [CrossRef]

66. Edwards, D.; Wehunt, E. Hosts of *Pratylenchus coffeae* with additions from central american banana-producing areas. *Plant Dis. Rep.* **1973**, *57*, 47.

67. Ayoub, S. *Pratylenchus zeae* found on corn, milo, and three suspected new hosts in California. *Plant Dis. Rep.* **1961**, *45*, 940.

68. Manuel, J.S.; Reynolds, D.; Bendixen, L.; Riedel, R. *Weeds as Hosts of Pratylenchus*; Ohio Agricultural Research and Development Center: Wooster, OH, USA, 1980.

69. Vanstone, V.A.; Russ, M.H. Ability of weeds to host the root lesion nematodes Pratylenchus neglectus and P. thornei I. Grass weeds. *Australas. Plant Pathol.* **2001**, *30*, 245–250. [CrossRef]

70. Katoh, K.; Standley, D.M. MAFFT multiple sequence alignment software version 7: Improvements in performance and usability. *Mol. Biol. Evol.* **2013**, *30*, 772–780. [CrossRef] [PubMed]

71. Stamatakis, A.; Hoover, P.; Rougemont, J. A rapid bootstrap algorithm for the RAxML web servers. *Syst. Biol.* **2008**, *57*, 758–771. [CrossRef]

72. Ronquist, F.; Teslenko, M.; Van Der Mark, P.; Ayres, D.L.; Darling, A.; Höhna, S.; Larget, B.; Liu, L.; Suchard, M.A.; Huelsenbeck, J.P. MrBayes 3.2: Efficient Bayesian phylogenetic inference and model choice across a large model space. *Syst. Biol.* **2012**, *61*, 539–542. [CrossRef]

73. Gamliel, A.; Katan, J. Control of plant disease through soil solarization. In *Disease Control in Crops*; Wiley-Blackwell: Hoboken, NJ, USA, 2009.

74. Gamliel, A.; Katan, J. *Soil Solarization: Theory and Practice*; APS Press/The American Phytopathological Society: St. Paul, MN, USA, 2012.

75. Katan, J. Cultural approaches for disease management: Present status and future prospects. *J. Plant Pathol.* **2010**, *92*, S7–S9.

76. Katan, J.; Gamliel, A. Soilborne diseases, control by physical methods. In *Encyclopedia of Agrophysics*; Springer: Dordrecht, The Netherlands, 2011; pp. 813–816.

77. Triky-Dotan, S.; Austerweil, M.; Steiner, B.; Peretz-Alon, Y.; Katan, J.; Gamliel, A. Accelerated degradation of metam-sodium in soil and consequences for root-disease management. *Phytopathology* **2009**, *99*, 362–368. [CrossRef]

78. Di Primo, P.; Gamliel, A.; Austerweil, M.; Steiner, B.; Beniches, M.; Peretz-Alon, I.; Katan, J. Accelerated degradation of metam-sodium and dazomet in soil: Characterization and consequences for pathogen control. *Crop Prot.* **2003**, *22*, 635–646. [CrossRef]

79. Triky-Dotan, S.; Austerweil, M.; Steiner, B.; Peretz-Alon, Y.; Katan, J.; Gamliel, A. Generation and dissipation of methyl isothiocyanate in soils following metam sodium fumigation: Impact on verticillium control and potato yield. *Plant Dis.* **2007**, *91*, 497–503. [CrossRef]

80. Hafez, S.; Sundararaj, P. Evaluation of chemical nematicides for the suppression of *Meloidogyne chitwoodi* and *M. hapla* on potato. *Int. J. Nematol.* **2001**, *11*, 192–194.

81. Ajwa, H.; Trout, T.; Mueller, J.; Wilhelm, S.; Nelson, S.; Soppe, R.; Shatley, D. Application of alternative fumigants through drip irrigation systems. *Phytopathology* **2002**, *92*, 1349–1355. [CrossRef]

82. Ruzo, L.O. Physical, chemical and environmental properties of selected chemical alternatives for the pre-plant use of methyl bromide as soil fumigant. *Pest Manag. Sci. Former. Pestic. Sci.* **2006**, *62*, 99–113. [CrossRef]

83. Klose, S.; Ajwa, H.A.; Browne, G.T.; Subbarao, K.V.; Martin, F.N.; Fennimore, S.A.; Westerdahl, B.B. Dose response of weed seeds, plant-parasitic nematodes, and pathogens to twelve rates of metam sodium in a California soil. *Plant Dis.* **2008**, *92*, 1537–1546. [CrossRef] [PubMed]

84. Gamliel, A. *Soil and substrate health. In Integrated Pest and Disease Management in Greenhouse Crops*; Springer: Dordrecht,TheNetherlands,2020;pp.355–383.

85. Talavera, M.; Miranda, L.; Gómez-Mora, J.A.; Vela, M.D.; Verdejo-Lucas, S. Nematode Management in the Strawberry Fields of Southern Spain. *Agronomy* **2019**, *9*, 252. [CrossRef]

86. Oka, Y.; Pivonia, S. Use of ammonia-releasing compounds for control of the root-knot nematode *Meloidogyne javanica*. *Nematology* **2002**,*4*,65–71. [CrossRef]

87. Rodriguez-Kabana, R.; Morgan-Jones, G.; Chet, I. Biological control of nematodes: Soil amendmentsand microbialantagonists. *Plant Soil* **1987**,*100*,237–247. [CrossRef]

88. Gamliel, A.; Stapleton, J. Effect of chicken compost or ammonium phosphate and solarizationonpathogen control,rhizospheremicroorganisms,andlettucegrowth. *Plant Dis.* **1993**,*77*,886–891. [CrossRef]

89. Oka, Y.; Tkachi, N.; Shuker, S.; Yerumiyahu, U. Enhanced nematicidal activity of organic and inorganic ammonia-releasing amendments by Azadirachta indica extracts. *J. Nematol.* **2007**,*39*,9.

90. Eleckcioglu, I.; Avci, M.; Nicol, J.; Meyveci, K.; Bolat, N.; Yorgancilar, A.; Sahim, E.; Kaplan, A. The Use of Crop Rotation as a Means to Control the Cyst and Lesion Nematode Under Rainfed Wheat Production Systems. In Proceedings of the Plant ProtectionCongressofTurkey,Samsun,Turkey,8–10September2004.

91. Storey, R.; Glazer, I.; Orion, D. Lipid Utilisation By Starved and Anhydrobiotic Individuals of Pratylenchus Thornei. *Nematologica* **1982**,*28*,373–378.

92. Abdel-Rahman, F.H.; Alaniz, N.M.; Saleh, M.A. Nematicidal activity of terpenoids. *J. Environ. Sci. Health Part B* **2013**,*48*,16–22. [CrossRef]

93. Thompson, J.; Brennan, P.; Clewett, T.; Sheedy, J.; Seymour, N. Progress in breeding wheat for tolerance and resistance to root-lesion nematode (Pratylenchus thornei). *Australas. Plant Pathol.* **1999**,*28*,45–52. [CrossRef]

94. Talavera, M.; Van stone, V.A. Monitoring *Pratylenchus thornei* densities in SOCL and roots under resistant (*Triticum turgidum durum*) and susceptible (*Triticum aestivum*) wheat cultivars. *Phytoparasitica* **2001**, *29*, 29–35. [CrossRef]

95. Farsi, M. Genetic Variation for Tolerance and Resistance to *Pratylenchusneglectus*/by Mohammed Farsi. Ph.D.Thesis,UniversityofAdelaide,Dept. ofPlantScience,Adelaide, Australia, 1996.

96. Linsell, K.J.; Riley, I.T.; Davies, K.A.; Oldach, K.H. Characterization of resistance to Pratylenchus thornei (Nematoda) in wheat (*Triticum aestivum*): Attraction, penetration, motility, and reproduction. *Phytopathology* **2014**, *104*, 174–187. [CrossRef]

97. Leibson, T.; Lifshitz, M. Organophosphate and carbamate poisoning: Review of the current literature and summaryofclinicalandlaboratoryexperienceinsouthernIsrael. *Isr. Med. Assoc. J.* **2008**,*10*,767.

98. Hague, N.; Gowen, S. Chemical control of nematodes. In *Principles and Practice of Nematode Control in Crops*; Brown, H., Kerry, B., Eds.; Academic Press: New York, NY, USA;Sydney,Australia,1987;pp.131–178.

99. Oka, Y. Nematicidal activity of fluensulfone compared to that of organophosphate and carbamate nematicides againstXiphinema indexand Longidorus vineacola. *Eur. J. Plant Pathol.* **2019**, *154*, 565–574. [CrossRef]

100. Al-Banna, L.; Williamson, V.; Gardner, S.L. Phylogenetic Analysis of Nematodes of the GenusPratylenchusUsing Nuclear 26S rDNA. *Mol. Phylogenet. Evol.* **1997**, *7*, 94–102. [CrossRef]

101. Andrés, M.; Pinochet, J.; Hernández-Dorrego, A.; Delibes, A. Detection and analysis of inter-and intraspecific diversity of Pratylenchus spp. using isozyme markers. *Plant Pathol.* **2000**, *49*, 640–649. [CrossRef]

102. Duncan, L.; Inserra, R.; Thomas, W.; Dunn, D.; Mustika, I.; Frisse, L.; Mendes, M.; Morris, K.; Kaplan, D. Molecular and morphological analysis of isolates of Pratylenchus coffeae and closely related species. *Nematropica* **1999**, *29*, 61–80.

103. Loof, P.A. Taxonomic studies on the genus Pratylenchus (Nematoda). Tijdschr. Over Plantenziekten 1960, 66, 29–90.

Chapter-5

Pratylenchus penetrans Parasitizing Potato Crops: Morphometric and Genetic Variability of Portuguese Isolates

1. Introduction

The root lesion nematode (RLN) *Pratylenchus penetrans* (Cobb, 1917) Filipjev and Schuurmans Stekhoven, 1941 is an important migratory endoparasite, often reported as a limiting factor of several herbaceous and fruit crops [1–3]. On potato (*Solanum tuberosum* L.), the nematode causes necrotic lesions on tubers and roots due to migration and feeding, and its presence increases the severity of the "potato early dying" disease caused by *Verticillium dahliae* Kleb. [3,4]. Damage of roots by *P. penetrans* diminishes water and uptake of nutrients, resulting in poor plant growth and consequent crop losses. In Europe and North America, *P. penetrans* has been considered a damaging species associated to potato crop [5–11]. In Portugal, *P. crenatus* Loof, 1960, *P. neglectus* Filipjev and S. Stekhoven, 1941, *P. penetrans* and *P. thornei* Sher and Allen, 1953 have been found parasitizing potato, coexisting frequently in soil with other plant-parasitic nematodes [12]. The correct iden- tification and characterization of *Pratylenchus* species are thus important, for example, to inform and advise farmers on the application of suitable pest management strategies. *Pratylenchus* species can be identified by means of morphological and morphometrical characters but requires specialized expertise since a considerable number of species share many morphological features and most specific differences can only be observed using high magnifications [3,13]. In addition, intraspecific morphological variability has been demon- strated in *P. penetrans* isolates in populations from different geographical locations [14]. To overcome the issues of overlapping morphological and morphometrical characters, identification of RLN should be complemented with the molecular analysis for accurate di- agnosis of this group of nematodes [15]. Molecular methods based on restriction fragment length polymorphism (RFLP) analysis of the ribosomal ribonucleic acid (rRNA) genes [16] and sequencing of different fragments of the rDNA cluster, including internal transcribed spacers (ITS) [17,18], 18S [19,20] and 26S [19,21,22] have been used for diagnostics of RLN species [15]. Moreover, sequencing of the mitochondrial DNA (mtDNA), cytochrome c oxidase subunit 1 (COI) gene [15,23–27] and the nuclear hsp90 gene [23,24,27–29] have also been largely used in the molecular characterization of RLN species. *Pratylenchus fallax* Seinhorst, 1968, and *P. convallariae* Seinhorst, 1959, were shown to be closely related to *P. penetrans* (96–97% similarity) after sequence analysis of the D2–D3 region of the 28S rRNA gene [22,30]. Phylogenetic analyses of sequences of D2-D3 of 28S rDNA or partial 18S rDNA conducted by Subbotin et al. [19] grouped *P. penetrans* with *P. arlingtoni* Carta and Skantar, 2001, *P. convallariae*, *P. dunensis* de la Pena, van Aelst, Karssen and Moens, 2006, *P. fallax* and *P. pinguicaudatus* Corbett, 1969, in clade IV, forming the Penetrans group. Later, Palomares-Rius et al. [23] added *P. brachyurus* Filipjev and Schuurmans Stekhoven, 1941, and *P. oleae* n. sp. into this clade. Using a combination of phylogenetic data with molecular species delineation analysis, population genetics, morphometric information and sequences, Janssen et al. [15] reconstructed a multi-gene phylogeny of the Penetrans group using the ITS, D2-D3 of the 28S rDNA regions from nuclear rDNA and the COI gene from mtDNA. The authors were able to confirm the taxonomic status of *P. penetrans*, *P. fallax* and *P. convallariae*, clarifying the boundaries within the Penetrans group. In the same study, *P. fallax* populations demonstrated low intraspecific variability whereas *P. penetrans* showed diverse haplotypes, with extremely variable intraspecific variability. Nonetheless, identical *P. penetrans* haplotypes were found to be geographically widespread, suggesting that *P. penetrans* could have spread anthropogenically through agricultural development and crop exchange [15]. Despite *P. penetrans* has already been detected in Portuguese potato crops, little information is still available on the morphometric and molecular variability of these *P. penetrans* isolates. The knowledge acquired in this study can be valuable in help defining effective strategies for RLN management in this crop. Therefore, the objectives of this research were to evaluate the morphometric variability of *P. penetrans* Portuguese isolates and to assess their genetic diversity, geographical and host relations. Information on intra- and interspecific variation of *P. penetrans* parasitizing potato increase the awareness of the genetic diversity of this species, and relationships with other *P. penetrans* isolates in other countries and hosts.

2. Results

2.1 Morphology of P. penetrans Portuguese Isolates

2.1.1 Female

Body moderately slender, almost straight when heat relaxed, with body length 672.5 (522.1–869.5) µm long (Figure 1A and Table 1). Lip region slightly offset from body, body annules distinct, lip with three annules low flat anteriorly with rounded margins. Stylet stout 16.5 (15.2–17.9) µm long, with knobs varying from rounded to cupped anteriorly (Figure 1C–E). Lateral fields with four straight lines (Figure 1F). Pharyngeal glands over- lapping intestine ventrally and slightly laterally. Excretory pore at from anterior extremity located opposite to pharyngo-intestinal junction. Spermatheca rounded, filled with sperm. Post uterine sac 1–1.5 times longer than vulval body diameter. Vulva located at 81.1 (75.7–83.9) % of body length (Figure 1G and Table 1). Tail cylindrical, 30.34 (19.0–41.5) µm long with smooth tip (Figure 1H,I and Table 1).

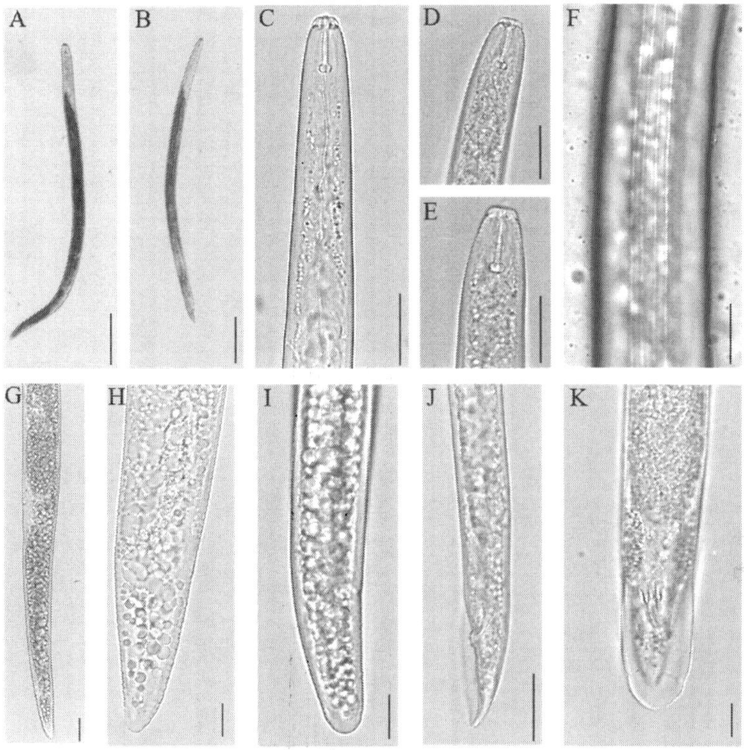

Figure 1. Pratylenchus penetrans specimens from Portuguese isolates: A—Entire female body, B— entire male body; C–E—anterior region; F—lateral field; G—female posterior region showing vulva; H,I—female tail variability; J—male posterior region; K—male tail (ventral side). Scale bars A,B: 100 μm; C–E, J: 20 μm; F–I, K:10 μm.

Table 1. Morphometric characters of Pratylenchus penetrans females from Portugal. All measurements are in μm. Data are means of 10 nematodes ± standard deviation (range). In each row means followed by the same letters do not differ significantly at p > 0.05, according to the Fisher Least Significant Difference test.

Character	Isolate					Loof (1960)
	PpA21L2	PpA24L1	PpA34L3	PpA44L2	PpA44L4	
Tail	$29.4 \pm 4.9^{a,b}$ (19.0–35.8)	$31.8 \pm 2.6^{a,b,c}$ (27.7–35.6)	23.7 ± 3.9^{d} (19.5–31.4)	$34.0 \pm 4.2^{b,c}$ (26.6–41.0)	$32.8 \pm 3.9^{a,b,c}$ (25.9–38.4)	
V *	81.1 ± 2.2^{a} (75.7–83.7)	81.0 ± 1.7^{a} (77.7–83.9)	82.5 ± 2.2^{a} (78.3–83.9)	80.6 ± 2.0^{a} (77.0–83.5)	80.3 ± 4.0^{a} (79.8–83.4)	75–84
a *	22.8 ± 3.5^{a} (17.4–28.3)	21.6 ± 3.0^{a} (18.5–26.9)	20.1 ± 2.3^{a} (16.1–23.8)	23.1 ± 3.1^{a} (20.1–29.3)	21.4 ± 2.4^{a} (16.5–25.0)	19–32
b′ *	$5.1 \pm 0.6^{a,b,c}$ (4.3–6.4)	5.3 ± 0.5^{c} (4.8–6.3)	4.4 ± 0.4^{a} (3.8–5.2)	5.3 ± 0.5^{c} (4.6–6.0)	5.9 ± 2.0^{c} (4.3–10.5)	
c *	21.9 ± 3.3^{a} (18.1–29.6)	22.9 ± 3.2^{a} (17.6–27.2)	25.6 ± 2.8^{b} (21.1–28.9)	21.2 ± 2.3^{a} (18.5–25.6)	21.3 ± 2.3^{a} (18.2–24.6)	15–24
c′ *	2.0 ± 0.3^{a} (1.4–2.5)	1.9 ± 0.4^{a} (1.5–2.7)	1.5 ± 0.3^{b} (1.2–1.9)	1.9 ± 0.4^{a} (1.4–2.4)	2.0 ± 0.3^{a} (1.5–2.4)	

Table 1. Cont.

Character	Isolate					Loof (1960)
	PpA21L2	PpA24L1	PpA34L3	PpA44L2	PpA44L4	
Tail	29.4 ± 4.9 [a,b] (19.0–35.8)	31.8 ± 2.6 [a,b,c] (27.7–35.6)	23.7 ± 3.9 [d] (19.5–31.4)	34.0 ± 4.2 [b,c] (26.6–41.0)	32.8 ± 3.9 [a,b,c] (25.9–38.4)	
V *	81.1 ± 2.2 [a] (75.7–83.7)	81.0 ± 1.7 [a] (77.7–83.9)	82.5 ± 2.2 [a] (78.3–83.9)	80.6 ± 2.0 [a] (77.0–83.5)	80.3 ± 4.0 [a] (79.8–83.4)	75–84
a *	22.8 ± 3.5 [a] (17.4–28.3)	21.6 ± 3.0 [a] (18.5–26.9)	20.1 ± 2.3 [a] (16.1–23.8)	23.1 ± 3.1 [a] (20.1–29.3)	21.4 ± 2.4 [a] (16.5–25.0)	19–32
b′ *	5.1 ± 0.6 [a,b,c] (4.3–6.4)	5.3 ± 0.5 [c] (4.8–6.3)	4.4 ± 0.4 [a] (3.8–5.2)	5.3 ± 0.5 [c] (4.6–6.0)	5.9 ± 2.0 [c] (4.3–10.5)	
c *	21.9 ± 3.3 [a] (18.1–29.6)	22.9 ± 3.2 [a] (17.6–27.2)	25.6 ± 2.8 [b] (21.1–28.9)	21.2 ± 2.3 [a] (18.5–25.6)	21.3 ± 2.3 [a] (18.2–24.6)	15–24
c′ *	2.0 ± 0.3 [a] (1.4–2.5)	1.9 ± 0.4 [a] (1.5–2.7)	1.5 ± 0.3 [b] (1.2–1.9)	1.9 ± 0.4 [a] (1.4–2.4)	2.0 ± 0.3 [a] (1.5–2.4)	

** L—body length; V—position of vulva from the anterior end expressed as the percentageof body length; a—body length/maximum body width; b'—body length/distance from the anterior end to the base of esophageal glands; c—body length/tail length; c'—tail length/tail diameter at the anus.*

2.1.2 Male

Males were common in all the isolates, morphologically similar to females for all non-sexual characters but smaller, with body length 555.64 (470.5–670.1) μm long (Figure 1B and Table 2). Lateral field with four lines ending on bursa, spicules slender, gubernaculum ventrally curved. Bursa irregularly crenate along margin, enveloping the tail tip (Figure 1J,K).

Table 2. Morphometric characters of Pratylenchus penetrans males from Portugal. All measurements are in μm. Data are means of 10 nematodes ± standard deviation (range). In each row, means followed by the same letters do not differ significantly at p > 0.05, according to the Fisher Least Significant Difference test.

Character	Isolate					Loof (1960)
	PpA21L2	PpA24L1	PpA34L3	PpA44L2	PpA44L4	
L *	535.0 ± 30.7 [a] (483.3–582.4)	582.6 ± 35.0 [b] (529.0–629.4)	522.0 ± 32.2 [a] (470.5–580.5)	602.6 ± 36.0 [b] (531.5–670.1)	536.0 ± 17.9 [a] (508.2–570.0)	305–574
Stylet length	15.8 ± 0.4 [a] (15.3–16.7)	15.3 ± 0.4 [a] (14.9–15.7)	15.7 ± 0.6 [a] (15.0–16.5)	15.6 ± 0.6 [a] (14.8–16.7)	15.9 ± 1.1 [a] (14.4–17.5)	
Anterior end to medium bulb	57.5 ± 3.1 [a,b] (52.7–62.9)	53.6 ± 4.6 [b,c] (45.7–59.2)	59.5 ± 3.4 [a,b] (54.6–63.7)	63.7 ± 4.4 [d] (54.9–71.5)	56.0 ± 3.9 [a,b,c] (49.5–60-5)	
Anterior end to esophageal gland lobe	124.0 ± 12.9 [a] (105.2–149.2)	122.3 ± 6.6 [a] (110.9–131.2)	122.5 ± 8.0 [a] (110.7–137.9)	127.7 ± 10.2 [a] (108.5–144.0)	127.9 ± 9.5 [a] (112.0–146.2)	
Anterior end to excretory pore	85.0 ± 5.0 [a] (79.0–92.6)	83.2 ± 6.9 [a] (71.7–94.2)	84.5 ± 7.7 [a] (72.3–97.6)	94.9 ± 6.0 [b] (81.7–104.1)	87.1 ± 5.7 [a] (77.0–95.9)	
Maximum body width	21.9 ± 2.4 [a] (18.2–27.1)	20.8 ± 1.7 [a] (17.7–23.5)	21.3 ± 1.8 [a] (19.1–25.0)	21.2 ± 1.7 [a] (19.2–24.4)	19.8 ± 4.3 [a] (8.6–23.2)	
Body width at anus	14.0 ± 1.4 [a] (12.0–15.7)	13.3 ± 0.7 [a] (12.4–14.6)	13.3 ± 0.7 [a] (12.2–14.9)	14.3 ± 1.2 [a] (12.3–15.8)	13.8 ± 0.7 [a] (12.9–14.8)	
Spicule	16.2 ± 0.8 [a] (15.2–17.5)	15.8 ± 1.5 [a] (13.8–18.0)	19.1 ± 1.6 [b] (16.6–21.6)	18.6 ± 1.7 [b] (15.9–21.9)	16.7 ± 1.4 [a] (14.4–18.5)	14–17
Tail	25.6 ± 2.9 [a,b] (22.4–32.7)	29.0 ± 4.1 [b,c] (21.7–35.4)	21.9 ± 2.6 [d] (16.1–24.9)	28.2 ± 3.8 [a,b,c] (22.5–36.0)	27.4 ± 2.8 [a,b,c] (23.6–31.8)	
a *	24.6 ± 2.5 [a] (19.8–29.4)	28.2 ± 2.6 [b] (25.3–34.4)	24.6 ± 2.4 [a] (19.9–27.1)	28.5 ± 1.6 [b] (26.8–30.5)	29.3 ± 11.9 [b] (22.8–62.6)	23–34
b′ *	4.4 ± 0.5 [a] (3.4–4.9)	4.8 ± 0.4 [b] (4.2–5.5)	4.3 ± 0.2 [a] (4.0–4.6)	4.7 ± 0.2 [b] (4.4–5.1)	4.2 ± 0.4 [a] (3.7–4.8)	
c *	21.1 ± 2.5 [a] (16.4–24.3)	20.4 ± 2.5 [a] (15.6–24.4)	24.2 ± 3.4 [b] (20.7–31.4)	21.6 ± 1.9 [a] (18.6–24.1)	19.7 ± 1.9 [a] (16.9–22.8)	16–22
c′ *	1.8 ± 0.2 [a,b] (1.6–2.2)	2.2 ± 0.4 [b,c] (1.5–2.7)	1.6 ± 0.2 [a] (1.2–2.0)	2.0 ± 0.3 [a,b,c] (1.5–2.4)	2.0 ± 0.3 [a,b,c] (1.7–2.5)	

**L—body length; V—position of vulva from anterior end expressed as percentage of body length; a—body length/maximum body width; b'—body length/distance from anterior end to base of esophageal glands; c—body length/tail length; c'—tail length/tail diameter at anus.*

2.2 Morphometry of P. penetrans Portuguese Isolates

The morphometric measurements of *P. penetrans* isolates of Portugal were, in average, within the range described by Loof after [31], except for the c ratio in both PpA34L3 females and males, overall body length of PpA24L1 and PpA44L2 males and spicule length of PpA34L3 and PpA44L2 (Tables 1 and 2). Morphometric comparisons using ANOVA revealed a significant degree of intra- and inter-isolate variability on most studied characters. Nine out of fifteen morphometric characters studied in *P. penetrans* females, varied significantly among isolates ($p < 0.05$) (Table 1). Inter-isolate variability was high for the overall body length, anterior end to excretory pore, anterior end to vulva, body width at anus, vulva–anus distance, tail length and ratios b', c and c', whereas the stylet length, distance of anterior end to median bulb, anterior end to pharyngeal gland lobe, maximum body width, V% and ratio a did not vary significantly among isolates ($p > 0.05$). The stylet length and V% had the lowest CV intra and inter-isolates of females and the highest values of CV were found in tail length, vulva–anus distance and b' ratio (Table 3). In males, inter-isolate variability was found in nine out of thirteen morphometric characters: overall body length, anterior end to median bulb, anterior end to excretory pore, spicule length, tail and a, b', c and c' ratios ($p < 0.05$). The stylet length, distance from the anterior end to the tip of esophageal glands, body width at the anus and maximum body width were similar among isolates ($p > 0.05$) (Table 2). The stylet was the least variable character, whereas tail, c' ratio and spicule length were the most variable among isolates, supporting the results given by the ANOVA (Table 4).

Table 3. Intra- and inter-isolate coefficient of variability (%) of *Pratylenchus penetrans* females from Portugal.

Character			Isolate			Inter-Isolate Coefficient of Variability (%)
	PpA21L2	PpA24L1	PpA34L3	PpA44L2	PpA44L4	
L *	9.4	13.0	14.4	8.6	11.8	8.0
Stylet length	4.6	3.9	7.2	3.9	4.5	2.2
Anterior end to medium bulb	6.4	11.9	8.2	4.4	8.7	3.8
Anterior end to esophageal gland lobe	12.5	11.2	7.6	6.9	17.4	4.6
Anterior end to excretory pore	10.3	9.5	10.6	4.1	16.5	6.6
Anterior end to vulva	8.9	13.2	15.7	9.4	13.9	7.3
Maximum body width	18.2	19.8	12.4	14.4	17.9	7.5
Body width at anus	11.2	14.5	11.3	15.3	13.4	8.4
Vulva-anus	18.7	16.6	18.3	17.2	12.1	11.3
Tail	16.6	8.3	16.5	12.4	12.0	13.5
V *	2.8	2.0	2.7	2.5	5.0	1.0
a *	15.3	13.8	11.3	13.2	11.4	5.5
b' *	12.7	8.5	9.1	9.6	33.5	10.5
c *	14.9	13.9	10.8	10.9	10.6	8.1
c' *	16.2	19.2	17.1	19.2	14.3	10.1

**L—body length; V—position of vulva from anterior end expressed as percentage of body length; a—body length/maximum body width; b'—body length/distance from anterior end to base of esophageal glands; c—body length/tail length; c'—tail length/tail diameter at the anus.*

Table 4. Intra- and inter-isolate coefficient of variability (%) of *Pratylenchus penetrans* males from Portugal.

Character	Isolate					Inter-Isolate Coefficient of Variability (%)
	PpA21L2	PpA24L1	PpA34L3	PpA44L2	PpA44L4	
L*	5.7	6.0	6.2	6.0	3.3	6.3
Stylet length	2.7	2.8	3.6	3.9	6.7	1.4
Anterior end to medium bulb	5.4	8.5	5.7	6.9	7.0	6.6
Anterior end to esophageal gland lobe	10.4	5.4	6.5	8.0	7.5	2.2
Anterior end to excretory pore	5.8	8.3	9.1	6.3	6.5	5.4
Maximum body width	11.0	8.1	8.3	8.1	21.7	3.7
Body width at anus	9.9	5.6	5.3	8.4	5.0	3.3
Spicule	4.9	9.3	8.3	8.9	8.4	8.7
Tail	11.3	14.2	12.0	13.4	10.2	10.7
a*	10.3	9.2	9.6	5.6	40.7	8.3
b'*	10.7	8.1	4.5	4.4	8.7	5.9
c*	11.6	12.2	13.9	8.9	9.6	8.0
c'*	10.2	16.4	13.3	13.6	12.5	10.5

**L—body length; V—position of vulva from anterior end expressed as percentage of body length; a—body length/maximum body width; b'—body length/distance from anterior end to base of esophageal glands; c—body length/tail length; c'—tail length/tail diameter at the anus.*

2.3 Genetic Diversity of P. penetrans Portuguese Isolates

ITS sequences of three clones from each isolate were determined and submitted in GenBank database under the accession numbers MW633839-MW633853. For the COI gene, sequences of two clones from isolate PpA21L2 and three clones from the other isolates were determined and also submitted under the accession numbers MW660605-MW660618. A BLAST search against NCBI database of the determined ITS and COI sequences con- firmed the species identity, with sequences homologies ranging from 94.7% to 98.4%, and 99.2% to 100.00%, to other *P. penetrans* ITS and COI sequences, respectively, present in the database. The length variation on ITS region of all clones (671–683 bp) and the sequence analysis revealed high variability, not only between isolates but also within isolates, with a high number of polymorphic (S), mutation (Eta) sites and average number of nucleotide differences (k) then the ones found for COI region. All 15 ITS sequences and 14 COI se- quences corresponded, respectively, to 15 and 7 different haplotypes (Table 5). Intra-isolate nucleotide diversity (Pi) for the ITS region was lower for the PpA34L3 isolate (Pi = 0.00997) and higher for the PpA44L2 isolate (Pi = 0.06115). For the COI gene, a low number of polymorphic and mutation sites were found considering each isolate or even considering all isolates. The COI intra-isolate Pi was lower for PpA44L2 isolate (Pi = 0.00000), with all three clones being identical, and higher for PpA21L2 isolate (Pi = 0.00509). Considering all isolates, a higher Pi was found for the ITS region (Pi = 0.03350) than for the COI gene (Pi = 0.00587) (Table 5).

Table 5. Genetic diversity of cloned ITS and COI regions of five *Pratylenchus penetrans* isolates from Portugal.

Isolate	Genomic Region	No. of Clones	Sequences Length (bp)	S *	Eta *	No. of Haplotypes	Hd * (Standard Deviation)	Pi * (Standard Deviation)	K *
PpA21L2	ITS	3	677; 683; 673	35	35	3	1.000 (0.272)	0.03488 (0.00254)	23.333
	COI	2	393	2	2	2	1.000 (0.500)	0.00509 (0.00964)	2.000
PpA24L1	ITS	3	676; 673; 671	38	42	3	1.000 (0.272)	0.03992 (0.01097)	26.667
	COI	3	393	1	1	2	0.667 (0.314)	0.00170 (0.00080)	0.667
PpA34L3	ITS	3	671; 671; 673	10	10	3	0.667 (0.314)	0.00997 (0.00425)	6.667
	COI	3	393	1	1	2	1.000 (0.272)	0.00170 (0.00080)	0.667
PpA44L2	ITS	3	673; 675; 678	60	62	3	1.000 (0.272)	0.06115 (0.01759)	40.667
	COI	3	393	0	0	1	0.000 (0.000)	0.0000 (0.0000)	0.000
PpA44L4	ITS	3	676; 674; 674	20	20	3	1.000 (0.272)	0.00339 (0.00160)	13.333
	COI	3	393	2	2	2	0.667 (0.314)	0.01990 (0.00765)	1.333
All 5 isolates	ITS	15	-	99	109	15	1.000 (0.024)	0.03350 (0.00414)	21.743
	COI	14	-	10	10	7	0.758 (0.116)	0.00587 (0.00164)	2.308

**S—number of polymorphic sites; Eta—total number of mutations; Hd—haplotype diversity; Pi—nucleotide diversity; k—average number of nucleotide differen*

2.4 Phylogenetic and Molecular Evolution Relationships

Phylogenetic analysis was performed with the alignment of the 15 sequences obtained in this study and other ITS sequences from *P. penetrans*, *P. fallax*, *P. pinguicaudatus* and *P. thornei* present in the GenBank database. Results showed that *P. penetrans* isolates from Portugal clearly group up with other *P. penetrans* isolates but ITS sequences from the same isolates do not group together, reflecting the high intra- and inter-isolate estimated ITS divergence. Additionally, no grouping of isolates belonging to the same country or origi- nated from the same host was found (Figure 2). On the other hand, phylogenetic analysis based on COI sequences revealed lower divergence between sequences from the same isolate and also from different isolates, comparing to the ITS region phylogenetic analysis. All Portuguese *P. penetrans* COI sequences grouped together and with other *P. penetrans* isolates, revealing a closer relationship with one Dutch isolate from apple (KY816941), one African isolate from onion (KY817013) and five American isolates from potato (MK877988; MK877989; MK877990; MK877991 and MK877992) (Figure 3). The differences between the *P. pinguicaudatus*, *P. fallax* and *P. thornei*, included in the phylogenetic analysis, were visible on both trees, as they did not group together (Figures 2 and 3).

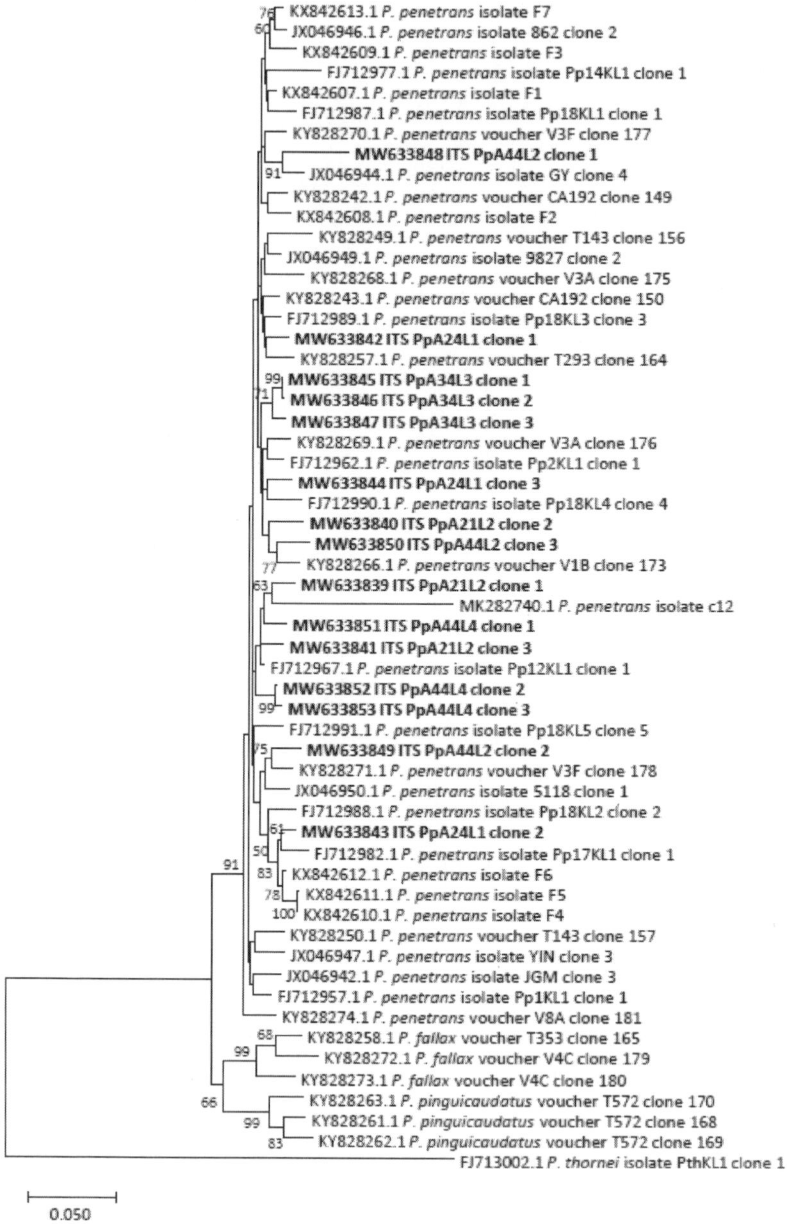

Figure 2. Neighbor-joining phylogenetic tree based on ITS nucleotide sequences of ***Pratylenchus penetrans, P. pinguicaudatus*** and ***P. fallax***. ITS sequence from *P. thornei* was used as an outgroup. Bootstrap values are shown next to the branches and values with less than 50% confidence were not shown. Scale bar represents nucleotide substitutions per site.

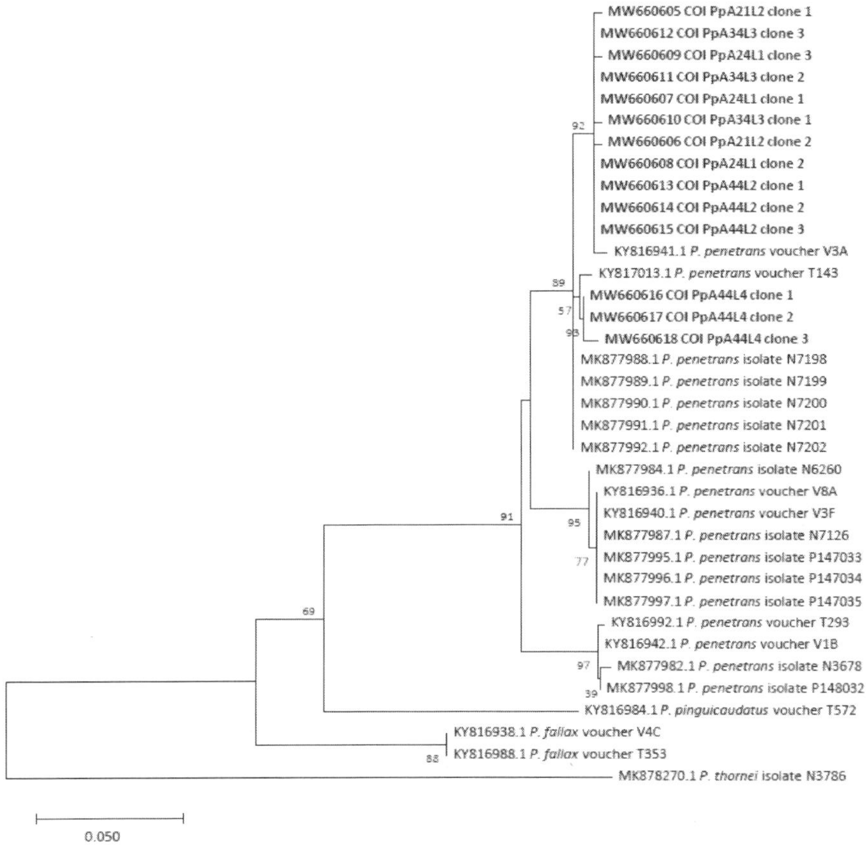

Figure 3. Neighbor-joining phylogenetic tree based on COI nucleotide sequences of *Pratylenchus penetrans*, *P. pinguicaudatus* and *P. fallax*. COI sequence from *P. thornei* was used as the outgroup. Bootstrap values are shown next to the branches and values with less than 50% confidence are not shown. Scale bar represents nucleotide substitutions per site.

The estimate of evolutionary divergence between sequences of P. penetrans Portuguese isolates showed that ITS region diverges by at least 0.01513 base substitutions per site (± 0.00487), considering different isolates and that value decreases for 0.00149 (± 0.00149), considering intra-isolate divergence (isolate PpA34L3). However, there were ITS sequences from clones from the same isolate with an estimated divergence higher than from distinct isolates. The higher value of base substitutions per site, 0.08121 (± 0.01253), was found between the PpA44L2 isolate, clone 1, and PpA24L1 isolate, clone 2 (Table S1). On the other hand, the COI gene revealed much lower nucleotide divergence with a minimum of 0.00000 base substitutions per site (±0.00000), considering both, intra and inter-isolate divergence. A maximum of 0.01821 (± 0.00667) base substitutions per base on the COI gene was found between PpA21L2 isolate, clone 1 and PpA44L4 isolate, clone 3 (Table S2).

From neutrality tests, estimated Tajima's D values, using the total number of mu- tations, were −1.54235 (p > 0.05) and 1.03620 (p > 0.05) for ITS and COI respectively, indicating that the changes were not significant and all sequences underwent neutral selection. Additionally, the mismatch distribution of both ITS and COI sequences revealed to be a multimodal distribution, with several peaks of pairwise differences, excluding the possibility of abrupt selection events (Figure 4).

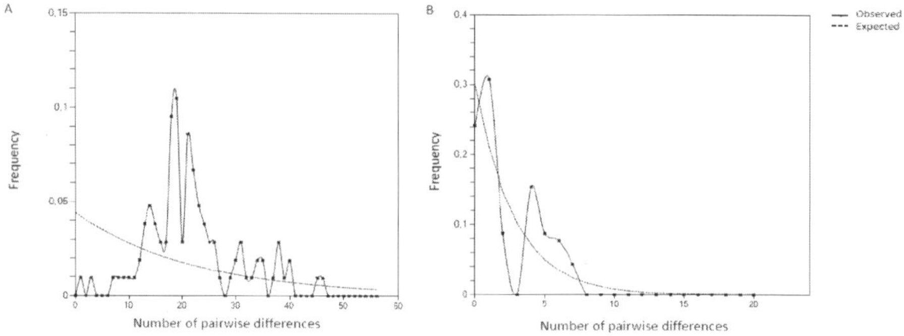

Figure 4. Mismatch distribution of ITS (A) and COI (B) sequences of five Portuguese isolates of *Pratylenchus penetrans*.

The possible correlation of genetic distance and geographic distance of the five *P. penetrans* isolates were also investigated, considering both ITS and COI gene, and there were no significant correlation between this two variables with a Kendall tau of 0.02458 ($p > 0.05$) for ITS region and a Kendall tau of 0.08254 ($p > 0.05$) for the COI gene, showing that geographical distance is not the main factor leading to *P. penetrans* isolates differentiation.

3. Discussion

In this study, *P. penetrans* isolates from potato in different geographic locations of Portugal were characterized for the first time, using both morphometric and molecular analyses. The comparative morphometrical analyses revealed the presence of substantial inter and intra variability between isolates, although differences fall within the range of the morphometrical variability described previously in *P. penetrans* [3,31]. The body size of these isolates appears to be larger than that described by Rusinque et al. in *P. penetrans* parasitizing amaryllis (*Hippeastrum × hybridum*), in Portugal [32]. Spicule size of males and overall body length of the Portuguese isolates were also greater than those observed by Mokrini et al. in populations associated to maize (*Zea mays* L.) in Morocco [33]. Variations in morphometric characters can be caused from differences in fixation methods or changes in environmental conditions [34]. The morphometric characters of Portuguese isolates were recorded on fresh mounted nematodes (not glycerin-infiltrated specimens) and compared with type specimens in permanent mounts, and therefore affected by "shrinkage" due to the fixation process. Environmental conditions, like the host plant, influence morphometric characters such as body length, width, esophagus length, stylet length, V value, a and b ratios and qualitative characters such as tail terminus, growth of ovary and shape of the median bulb [14]. Townshend [35] reported that morphometric variations existed between populations of *P. penetrans* associated with strawberry (*Fragaria ananassa*) and those asso- ciated with celery (*Apium graveolens* L.) in Ontario, Canada. Furthermore, variations in size were also found between *P. penetrans* populations recovered from strawberry collected at different geographical areas [35]. In our study, intra- and inter-isolate variability was found in most of the morphometric characters that were analyzed in females and males. However, the results obtained with ANOVA and the analysis of the CV allowed one to verify that the characters V and stylet length proved to be stable among isolates and between replicates within the same isolate. As previously noted by Roman and Hirschmann [13] and Tarte and Mai [14], the stability of these characters confirms its usefulness for discriminating this species. All other morphometrical characters, including those commonly used in nematode taxonomy (body length, body width, anterior end to esophageal glands and a, b', c and c' ratios), have shown relatively high coefficients of variation.

The ITS and COI genomic regions from the five Portuguese *P. penetrans* isolates were selected for sequencing to evaluate the intraspecific genetic diversity of this species. From the two regions, the ITS region revealed higher genetic diversity than the COI gene with 15 and 6 different haplotypes from the 15 ITS and 14 COI sequences, respectively. Besides, inter-isolate genetic diversity also intra-isolate genetic diversity was found in all isolates with exception for one isolate in the COI gene. Sequence comparisons performed by De Luca et al. [17] revealed high intraspecific variability in ITS sequences of several *Pratylenchus* species, including *P. penetrans*. Sequence analyses showed high sequence variability not only between populations or isolates but also within individuals. The same study concluded that ITS sequences allow a clear separation of the *Pratylenchus* species, despite the high intraspecific variability. Janssen et al. [15] also reported intraspecific variability of *P. penetrans* isolates based on sequence analysis and phylogenetic reconstruction of the ITS, D2-D3 regions of 28S rDNA and the COI gene. Furthermore, the phylogenetic analy- ses based on the sequences of the ITS and D2-D3 regions also confirmed high sequence variability among populations of *P. penetrans* [29].

Despite the high intraspecific diversity found for *P. penetrans* in our studies, phylogenetic analyses revealed that both ITS and COI genomic regions separate *P. penetrans* from other related species, such as *P. pinguicaudatus*, *P. fallax* and *P. thornei*, which is also in accordance with that previously reported [15,17,29]. Additionally, no grouping of isolates belonging to the same country or originated from the same host was found in phylogenetic analyses of both ITS and COI genomic regions. This is in agreement with the no correlation of genetic and geographic distance found among the Portuguese isolates, being the same COI haplotypes from isolates sampled in fields that are more than 90 km away, suggesting that geographical distance is not the main factor leading to the differentiation of isolates. Janssen et al. [15] referred that although the large intraspecific variability recovered in *P. penetrans*, identical haplotypes were found to be geographically widespread and this could be a result of the anthropogenic spread of *P. penetrans* through agriculture devel- opment and crop exchange. Our findings contribute to increase the understanding of *P. penetrans* variability.

4. Materials and Methods

4.1 Pratylenchus penetrans Isolates

Five *P. penetrans* isolates, obtained previously from potato roots sampled in the north and centre regions of mainland Portugal [12], were used in this study. The isolates were originated from a gravid female and propagated on carrot discs [36]. Isolates PpA21L2, PpA4L1 and PpA34L3 are from potato fields in different geographical locations, whereas PpA44L2 and PpA44L4 shared the same sampling geographic origin (Table 6).

4.2 Morphometrical Analyses

Twenty individual adults (10 females and 10 males), from each isolate, were mounted into a drop of water and used for the morphometric analyses. Before covering and sealing slides with the coverslips, nematodes were immobilized by gently heating the slide underneath, just enough to stop movement. Nematode measurements were made directly using a DM2500 microscope equipped with a ICC50HD digital camera (Leica Microsystems, Wetzlar, Germany) and LAS 4.8.0 software (Leica) and results compared with previous de- scriptions for this species [31]. Microscopic observations were made in nematodes without using a fixation method since the software used for nematode measurements allows the capture of the image and simultaneous measurement of specimens, without the need of a preservation method. All measurements were expressed in micrometers (µm). To assess the morphometric variation of the isolates, data was subjected to a one-way analysis of variance (ANOVA) using Statistica® V.7 (StatSoft, Tulsa, Germany), after ensuring that the assumptions of normality and constant variance were met, as checked by using the Shapiro–Wilk and Levene's tests, respectively. Logarithmic and square root transformations were applied to data whenever needed. Following ANOVA, to test differences between isolates Fisher Least Significant Difference test at the 95% confidence level was applied. The coefficients of variability (CV) were calculated to determine which characters were most stable and more variable among isolates.

Table 6. *Pratylenchus penetrans* isolates used in this study, respective geographical origin and GenBank accession numbers.

Isolate	GPS Coordinates	Locality	Accession (ITS)	Accession (COI)
PpA21L2	41°16'18"N 8°41'23" W	Aveleda, Maia, Portugal	MW633839 MW633840 MW633841	MW660605 MW660606 -
PpA24L1	41°15'27"N 8°40'30" W	Vila Nova da Telha, Maia, Portugal	MW633842 MW633843 MW633844	MW660607 MW660608 MW660609
PpA34L3	40°37'28"N 8°38'19" W	Aveiro, Portugal	MW633845 MW633846 MW633847	MW660610 MW660611 MW660612
PpA44L2	40°23'25.2" N 8°30'07.7" W	Coimbra, Portugal	MW633848 MW633849 MW633850	MW660613 MW660614 MW660615
PpA44L4	40°23'25.2" N 8°30'07.7" W	Coimbra, Portugal	MW633851 MW633852 MW633853	MW660616 MW660617 MW660618

4.3 DNA Extraction, PCR, Cloning and Sequencing

Nematode DNA was extracted from 50 to 100 mix developmental stages of *P. penetrans* PpA21L2, PpA24L1, PpA34L3, PpA44L2 and PpA44L4 isolates (Table 6) using the DNeasy® Blood and Tissue Mini kit (Qiagen, Hilden, Germany) following the manufac- turer's instructions.

Two genomic regions were selected to evaluate the intraspecific genetic diversity of this species, the internal transcribed spacers (ITS) rDNA region containing partial 18S and 28S and complete ITS1, 5.8S and ITS2 sequences and partial cytochrome c oxidase subunit I (COI) gene.

PCR amplifications were carried out using 20–50 ng extracted DNA and2U of BioTaq DNA polymerase (Meridian Bioscience, Memphis, TN, USA) in the 1 ×reaction buffer, 0.2 mM each dNTPs, 1.25 mM $MgCl_2$ and 2.0 µM of each primer, PRATTW81 (5′GTAGGTGAACCTGCTGCTG3′) and AB28 (5′ATATGCTTAAGTTCAGCGGGT3′) for ITS region [16] and JB3 (5′TTTTTTGGGCATCCTGAGGTTTAT3′) and JB4.5 (5′TAAAGAAA GAACATAATGAAAATG3′) for the COI gene [37]. Reactions were carried out in a Thermal Cycler (Bio-Rad, California, USA) with an initial denaturation step of 95 °C for 3 min followed by 35 reaction cycles of 94 °C for 30 s, annealing for 30 s at 60 °C and 54 °C for ITS region and COI region, respectively, extension at 72 °C for 30 s and a final extension at 72 °C for 7 min. The PCR products were purified using the NucleoSpin® Gel and PCR Clean-up kit (Macherey-Nagel, Duren, Germany) according to the manufacturer's instructions and cloned. Purified ITS and COI amplified products were ligated into pGEM®-T Easy Vector (Promega, Madison, USA using 50 ng vector in a 10 µL reaction with 3 U T4 DNA Ligase (Promega) and 36 ng purified ITS or 22 ng of COI products in the 1 ×Rapid Ligation Buffer (Promega). Ligation reactions were incubated for 1 h at room temperature. Then, 2 µL of the ligation product was used to transform *Escherichia coli* JM109 high efficiency competent cells (Promega) following the manufacturer's instructions. Plasmid DNA was extracted from *E. coli* cells using the Nzymini Prep kit (Nzytech, Lisbon, Portugal and three selected positive clones for each genomic region and each *P. penetrans* isolate were fully sequenced in both strands in an Automatic Sequencer 3730xl under BigDyeTM terminator cycling conditions at Macrogen Company (Madrid, Spain).

4.4 Sequence Analysis

Sequence analysis and alignments were achieved using BioEdit [38]. The region containing primers sequence was removed from all sequence analyses. Homologous sequences in the databases were searched using the Basic Local Alignment Search Tool [39]. Sequence statistics such as number of polymorphic (S) and mutation (Eta) sites, nucleotide diversity (Pi), haplotype diversity (Hd), average number of nucleotide differences (k) and mismatch distributions were estimated using DnaSP 6.12.03 software [40]. Intra-isolate sequence analyses were performed from the alignments obtained with sequences of each isolate and overall sequence diversity with the alignment obtained with all sequences of the five isolates.

4.5 Phylogenetic and Molecular Evolutionary Analyses

Phylogenetic and molecular evolutionary analyses were conducted in MEGA v10.1.8 software [41]. Phylogenetic trees were constructed by the neighbor-joining method [42] with 1000 replications of bootstrap, with the evolutionary distances computed using the maximum composite likelihood [43] model and ambiguous positions removed for each sequence pair (pairwise deletion option), using the ITS and COI nucleotide sequence alignments of the five isolates used in this study and homologous sequences retrieved from the GenBank database (Table 7). Genetic distance between sequences from the five Portuguese isolates were accomplished by pairwise distance using the maximum composite likelihood model with pairwise deletion option and standard error estimate by a bootstrap procedure (1000 replicates), using the alignments of ITS and COI nucleotide sequences determined in this study. Additionally, Tajima's D neutrality tests [44], which distinguish between a DNA sequence evolving randomly (or neutrally) and one evolving under a non-random process, and mismatch distribution of ITS and COI sequences of Portuguese *P. penetrans* isolates were performed in DnaSP v6.12.03 software.

Table 7. Sequences used in this study.

				Acession Number	
Species	Isolate	Region	Host	ITS	COI
Pratylenchus fallax	T353	The Netherlands, Doornenburg	*Malus pumila*	KY828258	KY816988
P. fallax	V4 C	The Netherlands, Ysbrechtum	*Vitis vinifera*	KY828272;KY828273	KY816938
P. penetrans	V3 A	The Netherlands, Baarlo	M. pumila	KY828268;KY828269	KY816941
P. penetrans	V8 A	The Netherlands, Baarlo	M. pumila	KY828274	KY816936
P. penetrans	V1B	The Netherlands, Meijel	M. pumila	KY828266	KY816942
P. penetrans	V3 F	The Netherlands, Nagele	M. pumila	KY828270; KY828271	KY816940
P. penetrans	N3678	USA, Minnesota	Zea mays	-	MK877982
P. penetrans	N6260	USA, Fairbanks County	*Paeonia* sp.	-	MK877984
P. penetrans	N7126	USA, Otoe County	*Malus* sp.	-	MK877987
P. penetrans	N7198	USA, Idaho	Solanum tuberosum	-	MK877988
P. penetrans	N7199	USA, Idaho	S. tuberosum	-	MK877989
P. penetrans	N7200	USA, Idaho	S. tuberosum	-	MK877990
P. penetrans	N7201	USA, Idaho	S. tuberosum	-	MK877991
P. penetrans	N7202	USA, Idaho	S. tuberosum	-	MK877992
P. penetrans	P148032	USA, Portage County	S. tuberosum	-	MK877998
P. penetrans	P147033	USA, Portage County	S. tuberosum	-	MK877995
P. penetrans	P147034	USA, Portage County	S. tuberosum	-	MK877996
P. penetrans	P147035	USA, Portage County	S. tuberosum	-	MK877997
P. penetrans	c12	Canada, Kentville	*Prunus* sp.	MK282740	-
P. penetrans	862	Chile	*Lillium* sp.	JX046946	-
P. penetrans	GY	France	*Prunus* sp.	JX046944	-
P. penetrans	JGM	France	*Sambucus* sp.	JX046942	-
P. penetrans	CA192	France, Britany	M. pumila	KY828242;KY828243	-
P. penetrans	Pp18KL1	Long Island, USA	S. tuberosum	FJ712987	-

P. penetrans	Pp18KL2	Long Island, USA	S. tuberosum	FJ712988	-
P. penetrans	Pp18KL3	Long Island, USA	S. tuberosum	FJ712989	-
P. penetrans	Pp18KL4	Long Island, USA	S. tuberosum	FJ712990	-
P. penetrans	Pp18KL5	Long Island, USA	*S. tuberosum*	FJ712991	-
P. penetrans	F1	MN, USA	*S. tuberosum*	KX842607	-
P. penetrans	F2	MN, USA	*S. tuberosum*	KX842608	-
P. penetrans	F3	MN, USA	*S. tuberosum*	KX842609	-
P. penetrans	F4	MN, USA	*S. tuberosum*	KX842610	-
P. penetrans	F5	MN, USA	*S. tuberosum*	KX842611	-
P. penetrans	F6	MN, USA	*S. tuberosum*	KX842612	-
P. penetrans	F7	MN, USA	*S. tuberosum*	KX842613	-
P. penetrans	Pp17KL1	Monroe County, USA	*Prunus cerasus*	FJ712982	-
P. penetrans	Pp12KL1	Rennes, France	*Malus* sp.	FJ712967	-
P. penetrans	T143	Rwanda, Nyakiriba	*Allium. cepa*	KY828249;KY828250	KY817013
P. penetrans	Pp14KL1	Spain	*Malus* sp.	FJ712977	-
P. penetrans	9827	The Netherlands	*Iris* sp.	JX046949	-
P. penetrans	5118	The Netherlands	*Lillium* sp.	JX046950	-
P. penetrans	T293	The Netherlands, Apeldoorn	Pyrus sp.	KY828257	KY816992
P. penetrans	Pp1KL1	Tongeren, Belgium	Rubus sp.	FJ712957	-
P. penetrans	YIN	USA	Acer x freemanii	JX046947	-
P. penetrans	Pp2KL1	Zandhoven, Belgium	Z. mays	FJ712962	-
P. pinguicaudatus	T572	UK, England, Rothemstadt	Triticum sp.	KY828261;KY828262; KY828263KY816984	
P. thornei	N3786	California, USA	V. vinifera	-	MK878270
P. thornei	PthKL1	Santaella, Spain	Cicer arietinum	FJ713002	-

The correlation between genetic and geographic distance of Portuguese *P. penetrans* isolates was also evaluated computing the determined pairwise distance versus the distance between the sampling locations of each of the five isolates. Geographic distance between isolates was calculated using the script available at https://www.movable-type.co.uk/ scripts/latlong.html (accessed on 15 January 2021) with the GPS coordinates of each isolate sampling location (Table 6 and Table S3). The significance of genetic and geographic distance correlation was calculated using Kendall tau rank correlation in Free Statistics Software v1.2.1 [45].

5. Conclusions

In conclusion, morphometric and genetic diversity were found among *P. penetrans* isolates and this variability was not only a result of the diversity found between isolates but also due to the diversity within each isolate. The information gathered highlights the importance of the knowledge about this relevant plant–parasitic nematode in potato crops, and can be used further in larger genetic studies, focusing this nematode species. Future research should also be conducted to evaluate whether the differences in pathogenicity among *P. penetrans* isolates are related to the observed morphometric and molecular variability.

References

1. Cobb, N.A. A new parasitic nema found infesting cotton and potatoes. J. Agric. Res. 1917, 11, 27–33.
2. Filipjev, I.N.; Stekhoven, J.H.S. A Manual of Agricultural Helminthology; Brill: Leiden, The Netherlands, 1941.
3. Castillo, P.; Vovlas, N. Pratylenchus (Nematoda: Pratylenchidae): Diagnosis, Biology, Pathogenicity and Management; Brill: Leiden, The Netherlands, 2007.
4. Rowe, R.C.; Riedel, R.M.; Martin, M.J. Synergistic interactions between Verticillium dahliae and Pratylenchus penetrans in potato early dying disease. Phytopathology 1985, 75, 412–418. [CrossRef]
5. Kimpinski, J. Root lesion nematodes in potatoes. Am. Potato J. 1979, 56, 79–86. [CrossRef]
6. Brown, M.J.; Riedel, R.M.; Rowe, R.C. Species of Pratylenchus associated with Solanum tuberosum cv. Superior in Ohio. J. Nematol. 1980, 12, 189–192. [PubMed]
7. Olthof, T.H.A.; Anderson, R.V.; Squire, S. Plant-parasitic nematodes associated with potatoes (Solanum tuberosum L.) in Simcoe County, Ontario. Can. J. Plant Pathol. 1982, 4, 389–391. [CrossRef]
8. MacGuidwin, A.E.; Rouse, D.I. Role of Pratylenchus penetrans in potato early dying disease of Russet Burbank potato. Phytopathol- ogy 1990, 80, 1077–1082. [CrossRef]
9. Ingham, R.E.; Hamm, P.B.; Riga, E.; Merrifield, K.J. First report of stunting and root rot of potato associated with Pratylenchus penetrans in the Columbia Basin of Washington. Plant Dis. 2005, 89, 207. [CrossRef]
10. Holgado, R.; Skau, K.A.O.; Magnusson, C. Field damage in potato by lesion nematode Pratylenchus penetrans, its association with tuber symptoms and its survival in storage. Nematol. Mediterr. 2009, 37, 25–39.
11. Hafez, S.L.; Sundararaj, P.; Handoo, Z.A.; Siddiqi, M.R. Occurrence and distribution of nematodes in Idaho crops. Int. J. Nematol. 2010, 20, 91–98.
12. Esteves, I.; Maleita, C.; Abrantes, I. Root-lesion and root-knot nematodes parasitizing potato. Eur. J. Plant Pathol. 2015, 141, 397–406. [CrossRef]
13. Roman, J.; Hirschmann, H. Morphology and morphometrics of six species of Pratylenchus. J. Nematol. 1969, 1, 363–386. [PubMed]
14. Tarte, R.; Mai, W.F. Morphological variation in Pratylenchus penetrans. J. Nematol. 1976, 8, 185–195. [PubMed]

15. Janssen, T.; Karssen, G.; Orlando, V.; Subbotin, S.A.; Bert, W. Molecular characterization and species delimiting of plant-parasitic nematodes of the genus Pratylenchus from the Penetrans group (Nematoda: Pratylenchidae). Mol. Phylogenet. Evol. 2017, 117, 30–48. [CrossRef]

16. Waeyenberge, L.; Ryss, A.; Moens, M.; Pinochet, J.; Vrain, T. Molecular characterisation of 18 Pratylenchus species using rDNA restriction fragment length polymorphism. Nematology 2000, 2, 135–142. [CrossRef]

17. De Luca, F.; Reyes, A.; Troccoli, A.; Castillo, P. Molecular variability and phylogenetic relationships among different species and populations of Pratylenchus (Nematoda: Pratylenchidae) as inferred from the analysis of the ITS rDNA. Eur. J. Plant Pathol. 2011, 130, 415–426. [CrossRef]

18. Wang, J.; Wei, Y.; Gu, J.; Zhang, R.; Huang, G.; Wang, X.; Li, H.M.; Sun, J. Phylogenetic analysis of Pratylenchus (Nematoda, Pratylenchidae) based on ribosomal internal transcribed spacers (ITS) and D2-D3 expansion segments of 28S rRNA gene. Acta Zootaxonomica Sin. 2012, 37, 687–693.

19. Subbotin, S.A.; Ragsdale, E.J.; Mullens, T.; Roberts, P.A.; Mundo-Ocampo, M.; Baldwin, J.G. A phylogenetic framework for root lesion nematodes of the genus Pratylenchus (Nematoda): Evidence from 18S and D2–D3 expansion segments of 28S ribosomal RNA genes and morphological characters. Mol. Phylogenet. Evol. 2008, 48, 491–505. [CrossRef]

20. Van Megen, H.; van den Elsen, S.; Holterman, M.; Karssen, G.; Mooyman, P.; Bongers, T.; Holovachov, O.; Bakker, J.; Helder, J. A phylogenetic tree of nematodes based on about 1200 full-length small subunit ribosomal DNA sequences. Nematology 2009, 11, 927–950. [CrossRef]

21. Al-Banna, L.; Ploeg, A.T.; Williamson, V.M.; Kaloshian, I. Discrimination of six Pratylenchus species using PCR and species-specific primers. J. Nematol. 2004, 36, 142–146.

22. Handoo, Z.A.; Carta, L.K.; Skantar, A.M. Morphological and molecular characterisation of Pratylenchus arlingtoni n. sp., P. convallariae and P. fallax (Nematoda: Pratylenchidae). Nematol. 2001, 3, 607–618. [CrossRef]

23. Palomares-Rius, J.E.; Guesmi, I.; Horrigue-Raouani, N.; Cantalapiedra-Navarrete, C.; Liébanas, G.; Castillo, P. Morphological and molecular characterisation of Pratylenchus oleae n. sp. (Nematoda: Pratylenchidae) parasitizing wild and cultivated olives in Spain and Tunisia. Eur. J. Plant Pathol. 2014, 140, 53–67. [CrossRef]

24. Troccoli, A.; Subbotin, S.A.; Chitambar, J.J.; Janssen, T.; Waeyenberge, L.; Stanley, J.D.; Duncan, L.; Agudelo, P.; Uribe, G.; Franco, J.; et al. Characterisation of amphimictic and parthenogenetic populations of Pratylenchus bolivianus Corbett, 1983 (Nematoda: Pratylenchidae) and their phylogenetic relationships with closely related species. Nematology 2016, 18, 651–678. [CrossRef]

25. Singh, P.R.; Nyiragatare, A.; Janssen, T.; Couvreur, M.; Decraemer, W.; Bert, W. Morphological and molecular characterisation of Pratylenchus rwandae n. sp. (Tylenchida: Pratylenchidae) associated with maize in Rwanda. Nematology 2018, 20, 781–794. [CrossRef]

26. Divsalar, N.; Shokoohi, E.; Hoseinipour, A.; Mashela, P. Molecular and morphological variation of the root-lesion nematode Pratylenchus neglectus. Biologia 2019, 74, 257–267. [CrossRef]

27. Mirghasemi, S.N.; Fanelli, E.; Troccoli, A.; Jamali, S.; Sohani, M.M.; De Luca, F. Molecular variability of the root-lesion nematode, Pratylenchus loosi (Nematoda: Pratylenchidae), from tea in Iran. Eur. J. Plant Pathol. 2019, 155, 557–569. [CrossRef]

28. De Luca, F.; Troccoli, A.; Duncan, L.W.; Subbotin, S.A.; Waeyenberge, L.; Coyne, D.L.; Brentu, F.C.; Inserra, R.N. Pratylenchus speijeri n. sp. (Nematoda: Pratylenchidae), a new root-lesion nematode pest of plantain in West Africa. Nematology 2012, 14, 987–1004. [CrossRef]

29. Fanelli, E.; Troccoli, A.; Capriglia, F.; Lucarelli, G.; Vovlas, N.; Greco, N.; De Luca, F. Sequence variation in ribosomal DNA and in the nuclear hsp90 gene of Pratylenchus penetrans (Nematoda: Pratylenchidae) populations and phylogenetic analysis. Eur. J. Plant Pathol. 2018, 152, 355–365. [CrossRef]

30. Carta, L.K.; Skantar, A.M.; Handoo, Z.A. Molecular, morphological and thermal characters of 19 Pratylenchus spp. and relatives using the D3 segment of the nuclear LSU rRNA gene. Nematropica 2001, 31, 193–208.

31. Loof, P.A. Taxonomic studies on the genus Pratylenchus (Nematoda). Tijdschrift Plantenziekten 1960, 66, 29–90.

32. Rusinque, L.C.M.; Vicente, C.; Inácio, M.L.S.; Nóbrega, F.; Camacho, M.J.; Lima, A.; Ramos, A.P. First report of Pratylenchus penetrans (Nematoda: Pratylenchida) associated with Amaryllis (Hippeastrum hybridum), in Portugal. Plant Dis. 2020, 1–5. [CrossRef]

33. Mokrini, F.; Waeyenberge, L.; Viaene, N.; Andaloussi, F.A.; Moens, M. Diversity of root-lesion nematodes (Pratylenchus spp.) associated with wheat (Triticum aestivum and T. durum) in Morocco. Nematology 2016, 18, 781–801. [CrossRef]

34. Machado, A.C.Z.; Siqueira, K.M.S.; Ferraz, L.C.C.B.; Inomoto, M.M.; Bessi, R.; Harakava, R.; Oliveira, C.M.G. Characterization of Brazilian populations of Pratylenchus brachyurus using morphological and molecular analyses. Trop. Plant Pathol. 2015, 40, 102–110. [CrossRef]

35. Townshend, J.L. Morphological observations of Pratylenchus penetrans from celery and strawberry in southern Ontario. J. Nematol. 1991, 23, 205. [PubMed]

36. Boisseau, M.; Sarah, J.L. In vitro rearing of Pratylenchidae nematodes on carrot discs. Fruits 2008, 63, 307–310. [CrossRef]

37. Derycke, S.; Remerie, T.; Vierstraete, A.; Backeljau, T.; Vanfleteren, J.; Vincx, M.; Moens, T. Mitochondrial DNA variation and cryptic speciation within the free-living marine nematode Pellioditis marina. Mar. Ecol. Prog. Ser. 2005, 300, 91–103. [CrossRef]

38. Hall, T.A. BioEdit: A user-friendly biological sequence alignment editor and analysis program for Windows 95/98/NT. Nucleic Acids Symp. Ser. 1999, 41, 95–98.

39. Altschul, S.F.; Madden, T.L.; Schaffer, A.A.; Zhang, J.; Zhang, Z.; Miller, W.; Lipman, D.J. Gapped BLAST and PSI-BLAST: A new generation of protein database search programs. Nucleic Acids Res. 1997, 25, 3389–3402. [CrossRef]

40. Rozas, J.; Ferrer-Mata, A.; Sánchez-DelBarrio, J.C.; Guirao-Rico, S.; Librado, P.; Ramos-Onsins, S.E.; Sánchez-Gracia, A. DnaSP 6: DNA sequence polymorphism analysis of large data sets. Mol. Biol. Evol. 2017, 34, 3299–3302. [CrossRef] [PubMed]

41. Kumar, S.; Stecher, G.; Li, M.; Knyaz, C.; Tamura, K. MEGA X: Molecular evolutionary genetics analysis across computing platforms. Mol. Biol. Evol. 2018, 35, 1547–1549. [CrossRef]

42. Saitou, N.; Nei, M. The neighbor-joining method: A new method for reconstructing phylogenetic trees. Mol. Biol. Evol. 1987, 4, 406–425.

43. Tamura, K.; Nei, M.; Kumar, S. Prospects for inferring very large phylogenies by using the neighbor-joining method. Proc. Natl. Acad. Sci. USA 2004, 101, 11030–11035. [CrossRef] [PubMed]

44. Tajima, F. Statistical method for testing the neutral mutation hypothesis by DNA polymorphism. Genetics 1989, 123, 585–595. [CrossRef]

45. Wessa, P. Kendall tau Rank Correlation (v1.0.13) in Free Statistics Software (v1.2.1). Office for Research Development and Education, 2017. Available online: https://www.wessa.net/rwasp_kendall.wasp/ (accessed on 10 January 2021).

Chapter-6

Integrative Taxonomy Reveals Hidden Cryptic Diversity within Pin Nematodes of the Genus Paratylenchus (Nematoda: Tylenchulidae)

1. Introduction

Pin nematodes of the genus *Paratylenchus* Micoletzky, 1922 [1] are obligate plant-ectoparasitic nematodes of small body length (<600 μm) with variable stylet length (10–120 μm), widely dispersed in different natural environments and crops, and distributed worldwide [2–4].

The taxonomic consideration for several genera of pin nematodes *sensu lato* historically included in this group comprise *Gracilacus*, *Paratylenchoides*, *Gracilpaurus*, *Cacopaurus*, has been recently discussed by Singh et al. [3] concluding that all these genera were confirmed as synonyms with *Paratylenchus* since no clear separations were detected under phylogenetic relationships of ribosomal and mitochondrial genes [3]. Stylet drives the feeding habitand many species have a long stylet (>40 μm), becoming swollen and feeding from deeper layers in the root cortex as sedentary ectoparasites. Two stylet pattern shapes can be found in this genus: (a) long and flexible stylet > 40 μm with conus representing about more than 70% of the total stylet (m ratio), and juveniles with well-developed stylet, initially included in the genus *Gracilacus* Raski [5]; (b) short and rigid stylet < 40 μm with conus about 50% of the total stylet, and juveniles without well-developed stylet, initially included in the genus *Paratylenchus*. Nevertheless, these differences were not sufficiently supported by molecular analyses to separate and maintain both genera [3,4,6]. The genus *Paratylenchus* is a wide diverse group with about 130 nominal species, from which about 54 of them are molecularly characterized [2–4,6–8]. Consequently, about half of the nominal species of this genus are not yet linked to molecular data, and there is a need for completing that information. The conserved morphology that characterizes *Paratylenchus* species led to the development of molecular methods using different fragments of nuclear ribosomal and mitochondrial DNA gene sequences to be used in DNA barcoding [3,4,6–8]. Use of molecular markers in species identification of pin nematodes over the last years has indi- cated that many widespread species actually comprise multiple genetically divergent and morphologically similar cryptic species [3,4,6–8]. An emblematic example of these species complexes comprises the *Paratylenchus straeleni*-species complex, which

Singh et al. [3] distinguished among 4–9 putative species within this complex considering all the avail- able ribosomal and mitochondrial sequences. In 1988, Castillo and Gomez Barcina [9] identified a population of *P. straeleni* (De Coninck, 1931) Oostenbrink, 1960 from a natural environment (Portuguese oak forest, *Quercus faginea* Lam.) at southern Spain based on morphological and morphometric traits. This raises the possibility that this population was potentially misidentified and included under the common and widely distributed species *P. straeleni*. Consequently, this is an excellent opportunity which prompted us to apply integrative taxonomical approaches to unravel the cryptic diversity of this species complex. This study allowed us to verify if this species identification was correct or to prove if close morphology and morphometry with original description comprise some genetic diversity with recent molecularly studied *P. straeleni* populations from Belgium, USA, and Turkey [3,6,10,11].

Thirty species of *Paratylenchus* have been reported in Spain from cultivated and natural ecosystems including *P. aonli* Misra and Edward, 1971 [12], *P. arculatus* Luc and de Guiran, 1962 [13,14], *P. baldaccii* Raski, 1975 [4,15], *P. caravaquenus* Clavero Camacho, Cantalapiedra- Navarrete, Archidona-Yuste, Castillo and Palomares-Rius, 2021 [4], *P. ciccaronei* Raski, 1975 [16–18], *P. enatus* (Raski, 1976) Siddiqi, 1986 [19], *P. enigmaticus* Munawar, Yevtushenko, Palomares-Rius and Castillo, 2021 [4], *P. goodeyi* Oostenbrink, 1953 [20], *P. hamatus* Thorne and Allen, 1950 [4], *P. holdemani* Raski, 1975 [4], *P. indalus* Clavero Camacho, Cantalapiedra- Navarrete, Archidona-Yuste, Castillo and Palomares-Rius, 2021 [4], *P. israelensis* (Raski, 1973) Siddiqi, 1986 [4], *P. macrodorus* Brzeski, 1963 [18], *P. microdorus* Andrássy, 1959 [15–18], *P. minusculus* Tarjan, 1960 [21], *P. mirus* (Raski, 1962) Siddiqi and Goodey, 1964 [12], *P. nanus* Cobb, 1923 [18,22], *P. pedrami* Clavero-Camacho, Cantalapiedra-Navarrete, Archidona- Yuste, Castillo and Palomares-Rius, 2021 [4], *P. peraticus* (Raski, 1962) Siddiqi and Goodey, 1964 [20], *P. projectus* Jenkins, 1956 [19,23], *P. sheri* (Raski, 1973) Siddiqi, 1986 [16–18,22], *P. similis* Khan, Prasad and Mathur, 1967 [18,22], *P. steineri* Golden, 1961 [18,20], *P. straeleni* (De Coninck, 1931) Oostenbrink, 1960 [9], *P. tateae* Wu and Townshend (1973) [4], *P. tenuicau- datus* Wu, 1961 [12], *P. teres* (Raski, 1976) Siddiqi, 1986 [24], *P. vandenbrandei* de Grisse, 1962 [16,17], *P. veruculatus* Wu, 1962 [4], and *P. zurgenerus* Clavero-Camacho, Cantalapiedra- Navarrete, Archidona-Yuste, Castillo and Palomares-Rius, 2021 [4]. However, for the majority of these studies, except that of Clavero-Camacho et al. [4], no molecular analyses were carried out for their identification, and the cryptic biodiversity of these nematodes could be underexplored, including some species identifications for Spanish populations performed by our group some years ago. For this reason, the identification and reliable estimation of pin nematode diversity in Spain is needed. This paper is the second in a series deciphering the cryptic diversity of pin nematodes in Spain using integrative taxonomical approaches, with the final aim to disentangle the real biodiversity of these nematodes in cultivated and natural environments in Spain. The first one dealt with pin nematodes associated with cultivated *Prunus* spp. in Spain, including almond, apricot, cherry, nectarine and peach [4]. This study tries to understand the biodiversity of *Paratylenchus* spp. in some almond samples and additional new natural environments as well as re-analyzing some previous studies carried out by our laboratory 30 years ago based on morphology and morphometry only [9,17,20], but now using more accurate and precise integrative taxonomical approaches.

In the genus *Paratylenchus*, species display a particular resting-stage which accumu- lates in soil under adverse environmental conditions (*viz.* drought conditions) [4]. This state is non-feeding, molting to adults after stimulation by host-plant roots, and may provide some useful data for species identification [25,26]. Usually the resting stage is fourth-stage juvenile (J4), but third-stage (J3) appears in other species, recognized by granular body con- tents and presence/absence of stylet [2,26]. In *P. straeleni*, all juveniles had a well-developed stylet and pharynx, while the body of J4 contained numerous dark granules and this is considered the resting stage [26]. However, in close-related species such as *P. steineri*, stylet and pharynx are well-developed in second- and third-stage juveniles (J2 and J3), but J4 had no stylet and pharynx is much reduced. Morphological changes in stylet morphology in juveniles of some *Paratylenchus* species need to be studied with regard to adult state. In this research we study the stylet morphology of quiescent juvenile stages (J4) based on an integrative taxonomical approach [4].

The main objectives of this study were to (i) conduct identification with morphological and morphometrical approaches of some *Paratylenchus* species collected in several nematode surveys on almond and natural environments in Spain; (ii) provide molecu- lar characterization of several species using ribosomal (D2-D3 expansion segments of 28S rRNA, Internal Transcribed Spacer region (ITS) rRNA) and the mitochondrial region cytochrome c oxidase subunit 1 (COI); (iii) study phylogenetic relationships within *Paraty- lenchus* spp. using the obtained molecular markers.

2. Results

Eighteen species were identified from 24 isolates of *Paratylenchus* spp. from 15 soil samples in nine municipalities in Spain (Table 1). In these populations, females, males (when available) and juveniles were morphologically and morphometrically studied in detail and molecular markers for their identification were provided (Table 1). From these, one isolate was considered a new undescribed species and 17 were already known de- scribed species (Table 1). The new species include an isolate from the *P. straeleni*-complex and was described herein as *Paratylenchus parastraeleni* sp. nov. The already known species included *P. amundseni* Bernard, 1982, *P. aciculus* Brown, 1959, *P. baldaccii* (Oostenbrink, 1953) Raski, 1962, *P. enigmaticus* Munawar et al., 2021, *P. goodeyi* Oostenbrink, 1953, *P. holdemani* Raski, 1975, *P. macrodorus* Brzeski, 1963, *P. neoamblycephalus* Geraert, 1965, *P. pandatus* (Raski, 1976) Siddiqi, 1986, *P. pedrami* Clavero-Camacho et al., 2021, *P. recisus* Siddiqi, 1996, *P. sheri* (Raski, 1973) Siddiqi, 1986, *P. tateae* Wu and Townshend (1973), *P. variabilis* Raski, 1975, *P. veruculatus* Wu, 1962, *P. verus* (Brzeski, 1995) Brzeski, 1998, and *P. vitecus* (Pramodini et al., 2006) Ghaderi et al., 2014. Eight of these species need to be considered as first reports for Spain in this research (*viz. P. amundseni, P. aciculus, P. neoamblycephalus, P. pandatus, P. recisus, P. variabilis, P. verus* and *P. vitecus*) and measurements from females, males (if available) and juveniles, as well as molecular markers were provided for their unequivocal identification.

Table 1. Isolates sampled and sequenced for *Paratylenchus* spp. from several localities in Spain used in this study.

Species	Sample Code	Locality, Province, Host	D2-D3	ITS	*COI*
1. *P. parastraeleni* sp. nov.	CAZ_05	Arroyo Frío, Jaén, *Quercus faginea* Lam.	MZ265064-MZ265070	MZ265004-MZ265007	MZ262208-MZ262211
2. *P. aciculus* Brown, 1959	CAZ_07	Coto Rios, Jaén, *Pinus halepensis* Mill.	MZ265071-MZ265075	MZ265008-MZ265011	MZ262212-MZ262214
3. *P. amundseni* Bernard, 1982	CAZ_02	La Iruela, Jaén, *Pinus halepensis* Mill.	MZ265076-MZ265078	MZ265012-MZ265014	MZ262215-MZ262219
4. *P. baldaccii* (Oostenbrink, 1953) Raski, 1962	CAZ_04	Arroyo Frío, Jaén, grasses	MZ265079	MZ265015-MZ265016	MZ262220-MZ262221
5. *P. enigmaticus* Munawar et al., 2021	IAS_21	Córdoba, Córdoba, grasses	MZ265080-MZ265083	MZ265017-MZ265019	MZ262222-MZ262226
6. *P. goodeyi* Oostenbrink, 1953	EP_ACA	Córdoba, Córdoba, wild olive	MZ265084-MZ265091	MZ265020-MZ265024	MZ262227-MZ262230
	AR_097	Santa Mª de Trasierra, Córdoba, wild olive	MZ265092-MZ265097	MZ265025-MZ265026	MZ262231-MZ262233
	PR_050	Montalbán, Córdoba, almond	MZ265098-MZ265099	MZ265027	MZ262234-MZ262235
	PR_017	Córdoba, Córdoba, almond	MZ265100	MZ265028	MZ262236
	PR_076	Marmolejo, Jaén, almond	MZ265101-MZ265102	MZ265029-MZ265031	-*
	PR_019	Córdoba, Córdoba, almond	MZ265103-MZ265105	MZ265032-MZ265033	MZ262237-MZ262238
7. *P. holdemani* Raski, 1975	AR_102	Santa Mª de Trasierra, Córdoba, wild olive	MZ265106-MZ265107	-	-
8. *P. macrodorus* Brzeski, 1963	AR_102	Santa Mª de Trasierra, Córdoba, wild olive	MZ265108-MZ265113	MZ265034-MZ265038	MZ262239-MZ262244
9. *P. neoamblycephalus* Geraert, 1965	CAZ_05	Arroyo Frío, Jaén, *Quercus faginea* Lam.	MZ265114-MZ265115	MZ265039-MZ265040	MZ262245-MZ262246
10. *P. pandatus* (Raski, 1976) Siddiqi, 1986	PIN_AR	Caravaca, Murcia, *Pinus halepensis* Mill.	MZ265116-MZ265117	MZ265041-MZ265042	MZ262247-MZ262251
11. *P. pedrami* Clavero-Camacho et al., 2021	AR_102	Santa Mª de Trasierra Córdoba, wild olive	MZ265118	-	-
12. *P. recisus* Siddiqi, 1996	CAZ_06	Arroyo Frío, Jaén, *Quercus faginea* Lam.	MZ265119-MZ265120	MZ265043	MZ262252
13. *P. sheri* (Raski, 1973) Siddiqi, 1986	CAZ_04	Arroyo Frío, Jaén, grasses	MZ265121-MZ265124	MZ265044-MZ265048	MZ262253-MZ262259
	CAZ_07	Coto Rios, Jaén, *Pinus halepensis* Mill.	MZ265125-MZ265126	MZ265049-MZ265050	MZ262260-MZ262261
14. *P. tateae* Wu and Townshend (1973)	PR_187	Ariza, Zaragoza, almond	MW282754-MW282759	MW282766-MW282771	MZ262262-MZ262264
15. *P. variabilis* Raski, 1975	EP_ACA	Córdoba, Córdoba, wild olive	MZ265127-MZ265129	MZ265051-MZ265053	MZ262265-MZ262267
16. *P. veruculatus* Wu, 1962	AR_102	Santa Mª de Trasierra Córdoba, wild olive	MZ265134-MZ265135	-	-
17. *P. verus* (Brzeski, 1995) Brzeski, 1998	AR_097	Santa Mª de Trasierra, Córdoba, wild olive	MZ265130-MZ265133	MZ265054-MZ265058	MZ262268-MZ262271
18. *P. vitecus* (Pramodini et al., 2006) Ghaderi et al., 2014	EP_ACA	Córdoba, Córdoba, wild olive	MZ265136-MZ265141	MZ265059-MZ265062	MZ262272-MZ262274

* = not sequenced.

** = not sequenced.*

2.1 Systematics

2.1.1 Description of *Paratylenchus parastraeleni* sp. nov.

(Figures 1–3, Table 2) http://zoobank.org/urn:lsid:zoobank.org:act:61B40ACF-177F- 4D92-A16F-0A3CAF78FD4A (accessed on 8 July 2021).

Female: body slender, ventrally arcuate to form an open, C-shaped body habitus when heat relaxed; cuticle finely annulated; lateral field equidistant with four distinct smooth lines. Lip region rounded, truncate, submedian lobes almost indistinct; with very slight sclerotization. Stylet flexible, 11.3–14.6% of body length. Conus of stylet 2.4–3.5 times longer than shaft, 73–80% of total stylet length. Stylet knobs small, 2.5–3.0 µm across, laterally directed. Procorpus cylindrical, about 50 µm long. Excretory pore situated at distal end of basal pharyngeal bulb. Hemizonid conspicuous, located two annuli anterior to excretory pore. Valvular apparatus in metacorpus 6.0–7.0 µm long, at 58–70% of pharynx length from anterior end. Basal pharyngeal bulb pyriform. Ovary outstretched, spermatheca almost spherical, 21 (19–28) µm wide, filled with rounded sperm 1.0–1.5 µm in diameter. Lateral vulval membranes, 5.5–6.0 µm long. Tail elongate-conoid gradually tapering to form a rounded terminus, 0.5–0.8 times as long as vulva–anus distance.

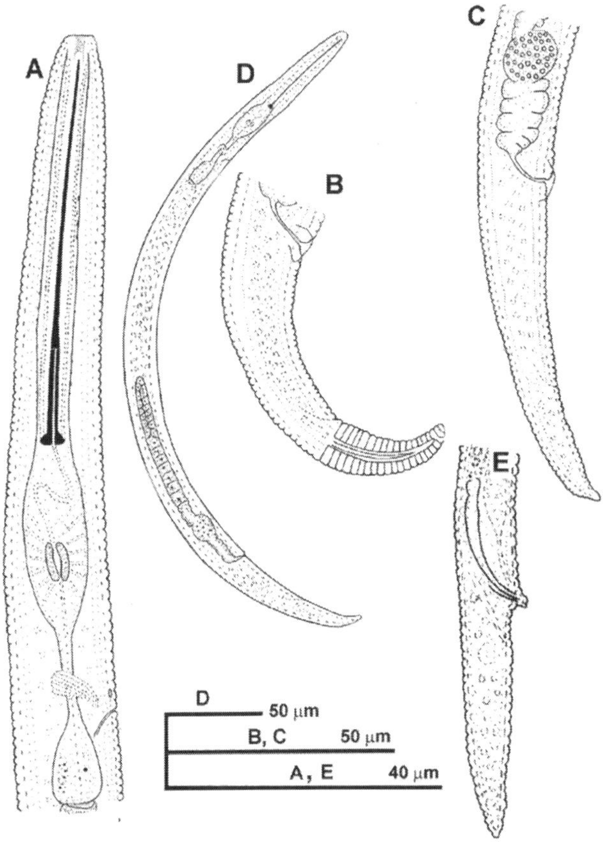

Figure 1. **Line drawings of** ***Paratylenchus parastraeleni*** **sp. nov. (A): Female pharyngeal region; (B,C): Female posterior region; (D): Entire female; (E): Male posterior region.**

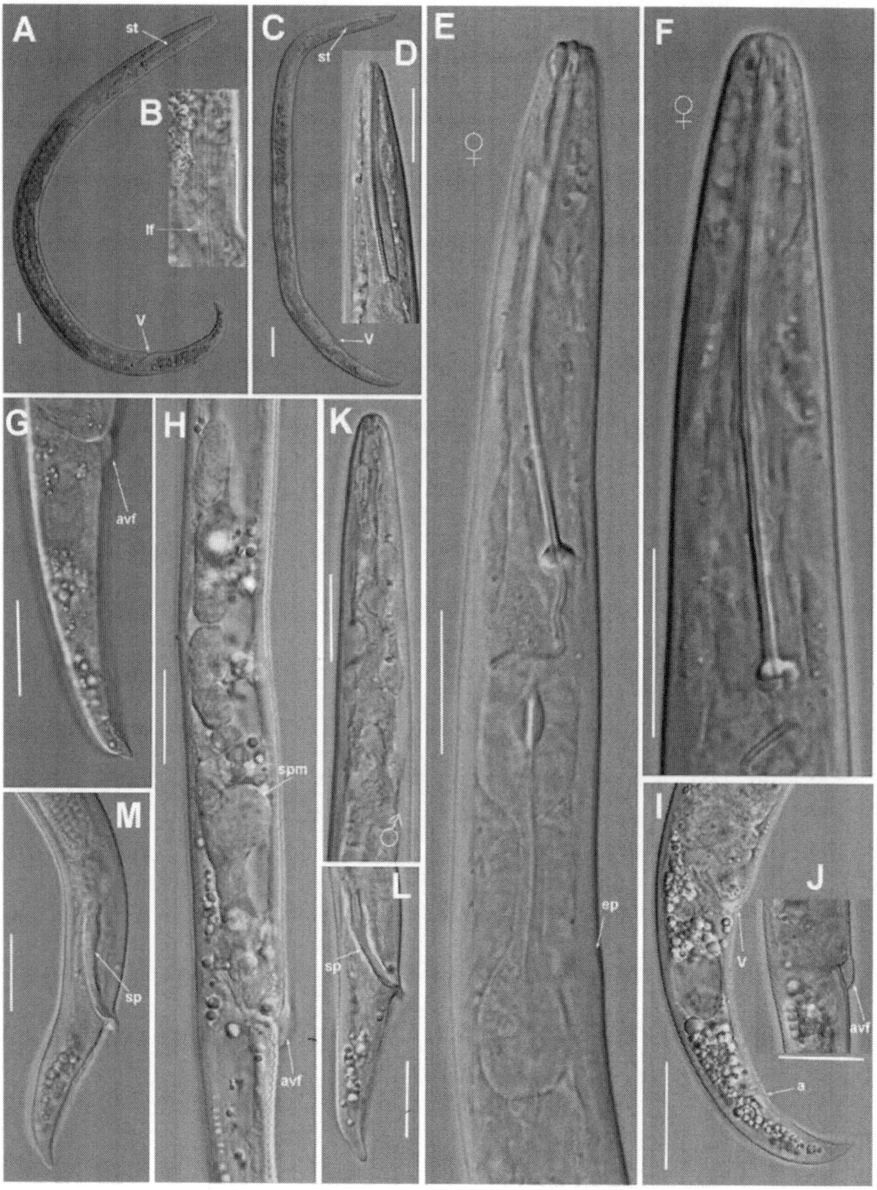

Figure 2. Light photomicrographs of *Paratylenchus parastraeleni* sp. nov. female and male. (A,C) Entire female with vulva arrowed; (B) detail of lateral fields; (D,F) detail of female stylet region; (E) female pharyngeal region; (G–J) female posterior region with vulva and anus (arrowed) and detail of vulva showing advulval flap (arrowed); (K) male pharyngeal region showing absence of stylet; (L,M) male posterior region showing spicules (arrowed). Scale bars (A–M = 20 μm). (Abbreviations: a = anus; avf = advulval flap; ep = excretory pore; lf = lateral field; sp = spicules; spm = spermatheca; V = vulva).

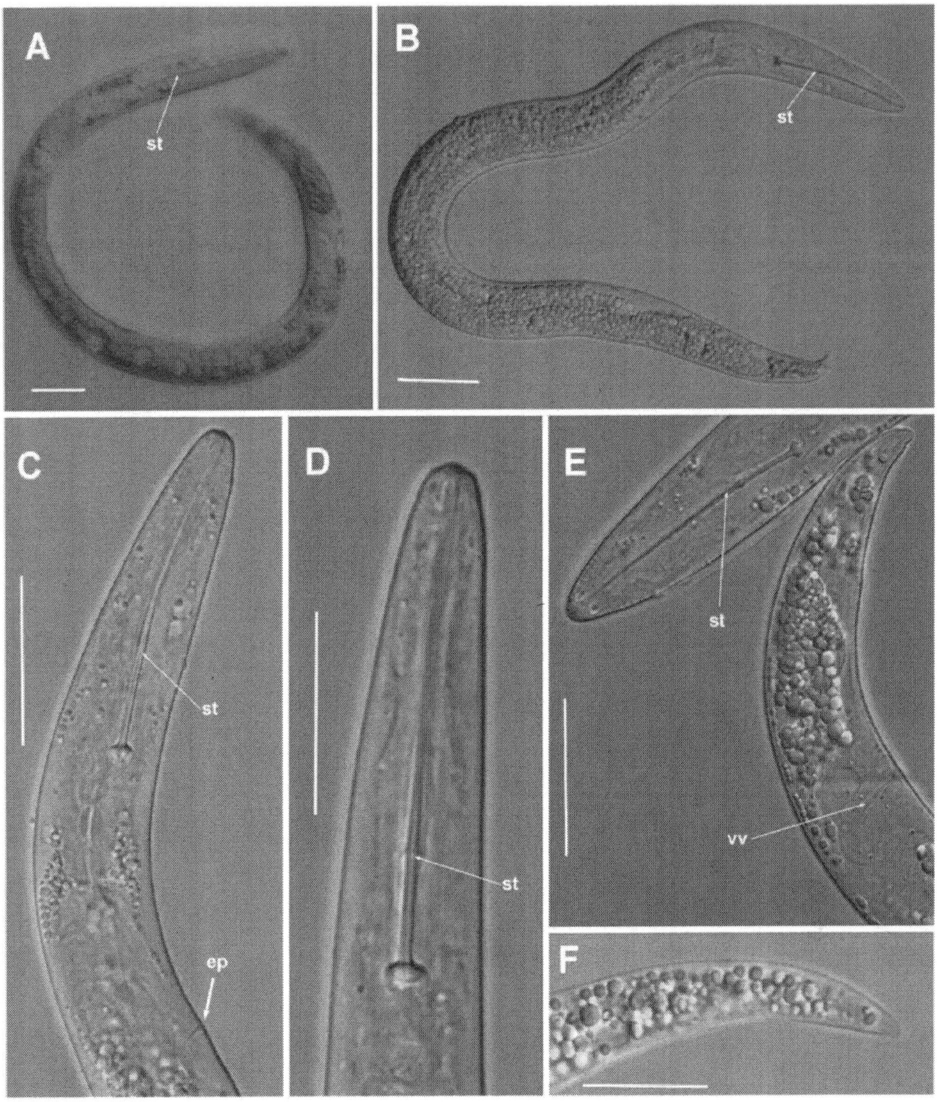

Figure 3. Light photomicrographs of ***Paratylenchus parastraeleni*** sp. nov. fourth-stage juveniles. (A,B) Entire fourth-stage juveniles showing stylet (arrowed); (C,D) fourth-stage juvenile pharyngeal region showing stylet; (E) fourth-stage juvenile anterior and posterior region showing stylet and initial vestigium of vagina (arrowed); (F) fourth-stage juvenile tail. Scale bars (A–F = 20 μm). (Abbreviations: ep = excretory pore; st = stylet; vv = vaginal vestigium).

Table 2. Morphometrics of *Paratylenchus parastraeleni* sp. nov. paratype females, males and fourth-stage juveniles. All measurements are in μm and in the form: mean ± s.d. (range).

	Holotype	Paratypes		
	Female	Females	Males	Juveniles (J4)
Sample Code	CAZ_05	CAZ_05	CAZ_05	CAZ_05
Locality	Arroyo Frío, Jaén			
n	1	19	4	5
L	425	417.7 ± 35.2 (363–467)	389.5 ± 19.3 (369–414)	337.6 ± 37.5 (302–382)
a *	23.6	20.6 ± 3.4 (16.3–26.4)	28.9 ± 1.3 (27.3–30.3)	22.2 ± 1.2 (20.8–23.9)
b	3.5	3.6 ± 0.3 (3.1–4.2)	4.2 ± 0.7 (3.6–5.2)	3.3 ± 0.4 (2.9–3.8)
c	14.7	13.4 ± 1.5 (11.4–16.5)	12.7 ± 1.2 (11.9–14.5)	12.9 ± 1.9 (10.7–15.9)
c'	2.6	2.9 ± 0.2 (2.5–3.4)	2.9 ± 0.1 (2.7–3.0)	2.8 ± 0.3 (2.4–3.2)
V or T	80.9	82.1 ± 0.9 (80.2–83.5)	-	-
G1	35.5	44.4 ± 4.4 (35.5–50.3)	-	-
Stylet length	54.0	53.5 ± 1.5 (52.0–56.0)	-	45.8 ± 2.2 (43.0–48.0)
(Stylet length/body length) × 100	12.7	12.9 ± 0.9 (11.3–14.6)	-	13.7 ± 1.0 (12.6–14.7)
Conus length	43.0	41.2 ± 1.4 (38.0–43.0)	-	36.0 ± 3.2 (31.0–39.0)
m	79.6	77.1 ± 1.9 (73.1–79.6)	-	78.6 ± 6.3 (72.1–88.6)
DGO	5.5	5.4 ± 0.5 (4.5–6.0)	-	4.0 ± 0.7 (3.0–5.0)
O	10.2	10.2 ± 1.0 (8.0–11.5)	-	8.7 ± 1.4 (6.8–10.6)
Lip width	5.5	4.9 ± 0.4 (4.0–5.5)	4.1 ± 0.3 (4.0–4.5)	4.8 ± 0.4 (4.5–5.0)
Median bulb length	24.0	24.0 ± 2.7 (19.0–29.0)	-	23.8 ± 2.8 (19.0–26.0)
Median bulb width	11.0	11.1 ± 0.9 (9.0–13.0)	-	9.4 ± 1.0 (8.5–11.0)
Anterior end to center median bulb	78	74.0 ± 3.8 (67.0–81.0)	-	62.4 ± 4.0 (58.0–68.0)
MB	65.0	64.2 ± 2.5 (58.5–70.0)	-	61.3 ± 4.0 (56.9–67.3)
Nerve ring to anterior end	88.0	89.4 ± 5.1 (77.0–98.0)	-	78.8 ± 4.2 (74.0–84.0)
Excretory pore to anterior end	94.0	98.2 ± 6.6 (87.0–114.0)	81.5 ± 5.7 (74.0–88.0)	86.4 ± 6.1 (77.0–94.0)
Pharynx length	120.0	115.2 ± 4.9 (107.0–123.0)	95.0 ± 11.1 (80.0–106.0)	101.8 ± 0.8 (101.0–103.0)
Maximum body diam.	18.0	20.9 ± 4.3 (14.0–28.0)	13.2 ± 0.3 (13.0–13.5)	15.2 ± 1.8 (14.0–18.0)
Tail length	29.0	31.4 ± 4.1 (25.5–39.0)	30.9 ± 3.7 (25.5–34.0)	26.8 ± 5.9 (19.5–35.0)
Anal body diam.	11.0	11.0 ± 1.4 (8.5–14.0)	10.6 ± 0.9 (9.5–11.5)	9.7 ± 1.7 (8.0–12.0)
Spicules	-	-	22.0 ± 1.1 (21.0–23.5)	-
Gubernaculum	-	-	4.0 ± 0.4 (3.5–4.5)	-

* Abbreviations: a = body length/greatest body diameter; b = body length/distance from anterior end to pharyngo-intestinal junction; DGO = distance between stylet base and orifice of dorsal pharyngeal gland; c = body length/tail length; c' = tail length/tail diameter at anus or cloaca; G1 = anterior genital branch length expressed as percentage (%) of the body length; L = overall body length; m = length of conus as percentage of total stylet length; MB = distance between anterior end of body and center of median pharyngeal bulb expressed as percentage (%) of the pharynx length; n = number of specimens on which measurements are based; O = DGO as percentage of stylet length; T = distance from cloacal aperture to anterior end of testis expressed as percentage (%) of the body length; V = distance from body anterior end to vulva expressed as percentage (%) of the body length.

Male: Less common than females (ratio ca. 1:4). Male body is slender than female body, tapering towards both ends, posterior region ventrally arcuate when heat relaxed. Cuticle apparently smooth with fine annulations; labial region similar to that of female but narrower and slightly truncated, continuous with body, sclerotization in labial region weak; stylet lacking. Pharynx rudimentary and non-functional, procorpus, metacorpus, and basal bulb inconspicuous; excretory pore located 81.5 μm away from anterior end. Testis outstretched, with small spermatozoa; spicule slender, slightly curved towards end; gubernaculum curved; bursa absent. Tail elongate-conoid, tapering gradually to a finely pointed tip.

Juveniles: J4 similar in morphology to adult females (Figures 2 and 3), bearing flexible stylet 45.8 (43.0–48.0) μm-long. Pharynx well developed, functional. Genital primordium underdeveloped, primordium of vagina discernible, anus indistinct, posterior region similar to female but slightly more rounded terminus.

Diagnosis and Relationships

The new species can be characterized by the presence of four lateral lines in lateral field, advulval flaps present, and a moderately long female stylet of 53.5 (52.0–56.0) μm. Lip region rounded, truncate, submedian lobes almost indistinct; with very slight sclerotization, continuous with the rest of the body. Spermatheca spherical. Tail elongate-conoid gradually tapering to form a rounded terminus. According to species grouping by Ghaderi et al. [2] belongs to group 10 characterized by stylet length more than 40 μm, four lateral lines and advulval flaps present.

Morphologically and morphometrically, the new species is very close to *P. straeleni*, and can be also similar to *P. goodeyi* and *P. ivorensis* Luc and de Guiran, 1962. In fact, the description of the Spanish population agrees well with original description by De Coninck [27], and other populations from The Netherlands, Poland, Italy, Czech Republic, Iran, USA, Turkey and Belgium [3,6,10,26,28–30], and no major differences in morphology or morphometry can be detected. Consequently, based on the molecular markers, this is an extraordinary example of cryptic species within the *P. straeleni*-complex species, and this can help to clarify the identity of other populations with similar morphology and morphometry. From *P. goodeyi* can be differentiated by lip region shape (conoid-rounded to truncate vs. conoid) [2], and from *P. ivorensis* in a posterior position of vulva (80.2–83.5 vs. 73–77).

Molecular Characterization

Seven D2-D3 of 28S rRNA (MZ265064-MZ265070), four ITS (MZ265004-MZ265007) and four COI gene sequences (MZ262208-MZ262211) were generated for this new species without intraspecific sequence variations, except for the ITS where only one variable posi- tion was detected. The closest species to *P. parastraeleni* sp. nov. was *P. straeleni*, being 95% similar for the D2-D3 region (MZ265064-MZ265070) (differing from 32 to 38 nucleotides and no indels) to several accessions deposited in GenBank. For the COI gene sequences (MZ262208-MZ262211), the similarity values were 93 and 94% (differing from 21 to 26 nu- cleotides and no indels) from *P. straeleni* sequences deposited in GenBank; finally, the similarity for the ITS region was 86–88% (differing from 89 to 111 nucleotides and 35 to 43 indels) from *P. straeleni* sequences deposited in GenBank. All molecular markers studied clearly separate both species. Due to the presence of more than one species in the same soil sample, J4 individual identification for morphological-morphometrical analysis was based on a molecular barcoding using the 28S rRNA markers, and nematodes with identical sequences as adults were considered as the same species, in this case, *P. parastraeleni* sp. nov.

Type Habitat and Locality

Paratylenchus parastraeleni sp. nov. was found in the rhizosphere of a *Quercus faginea* Lam., forest (coordinates 37°58′33.0" N 2°54′18.8" W); the municipal district of Arroyo Frío, Jaén province, Spain.

Etymology

The species epithet, *parastraeleni*, refers to Gr. prep. para, alongside of and resembling, because of its close resemblance to *Paratylenchus straeleni*.

Type Material

Holotype female, 17 paratypes females, 5 fourth-stage juveniles and 4 male paratypes (slide numbers CAZ_05-01 to CAZ_05-12) were deposited in the Nematode Collection of the Institute for Sustainable Agriculture, CSIC, Córdoba, Spain, and four females deposited at the USDA Nematode Collection (slides T-7511p and T-7512p).

2.1.2 Remarks of *Paratylenchus aciculus* Brown, 1959 (Figure 4, Table 3).

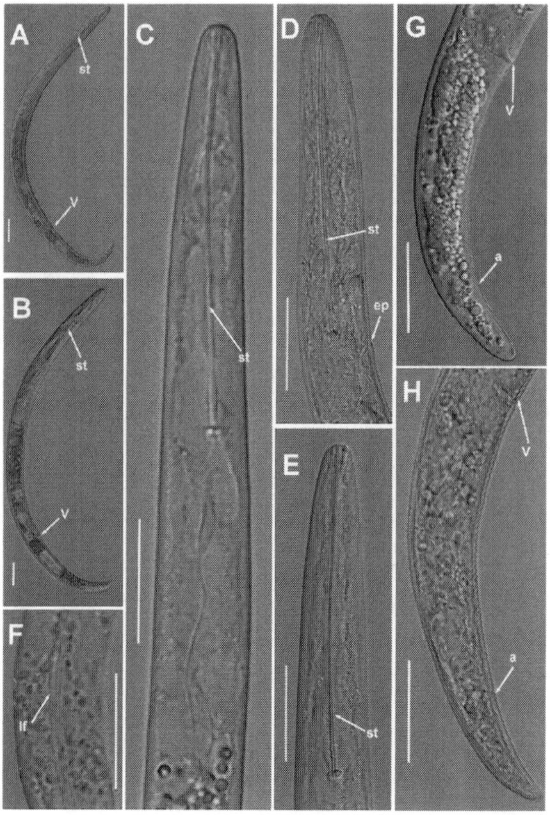

Figure 4. Light photomicrographs of Paratylenchus aciculus Brown, 1959. (A,B) Entire female with vulva arrowed; (C) female pharyngeal region; (D,E) female lip region; (F) detail of lateral field; (G,H) female posterior region with vulva and anus (arrowed). Scale bars (A–H = 20 µm). (Abbrevia- tions: a = anus; ep = excretory pore; lf = lateral field; st = stylet; V = vulva).

Table 3. Morphometrics of *Paratylenchus aciculus* Brown, 1959 from Coto Ríos, Jaén, Spain, type population, and *P. aculentus* from Belgium. All measurements are in µm and in the form: mean ± s.d. (range).

Species	*P. aciculus*	*P. aciculus*	*P. aculentus*
Life Stage	Females	Females	Females
Sample Code	CAZ_07	Type	Belgium
Locality, Province	Coto Ríos, Jaén	Population [31]	Singh et al. [3]
n	12	25	12
L	309.3 ± 16.2 (285–339)	240–310	266 ± 20.1 (233–03)
a *	19.1 ± 1.7 (17.0–21.5)	18–24	19.6 ± 2.0 (16.3–23.2)
b	2.5 ± 0.2 (2.2–2.8)	2.4–2.7	2.6 ± 0.1 (2.4–2.8)
c	12.4 ± 1.4 (10.3–14.8)	10–16	12.4 ± 1.5 (10.8–15.2)
c′	2.9 ± 0.2 (2.6–3.5)	3.0	2.8 ± 0.3 (2.4–3.1)
V	73.4 ± 0.8 (72.3–74.7)	68–74	72.5 ± 1.5 (70.8–75.7)
G1	34.8 ± 7.6 (23.5–42.5)	-	-
Stylet length	71.4 ± 2.8 (67.5–75.0)	61–69	56.0 ± 3.3 (52.4–61.2)
(Stylet length/body length) × 100	23.1 ± 1.0 (21.8–24.6)	-	-
Conus length	64.7 ± 3.5 (58.0–69.0)	-	49.1 ± 3.6 (43.0–54.9)
m	90.6 ± 2.7 (84.1–93.2)	-	-
DGO	5.3 ± 0.6 (4.5–6.5)	-	-
O	7.4 ± 0.7 (6.3–8.7)	-	-
Lip width	5.4 ± 0.5 (5.5–6.5)	-	-
Median bulb length	24.3 ± 1.9 (22.0–27.0)	-	-
Median bulb width	11.5 ± 0.5 (11.0–12.0)	-	-
Anterior end to center median bulb	85.0 ± 3.3 (79.0–89.0)	-	-
MB	66.2 ± 2.5 (62.3–69.0)	-	-
Nerve ring to anterior end	103.2 ± 4.7 (94.0–100.0)	-	-
Excretory pore to anterior end	83.8 ± 5.9 (72.5–91.0)	70	66.7 ± 5.2 (54.3–74.4)
Pharynx length	125.5 ± 8.2 (109.0–138.0)		101 ± 8.3 (87.0–113)
Maximum body diam.	16.4 ± 2.1 (14.0–20.0)	15	13.6 ± 1.3 (11.6–15.5)
Tail length	25.3 ± 3.4 (20.5–33.0)	50	20.9 ± 2.3 (18.1–25.1)
Anal body diam.	8.6 ± 0.7 (7.5–10.0)	-	7.6 ± 0.5 (7.0–8.3)

* *Abbreviations: a = body length/greatest body diameter; b = body length/distance from anterior end to pharyngo-intestinal junction; DGO = distance between stylet base and orifice of dorsal pharyngeal gland; c = body length/tail length; c′ = tail length/tail diameter at anus or cloaca; G1 = anterior genital branch length expressed as percentage (%) of the body length; L = overall body length; m = length of conus as percentage of total stylet length; MB = distance between anterior end of body and center of median pharyngeal bulb expressed as percentage (%) of the pharynx length; n = number of specimens on which measurements are based; O = DGO as percentage of stylet length; V = distance from body anterior end to vulva expressed as percentage (%) of the body length.*

According to species grouping by Ghaderi et al. [2] this species belongs to group 9 characterized by stylet length more than 40 µm, three lateral lines and advulval flap absent. The Spanish population from Coto Ríos, Jaén province, was characterized by long flexible stylet 67.5–75.0 µm, lip region rounded and continuous with body contour, female tail subacute to finely rounded, and spermatheca ellipsoid and filled with sperm, which indicates that males are required for reproduction but their numbers are lower than females. J4 not found. Morphometrics of the Spanish population agree well with original description as well as other populations with small differences in stylet length (67.5–75.0 µm vs. 61.0–69.0 µm), which may be due to geographical intraspecific variability [2]. This species was described from Canada and has been reported in USA, several European countries, including the recent integrative identification from Belgium [3], and this study comprises the first report from Spain. Although ribosomal markers (D2–D3 and ITS) between the Spanish population of *P. aciculus* and the Belgian population of *P. aculentus* are quite similar (see below), these species can be separated by COI (see below), and by clear differences in stylet length (67.5–75.0 µm vs. 52.4–61.2 µm), advulval flap (absent vs. small advulval flap present), and spermatheca shape (ellipsoid vs. rounded) [3].

Molecular Characterization

Five D2-D3 of 28S rRNA (MZ265071-MZ265075), four ITS sequences (MZ265008-MZ265011), and three COI sequences (MZ262212-MZ262214) were obtained for this species. In both ribosomal genes, no intraspecific variability was detected, however, one variable position was found between the three COI sequences included in this study (MZ262212-MZ262214). Ribosomal genes (MZ265071-MZ265075, MZ265008-MZ265011) showed a high similarity with *P. aculentus*, being 99% (2 out of 698 bp difference) and 98% (11–12 out of 742 bp difference) similar for the D2-D3 (MW413626- MW413628) and ITS region (MW413588- MW413589), respectively. However, the separation of both species is possible using the COI gene (MZ262212-MZ262214), since for this marker the similarity found was 89% (differing by 40–41 nucleotides and no indels) with the accessions belonging to *P. aculentus* (MW421639-MW421641).

2.1.3 Remarks of *Paratylenchus amundseni* Bernard, 1982 (Figure 5, Table 4).

According to species grouping by Ghaderi et al. [2] this species belongs to group 3 characterized by stylet length less than 40 µm, four lateral lines and advulval flaps present. The Spanish population from La Iruela, Jaén province, was characterized by a conoid-truncate lip region with submedian lobes indistinct, a female tail finely rounded to acute, and a rounded spermatheca filled with sperm, which indicates that males are required for reproduction but their numbers are lower than females. J4 bearing a delicate stylet. Some morphometric differences with original description include slightly larger body length (335–450 µm vs. 320–370 µm), slightly shorter stylet length (16.0–18.0 µm vs. 17.0–19.0 µm), and slightly posterior position of vulva (78.6–82.8 vs. 76.0–80.0), which may be considered as intraspecific variability. This species is very close morphologically and morphometrically to *P. tateae*, from which they can be separated by lip region (conoid-truncate and submedian lobes indistinct vs. conoid narrow, with anterior end flattened and protuberant submedian lips) (Figure 5), as well as by molecular markers (see below).

This species has only been reported from original description in the rhizosphere of grasses (*Leymus mollis* (Trin.) Pilg.) at Adak Island, Alaska (USA) [32], and this consists of the first report from Spain and the second written record.

Molecular Characterization

Three D2-D3 of 28S rRNA (MZ265076-MZ265078), three ITS (MZ265012-MZ265014), and five COI gene sequences (MZ262215-MZ262219) were generated herein for this species, including J4 and female adult sequences. All sequences showed no intraspecific variation. *Paratylenchus amundseni* was molecularly closely related with *P. tateae*, showing similarity values of 98% (differing from 11 to 14 nucleotides and no indels) for D2-D3 region. How- ever, for the ITS region, the similarity value was 95% (differing by 31 to 43 nucleotides and 7 to 11 indels) with *P. tateae* accessions (MW282766-MW282771) from Spain and Canada [8]. Finally, the similarity found for COI gene sequences was 90% (differing by 34–36 nu- cleotides) with the COI accessions of *P. tateae* (MZ262262-MZ262264) from Spain, newly obtained in the present study.

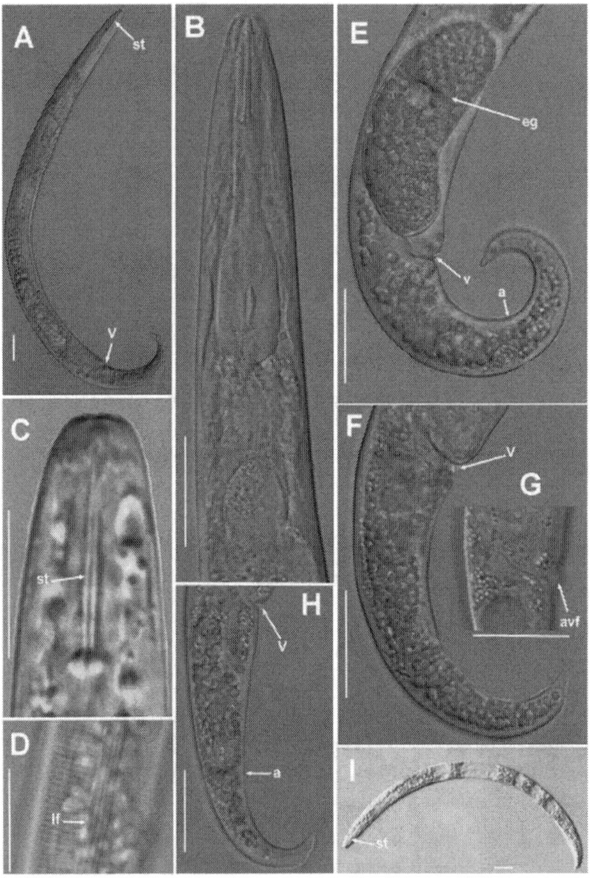

Figure 5. Light photomicrographs of Paratylenchus amundseni Bernard, 1982. (A) Entire female with vulva arrowed; (B) female pharyngeal region; (C) female lip region; (D) detail of lateral field; (E–H) female posterior region with vulva, anus, and

advulval flap (arrowed); (I) entire fourth-stage juvenile with stylet (arrowed). Scale bars (A–I = 20 μm). (Abbreviations: a = anus; avf = advulval flap; eg = egg; lf = lateral field; st = stylet; V = vulva).

Table 4. Morphometrics of Paratylenchus amundseni Bernard, 1982 from La Iruela, Jaén, Spain. All measurements are in μm and in the form: mean ± s.d. (range).

Life Stage	Females	Fourth-Stage Juveniles	Females
Sample Code	CAZ_02	CAZ_02	Type
Locality, Province	La Iruela, Jaén	La Iruela, Jaén	Population [32]
n	17	6	16
L	397.1 ± 35.0 (335–450)	358.7 ± 10.8 (340–369)	320–370
a *	18.7 ± 1.7 (16.0–22.5)	19.0 ± 2.7 (15.6–21.7)	19–25
b	4.2 ± 0.4 (3.6–5.2)	3.8 ± 0.2 (3.6–4.1)	3.6–4.6
c	11.5 ± 2.0 (7.9–16.0)	13.5 ± 2.0 (11.7–17.3)	9–14
c'	3.5 ± 0.4 (2.8–4.4)	2.7 ± 0.4 (2.3–3.2)	4.5
V	80.5 ± 1.4 (78.6–82.8)	-	76–80
G1	45.1 ± 5.6 (35.2–56.7)	-	-
Stylet length	17.0 ± 0.5 (16.0–18.0)	15.7 ± 0.8 (14.0–16.0)	17–19
(Stylet length/body length) × 100	4.3 ± 0.4 (3.8–5.1)	4.4 ± 0.1 (4.1–4.5)	-
Conus length	10.8 ± 0.7 (10.0–12.0)	10.5 ± 0.6 (9.5–11.0)	-
m	63.7 ± 3.4 (58.8–70.6)	67.0 ± 2.5 (62.5–68.8)	-
DGO	4.9 ± 0.6 (4.0–6.0)	4.3 ± 0.5 (3.5–5.0)	-
O	28.5 ± 3.2 (23.5–36.4)	27.6 ± 2.4 (25.0–31.3)	-
Lip width	4.4 ± 0.5 (4.0–5.5)	5.0 ± 0.6 (4.0–5.5)	-
Median bulb length	22.8 ± 2.0 (19.0–25.0)	20.4 ± 0.5 (20.0–21.0)	-
Median bulb width	11.1 ± 1.6 (9.5–14.0)	8.1 ± 0.2 (8.0–8.5)	-
Anterior end to center median bulb	51.1 ± 4.2 (43.0–59.0)	46.8 ± 1.4 (44.0–48.0)	-
MB	53.8 ± 3.1 (46.2–59.8)	50.1 ± 3.5 (44.9–53.9)	-
Nerve ring to anterior end	69.8 ± 5.4 (57.0–81.0)	67.5 ± 4.7 (62.0–74.0)	-
Excretory pore to anterior end	82.8 ± 5.2 (72.0–94.0)	85.0 ± 4.7 (79.0–91.0)	75
Pharynx length	93.8 ± 6.6 (76.0–104.0)	93.5 ± 5.0 (89.0–101.0)	-
Maximum body diam.	21.5 ± 2.9 (16.5–26.0)	19.2 ± 2.9 (16.0–23.0)	18
Tail length	35.5 ± 6.0 (22.0–43.0)	26.9 ± 3.1 (21.0–29.0)	34
Anal body diam.	10.3 ± 1.5 (8.0–13.5)	10.0 ± 1.5 (9.0–12.0)	-

* Abbreviations: a = body length/greatest body diameter; b = body length/distance from anterior end to pharyngo-intestinal junction; DGO = distance between stylet base and orifice of dorsal pharyngeal gland; c = body length/tail length; c' = tail length/tail diameter at anus or cloaca; G1 = anterior genital branch length expressed as percentage (%) of the body length; L = overall body length; m = length of conus as percentage of total stylet length; MB = distance between anterior end of body and center of median pharyngeal bulb expressed as percentage (%) of the pharynx length; n = number of specimens on which measurements are based; O = DGO as percentage of stylet length; V = distance from body anterior end to vulva expressed as percentage (%) of the body length.

2.1.4 Remarks on Paratylenchus baldaccii (Oostenbrink, 1953) Raski, 1962, Paratylenchus enigmaticus Munawar, Yevtushenko, Palomares-Rius and Castillo, 2021, Paratylenchus holdemani Raski, 1975, Paratylenchus neoamblycephalus Geraert, 1965, Paratylenchus pedrami Clavero-Camacho, Cantalapiedra-Navarrete, Archidona-Yuste, Castillo and Palomares-Rius, 2021, and Paratylenchus veruculatus Wu, 1962 (Table 5).

Paratylenchus baldaccii, *P. enigmaticus*, *P. holdemani*, *P. neoamblycephalus*, *P. pedrami*, and *P. veruculatus* have been previously recorded within recent studies of pin nematodes in Spain [4,15], and morphological and morphometrical data of them were coincident with previous reports. Consequently, only some morphometric data or D2-D3 sequences had been reported here for these nematode samples. *Paratylenchus baldaccii* was identified in grasses at Arroyo Frío, Jaén province, in the same sample that we previously identified a population of *P. vandenbrandei* [17]. These data suggest that most probably the previ- ous record of *P. vandenbrandei* [17] needs to be considered as *P. baldaccii*, as well as other reports [16,33], but additional studies need to be carried out to confirm these potential misidentifications on the basis of application of integrative taxonomy. *Paratylenchus baldaccii* has been reported in several localities at south and southeastern Spain, including Jaén, Granada and Murcia provinces [4,15,22,34]. *Paratylenchus enigmaticus* was detected in the rhizosphere of grasses at campus Alameda del Obispo, Córdoba; this report confirms a wider distribution than previously estimated, since it was detected only in the rhizosphere of cherry at Northeastern of Spain at La Almunia, Zaragoza province [4]. *Paratylenchus holdemani* has been recently reported in the rhizosphere of almond at Martos, Jaén province [4]. This new report under a natural environment (wild olive) at St. Maria de Trasierra, Córdoba province, also suggests that this species can be common in Andalucia (Southern part of the Iberian Peninsula). Finally, *P. neoamblycephalus* was confirmed by molecular and morphometrical data under a natural environment (Portuguese oak forest). Unfortunately, only a mature female was detected (Table 5), but morphometrics agree with original description [35] and recent data by Singh et al. [3]. Consequently, up to our knowledge, this is the first report of this species for Spain. Finally, the new findings of *P. pedrami* and *P. veruculatus* from natural environments (wild olive) at Córdoba province confirms also that these species are widely distributed in Spain [4].

Molecular Characterization

Several populations of species already molecularly characterized in previous works, such as *P. baldaccii*, *P. enigmaticus*, *P. holdemani*, *P. neoamblycephalus*, *P. pedrami*, and *P. veruculatus* have been sequenced herein. All sequences obtained for these species matched well with the accessions from the same species deposited in GenBank, showing similarity values from 99 to 100% [3,4].

2.1.5 Remarks on *Paratylenchus goodeyi* Oostenbrink, 1953 (Figure 6, Table 6).

Table 5. Morphometrics of Paratylenchus baldaccii (Oostenbrink, 1953) Raski, 1962 and Paratylenchus enigmaticus Munawar, Yevtushenko, Palomares-Rius and Castillo, 2021, Paratylenchus holdemani Raski, 1975, and Paratylenchus neoamblycephalus Geraert, 1965 from several localities in Spain. All measurements are in μm and in the form: mean ± s.d. (range).

	P. baldaccii	*P. enigmaticus*	*P. holdemani*	*P. neoamblycephalus*
Life Stage	Females	Females	Females	Female
Sample Code	CAZ_04	IAS_21	AR_102	CAZ_05
Locality, Province	Arroyo Frío, Jaén	Córdoba, Córdoba	St. Mª Trasierra, Córdoba	Arroyo Frío, Jaén
n	4	4	4	1
L	270.5 ± 5.2 (267–278)	367.0 ± 11.3 (358–383)	392.3 ± 30.2 (364–435)	363
a *	17.8 ± 2.1 (15.4–20.5)	21.5 ± 1.5 (20.2–23.2)	24.9 ± 1.6 (23.5–27.2)	18.2
b	3.4 ± 0.1 (3.3–3.5)	3.8 ± 0.2 (3.7–4.0)	3.9 ± 0.3 (3.6–4.1)	4.4
c	10.5 ± 1.3 (8.7–11.7)	15.0 ± 1.2 (13.2–16.0)	13.3 ± 1.0 (12.2–14.6)	14.0
c′	2.8 ± 0.1 (2.7–2.9)	2.5 ± 0.2 (2.3–2.8)	3.0 ± 0.4 (2.7–3.5)	2.2
V	81.2 ± 1.5 (79.5–83.0)	83.2 ± 1.2 (81.8–84.5)	81.0 ± 0.8 (79.9–81.9)	81.3
G1	40.1 ± 5.4 (34.5–46.4)	41.8 ± 9.3 (31.8–51.2)	28.4 ± 0.5 (28.0–28.7)	36.9
Stylet length	29.5 ± 1.2 (28.0–31.0)	26.8 ± 1.0 (26.0–28.0)	26.8 ± 0.6 (26.0–27.5)	33.0
(Stylet length/body length) × 100	10.9 ± 0.5 (10.5–11.6)	7.3 ± 0.1 (7.2–7.4)	6.8 ± 0.5 (6.3–7.4)	9.1
Conus length	21.4 ± 1.1 (20.5–23.0)	18.3 ± 0.5 (18.0–19.0)	16.0 ± 0.7 (15.0–16.5)	23
m	72.5 ± 4.8 (67.4–78.0)	68.3 ± 3.8 (64.3–73.1)	59.8 ± 1.5 (57.7–61.1)	69.7
DGO	5.0 ± 0.7 (4.0–5.5)	5.8 ± 0.9 (4.5–6.5)	6.3 ± 0.6 (5.5–7.0)	5.5
O	17.0 ± 2.5 (13.6–19.6)	21.5 ± 3.6 (16.7–25.0)	23.3 ± 2.0 (21.2–25.5)	16.7
Lip width	3.9 ± 0.3 (3.5–4.0)	6.3 ± 0.3 (6.0–6.5)	6.6 ± 0.5 (6.0–7.0)	6
Median bulb length	17.2 ± 0.8 (16.5–18.0)	23.3 ± 3.8 (20.0–27.0)	19.4 ± 3.1 (16.5–23.5)	26
Median bulb width	8.5 ± 0.5 (8.0–9.0)	10.0 ± 0.7 (9.5–11.0)	10.0 ± 0.4 (9.5–10.5)	11
Anterior end to center median bulb	44.8 ± 1.9 (42.0–46.0)	53.5 ± 1.3 (52.0–55.0)	56.5 ± 3.3 (52.5–60.5)	57
MB	56.4 ± 3.7 (52.1–59.7)	55.5 ± 1.1 (54.6–57.1)	56.4 ± 1.0 (55.2–57.6)	69.5
Nerve ring to anterior end	58.3 ± 1.5 (57.0–60.0)	73.3 ± 3.4 (70.0–78.0)	72.8 ± 3.6 (68.0–75.5)	66.0
Excretory pore to anterior end	68.3 ± 3.6 (63.0–71.0)	82.3 ± 4.3 (76.0–83.0)	85.5 ± 5.2 (79.5–82.0)	74.0
Pharynx length	79.5 ± 2.1 (77.0–82.0)	96.5 ± 3.9 (91.0–100.0)	100.3 ± 5.1 (93.0–105.0)	82.0
Maximum body diam.	15.4 ± 2.1 (13.0–18.0)	17.1 ± 1.7 (15.5–19.0)	15.8 ± 0.6 (15.0–16.5)	20.0
Tail length	26.3 ± 4.0 (23.0–32.0)	24.6 ± 2.9 (23.0–29.0)	29.6 ± 3.4 (26.5–33.5)	26.0
Anal body diam.	9.4 ± 1.1 (8.5–11.0)	9.8 ± 0.6 (9.0–10.5)	9.8 ± 0.3 (9.5–10.0)	12.0

* *Abbreviations: a = body length/greatest body diameter; b = body length/distance from anterior end to pharyngo-intestinal junction; DGO = distance between stylet base and orifice of dorsal pharyngeal gland; c = body length/tail length; c′ = tail length/tail diameter at anus or cloaca; G1 = anterior genital branch length expressed as percentage (%) of the body length; L = overall body length; m = length of conus as percentage of total stylet length; MB = distance between anterior end of body and center of median pharyngeal bulb expressed as percentage (%) of the pharynx length; n = number of specimens on which measurements are based; O = DGO as percentage of stylet length; V = distance from body anterior end to vulva expressed as percentage (%) of the body length.*

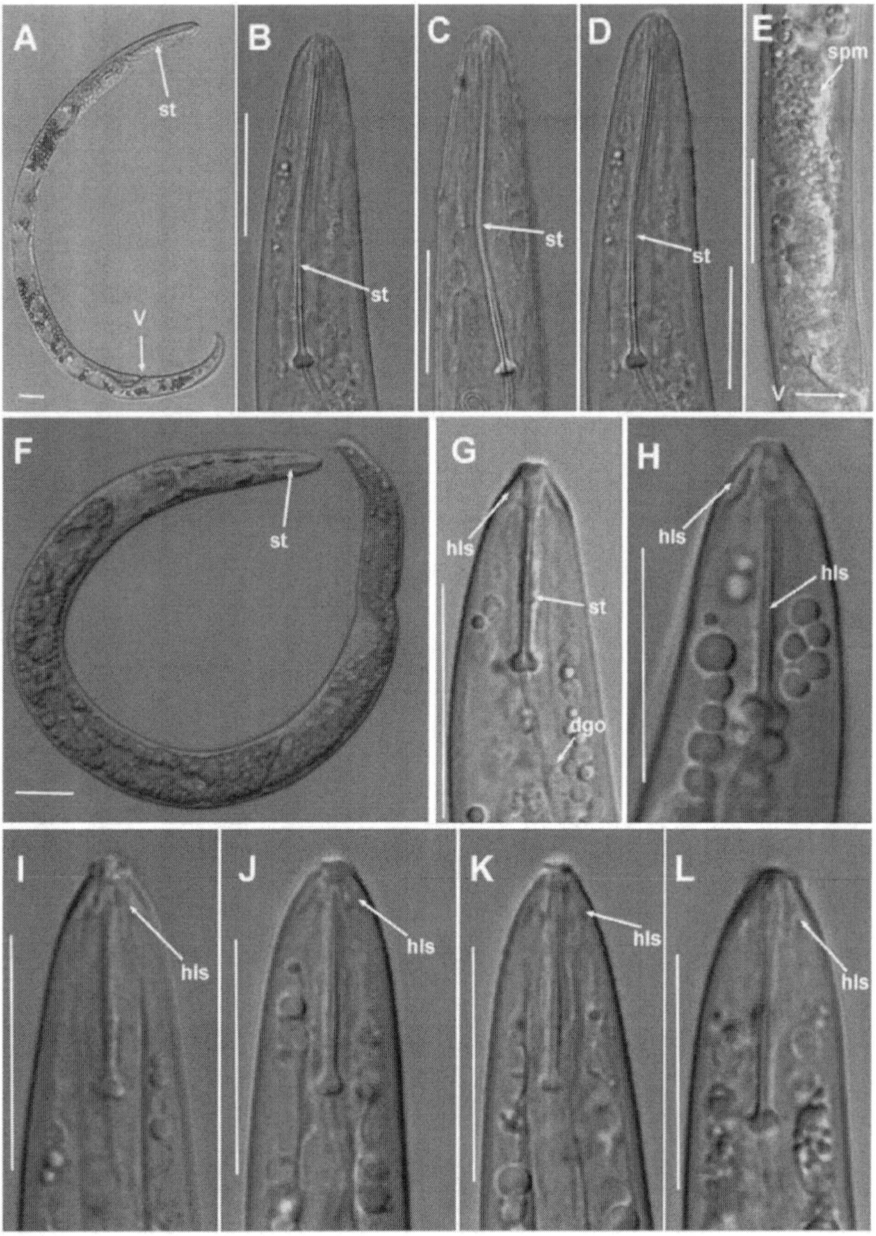

Figure 6. Light photomicrographs of *Paratylenchus goodeyi* Oostenbrink, 1953. (**A**) Entire female with vulva arrowed; (**B–D**) female lip region with stylet arrowed; (**E**) detail of vulval region showing spermatheca arrowed; (**F**) entire fourth- stage juvenile with short stylet arrowed; (**G–L**) fourth-stage juvenile lip regions showing labial sclerotization and short stylet (arrowed). Scale bars (**A–L** = 20 μm). (Abbreviations: dgo = pharyngeal dorsal gland orifice; hls = heavy labial sclerotization; spm = spermatheca; st = stylet; V = vulva).

Table 6. Morphometrics of *Paratylenchus goodeyi* Oostenbrink, 1953 from several localities in Spain. All measurements are in µm and in the form: mean ± s.d. (range).

Locality, Province	Córdoba, Córdoba		Sta. Mª Trasierra, Córdoba		Montalbán, Córdoba		Córdoba, Córdoba	
Life Stage	Females	Fourth-Stage Juveniles	Females	Fourth-Stage Juveniles	Females	Fourth-Stage Juveniles	Females	Fourth-Stage Juveniles
Sample Code	EP_ACA	EP_ACA	AR_097	AR_097	PR_050	PR_050	PR_017	PR_017
n	14	10	5	5	4	4	5	5
L	433.8 ± 38.3 (396–513)	440.8 ± 25.3 (409–486)	466.4 ± 35.1 (408–495)	432.8 ± 19.0 (412–461)	460.0 ± 34.6 (411–490)	395.8 ± 15.0 (375–409)	413.3 ± 8.2 (403–422)	395.8 ± 15.0 (375–409)
a *	22.6 ± 2.5 (18.8–26.8)	23.9 ± 1.5 (21.6–25.8)	22.3 ± 2.2 (19.6–24.3)	24.1 ± 1.6 (21.9–25.8)	22.5 ± 1.9 (20.6–24.3)	22.4 ± 1.9 (20.2–24.7)	20.3 ± 2.0 (17.5–22.2)	22.4 ± 1.8 (20.2–24.7)
b	3.9 ± 0.2 (3.5–4.3)	4.9 ± 0.3 (4.4–5.4)	3.9 ± 0.2 (3.6–4.3)	4.9 ± 0.3 (4.5–5.2)	4.0 ± 0.3 (3.6–4.3)	4.5 ± 0.4 (4.0–4.8)	3.6 ± 0.4 (3.3–4.2)	4.5 ± 0.4 (4.0–4.8)
c	12.6 ± 2.6 (10.7–20.9)	11.2 ± 1.0 (10.0–13.5)	13.7 ± 3.4 (11.3–19.3)	11.4 ± 1.1 (10.1–13.1)	13.4 ± 3.2 (11.6–18.1)	10.8 ± 1.3 (9.4–12.4)	12.0 ± 2.0 (10.6–14.9)	11.1 ± 1.5 (9.9–13.2)
c'	3.5 ± 0.5 (2.1–4.4)	3.2 ± 0.3 (2.6–3.5)	3.3 ± 0.7 (2.3–4.3)	3.0 ± 0.2 (2.7–3.3)	3.3 ± 0.4 (2.7–3.6)	3.1 ± 0.2 (2.8–3.3)	3.3 ± 0.4 (2.7–3.6)	3.0 ± 0.3 (2.6–3.3)
V	80.8 ± 1.3 (78.2–82.4)	-	79.7 ± 1.5 (77.8–81.2)	-	79.2 ± 1.4 (77.8–81.1)	-	80.9 ± 0.9 (79.7–81.8)	-
G1	30.6 ± 2.5 (25.7–33.9)	-	30.8 ± 3.4 (26.1–34.7)	-	29.8 ± 3.0 (26.1–33.1)	-	33.0 ± 2.3 (31.1–36.3)	-
Stylet length	50.9 ± 2.9 (46.0–56.0)	17.4 ± 1.2 (15.0–18.5)	52.8 ± 2.5 (51.0–56.0)	16.9 ± 0.9 (16.0–18.0)	51.8 ± 2.2 (50.0–55.0)	17.1 ± 0.9 (16.0–18.0)	51.8 ± 2.2 (50.0–55.0)	16.9 ± 0.6 (16.0–17.5)
(Stylet length/body length) × 100	11.8 ± 0.7 (10.4–12.7)	4.0 ± 0.3 (3.5–4.4)	11.4 ± 0.8 (10.4–12.5)	3.9 ± 0.3 (3.5–4.4)	11.3 ± 0.9 (10.2–12.4)	4.3 ± 0.3 (3.9–4.7)	12.5 ± 0.5 (12.1–13.2)	4.3 ± 0.3 (3.9–4.7)
Conus length	40.8 ± 2.4 (37.0–45.0)	12.4 ± 1.1 (10.0–14.0)	41.8 ± 1.9 (40.0–45.0)	12.2 ± 1.3 (10.0–13.0)	41.5 ± 1.7 (40.0–44.0)	12.8 ± 0.5 (12.0–13.0)	41.8 ± 2.2 (40.0–45.0)	12.8 ± 0.5 (12.0–13.0)
m	80.2 ± 2.1 (75.0–83.0)	71.3 ± 4.6 (62.5–80.0)	79.2 ± 2.6 (75.0–81.8)	72.1 ± 5.6 (62.5–76.5)	80.7 ± 0.8 (80.0–81.8)	74.4 ± 1.8 (72.2–76.5)	80.7 ± 0.8 (80.0–81.8)	75.6 ± 1.1 (74.3–76.5)
DGO	5.3 ± 0.4 (4.5–6.0)	4.1 ± 0.4 (3.5–5.0)	5.5 ± 0.5 (5.0–6.0)	4.2 ± 0.4 (4.0–5.0)	5.1 ± 0.5 (4.5–5.5)	4.3 ± 0.5 (4.0–5.0)	4.8 ± 0.3 (3.5–4.0)	4.1 ± 0.3 (4.0–4.5)
O	10.5 ± 1.0 (8.8–12.0)	23.6 ± 2.4 (20.0–27.8)	10.4 ± 1.0 (9.1–11.7)	24.9 ± 2.8 (22.2–29.4)	9.9 ± 1.1 (8.8–11.0)	24.9 ± 3.3 (22.2–24.9)	9.2 ± 0.6 (8.8–10.0)	24.5 ± 1.6 (22.9–26.5)
Lip width	4.8 ± 0.6 (4.0–5.5)	4.0 ± 0.2 (3.5–4.5)	4.2 ± 0.4 (4.0–5.0)	3.7 ± 0.3 (3.5–4.0)	4.1 ± 0.3 (4.0–4.5)	3.8 ± 0.3 (3.5–4.0)	4.1 ± 0.3 (4.0–4.5)	3.8 ± 0.3 (3.5–4.0)
Median bulb length	22.6 ± 2.9 (17.0–26.0)	24.3 ± 1.3 (22.0–27.0)	25.4 ± 0.5 (25.0–26.0)	24.4 ± 0.9 (24.0–26.0)	25.5 ± 0.6 (25.0–26.0)	24.5 ± 1.0 (24.0–26.0)	25.5 ± 0.6 (25.0–26.0)	23.8 ± 0.5 (23.0–24.0)

Table 6. *Cont.*

Locality, Province	Córdoba, Córdoba		Sta. M' Trasierra, Córdoba		Montalbán, Córdoba		Córdoba, Córdoba	
Life Stage	Females	Fourth-Stage Juveniles	Females	Fourth-Stage Juveniles	Females	Fourth-Stage Juveniles	Females	Fourth-Stage Juveniles
Sample Code	EP_ACA	EP_ACA	AR_097	AR_097	PR_050	PR_050	PR_017	PR_017
Median bulb width	11.2 ± 0.8 (10.0–13.0)	7.9 ± 0.4 (7.0–8.5)	11.3 ± 1.0 (10.0–12.5)	7.8 ± 0.6 (7.0–8.5)	11.3 ± 0.8 (10.2–12.4)	7.8 ± 0.6 (7.0–8.5)	11.3 ± 1.0 (10.0–12.0)	7.5 ± 0.4 (7.0–8.0)
Anterior end to center median bulb	74.8 ± 3.1 (70.0–81.0)	48.4 ± 1.5 (46.0–51.0)	77.0 ± 3.4 (73.0–81.0)	47.8 ± 0.8 (47.0–49.0)	76.3 ± 3.4 (73.0–81.0)	47.8 ± 1.0 (47.0–49.0)	75.8 ± 2.5 (73.0–79.0)	47.8 ± 1.0 (47.0–49.0)
MB	65.9 ± 4.8 (59.5–74.0)	53.5 ± 3.2 (48.5–57.5)	64.5 ± 6.6 (59.5–76.0)	53.7 ± 3.2 (53.9–56.6)	65.5 ± 8.2 (59.5–77.5)	54.2 ± 3.2 (50.0–57.3)	65.5 ± 8.2 (59.5–77.6)	54.2 ± 3.2 (51.0–57.3)
Nerve ring to anterior end	90.9 ± 7.1 (78.0–99.0)	65.0 ± 3.2 (61.0–71.0)	95.4 ± 2.7 (91.0–98.0)	64.2 ± 2.3 (62.0–69.0)	94.8 ± 2.6 (91.0–97.0)	64.8 ± 2.2 (63.0–68.0)	94.8 ± 2.6 (91.0–97.0)	64.8 ± 2.2 (63.0–68.0)
Excretory pore to anterior end	94.6 ± 6.8 (82.0–106.0)	83.1 ± 6.0 (74.0–91.0)	98.2 ± 3.3 (94.0–103.0)	81.6 ± 5.7 (75.0–88.0)	98.5 ± 3.7 (94.0–103.0)	81.8 ± 5.6 (76.0–87.0)	98.5 ± 3.7 (94.0–103.0)	81.8 ± 5.6 (76.0–87.0)
Pharynx length	112.2 ± 12.0 (97.0–132.0)	90.8 ± 6.6 (82.0–101.0)	119.4 ± 12.1 (100.0–129.0)	89.2 ± 5.3 (83.0–97.0)	116.5 ± 13.5 (98.0–127.0)	88.3 ± 5.3 (82.0–94.0)	116.5 ± 13.5 (98.0–127.0)	88.3 ± 5.3 (82.0–94.0)
Maximum body diam.	19.4 ± 2.8 (15.0–26.0)	18.5 ± 1.3 (16.0–20.0)	21.0 ± 2.3 (19.0–25.0)	18.0 ± 1.6 (16.0–20.0)	20.5 ± 1.7 (19.0–23.0)	17.8 ± 1.7 (16.0–20.0)	20.5 ± 1.7 (19.0–23.0)	17.8 ± 1.7 (16.0–20.0)
Tail length	35.2 ± 4.2 (23.5–42.0)	39.6 ± 4.3 (31.0–44.0)	35.2 ± 6.2 (25.0–41.0)	38.2 ± 4.3 (32.0–43.0)	35.3 ± 5.9 (27.0–40.0)	37.0 ± 3.2 (33.0–40.0)	35.3 ± 5.9 (27.0–40.0)	36.0 ± 3.6 (31.0–39.0)
Anal body diam.	10.2 ± 0.8 (9.0–11.0)	12.5 ± 0.9 (11.5–14.0)	10.7 ± 0.7 (9.5–11.0)	12.6 ± 0.8 (11.5–13.5)	10.8 ± 0.5 (10.0–11.0)	12.1 ± 0.9 (11.0–13.0)	10.8 ± 0.5 (10.0–11.0)	12.1 ± 0.9 (11.0–13.0)

* Abbreviations: a = body length/greatest body diameter; b = body length/distance from anterior end to pharyngo-intestinal junction; DGO = distance between stylet base and orifice of dorsal pharyngeal gland; c = body length/tail length; c' = tail length/tail diameter at anus or cloaca; G1 = anterior genital branch length expressed as percentage (%) of the body length; L = overall body length; m = length of conus as percentage of total stylet length; MB = distance between anterior end of body and center of median pharyngeal bulb expressed as percentage (%) of the pharynx length; n = number of specimens on which measurements are based; O = DGO as percentage of stylet length; V = distance from body anterior end to vulva expressed as percentage (%) of the body length.

This species has been detected in several samples of almond and natural environment (wild olive) in several localities of Córdoba and Jaén provinces (Table 1). Morphology and morphometrics of adult females are coincident with the original description and recent studies [3,4]. However, in none of the previous studies on this species J4 were studied under an integrative taxonomic point of view. In all of our populations, irrespective of cultivated almond fields or natural environments, all the J4 of this species were characterized by bearing a short rigid and straight stylet (15.0–18.5 μm), lip region-truncate with labial framework sclerotization strong; with numerous dark granules into the body (Figure 6, Table 6), and considered the resting-stage [26]. In the original description of *P. goodeyi* it is mentioned that "J3 and J4 from soil samples, which probably belonged to this species, on account of the typical shape of the lip region, all had a short spear below 20 μm" [36]. However, this is the first report documenting, by morphometric and molecular markers (see below), a clear stylet and lip region metamorphosis between J4 and adult female, from short rigid stylet and conoid-truncate lip region with strong labial sclerotization moving to a long and slender flexible stylet and a conoid-rounded lip region without labial sclerotization (Figure 6). These data suggest, that apart from the reserve dark granules for resting during adverse environmental conditions (such as the hard drought during the summer season in Mediterranean climates), J4 of *P. goodeyi* is ready for feeding on susceptible roots during the beginning of the next season. Except for the stylet and lip region, J4 showed similar morphology to adult females with a posterior body rounded terminus. The present reports extend the geographical distribution of this species in Spain which has been already reported in several provinces including Navarra [12], Jaén [20,21], Barcelona [19], and Córdoba [4].

Molecular Characterization

Twenty-two D2-D3 sequences of 28S rRNA (MZ265084-MZ265105), 14 ITS (MZ265020-MZ265033), and 12 COI gene sequences (MZ262227-MZ262238) of *P. goodeyi* were generated in this study, with an intraspecific sequence variation from 0 to 9 nucleotides for D2-D3 of 28S rRNA (MZ265084-MZ265105), 0 to 17 nucleotides for ITS region (MZ265020-MZ265033), and finally, 0 to 29 nucleotides for COI gene (MZ262227-MZ262238). Some intraspecific sequence variations were detected when comparing with the accessions of *P. goodeyi* deposited in GenBank, showing similarity values of 99% for the D2-D3 of 28S rRNA, from 96 to 99% for the ITS region and finally, from 96 to 98% for the COI gene [3,4]. Some accessions from the different populations, belonging to J4, and all of them, matched well, from 99 to 100% similarity, with the sequences obtained for adult females of the same population.

2.1.6 Remarks on *Paratylenchus macrodorus* Brzeski, 1963 (Figure 7, Table 7).

According to species grouping by Ghaderi et al. [2] this species belongs to group 11 characterized by stylet length more than 40 μm, four lateral lines and advulval flaps absent. The Spanish population from Santa Mª de Trasierra, Córdoba province, was characterized by long flexible stylet 70.0–84.0 μm, lip region continuous with body contour, tapering slightly to a blunt anterior end, submedian lobes fairly distinct, female tail tapering gradually to finely rounded terminus. Males without stylet, and J4 similar to female, except for shorter stylet (both stages confirmed belonging to this species by molecular markers).

Morphometrics of the Spanish population agree well with original description as well as other populations with small differences in stylet length (70.0–84.0 μm vs. 75.0–92.0 μm), which may be due to geographical intraspecific variability [2]. Molecularly *P. macrodorus* is close to *P. pandatus* and *P. wuae* (using D2-D3 region of 28S rRNA) from which can be mor- phological and morphometrically separated by submedian lobes (fairly distinct vs. clearly distinct, pronounced submedian lobes, respectively), body length (317–410 vs. 290–339, 300–360 μm, respectively), c and c' ratios (7.4–11.1 vs. 9.2–16.6, 10.5–11.3, and 3.5–4.9 vs. 2.2–3.0, 3.4–3.8, respectively), and J4 stylet (present vs. absent, absent, respectively) [2,37,38].

This species was described from vegetables from Poland [39] and has been reported from the Netherlands, Germany and Belgium [34], and New Caledonia [40]. This is the second report from Spain, the first being from natural environments in Almeria province [18].

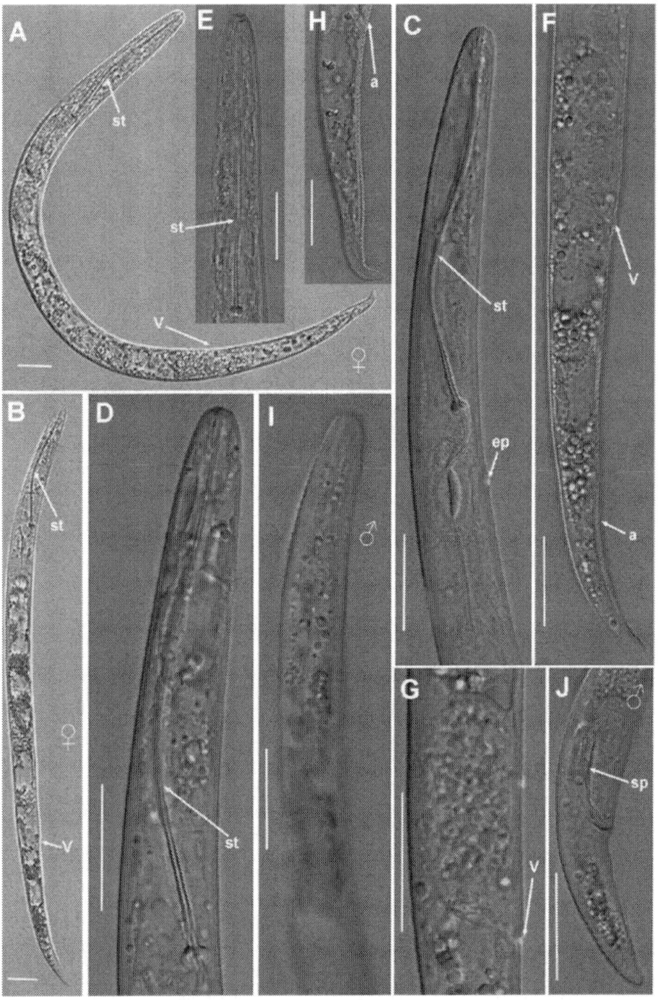

Figure 7. Light photomicrographs of *Paratylenchus macrodorus* Brzeski, 1963. (A,B) Entire female with stylet and vulva arrowed; (C) female pharyngeal region;

(D,E) female lip region; (F) female posterior region with vulva and anus (arrowed); (G) detail of vulva (arrowed); (H) female tail region with anus arrowed; (I) male pharyngeal region showing absence of stylet; (J) male posterior region showing spicules (arrowed). Scale bars (A–J = 20 μm). (Abbreviations: a = anus; ep = excretory pore; sp = spicules; st = stylet; V = vulva).

Table 7. Morphometrics of Paratylenchus macrodorus Brzeski, 1963 from Santa María de Trasierra, Córdoba, Spain. All measurements are in μm and in the form: mean ± s.d. (range).

Life Stage	Females	Male	Fourth-Stage Juveniles
Sample Code		AR_102	
Locality, province		St. Mª Trasierra, Córdoba	
n	10	1	3
L	365.8.3 ± 26.6 (317–410)	395	304.3 ± 6.1 (299–311)
a *	23.0 ± 2.6 (19.9–28.3)	31.6	20.8 ± 0.6 (20.2–21.4)
b	2.8 ± 0.2 (2.5–3.0)	4.0	2.8 ± 0.1 (2.7–2.8)
c	8.8 ± 1.1 (7.4–11.1)	12.3	13.6 ± 0.5 (13.2–14.1)
c′	4.2 ± 0.5 (3.5–4.9)	2.9	2.7 ± 0.03 (2.7–2.8)
V or T	75.1 ± 0.8 (73.9–76.3)	48.1	-
G1	29.7 ± 3.1 (24.1–33.8)	-	-
Stylet length	76.2 ± 3.9 (70.0–84.0)	-	62.3 ± 1.5 (61.0–64.0)
(Stylet length/body length) × 100	20.9 ± 1.6 (18.8–24.0)	-	20.5 ± 0.3 (20.1–20.7)
Conus length	68.9 ± 4.3 (61.5–77.0)	-	53.0 ± 1.0 (52.0–54.0)
m	90.4 ± 1.6 (87.9–93.7)	-	85.0 ± 1.6 (83.9–86.9)
DGO	5.6 ± 0.6 (5.0–6.5)	-	5.5 ± 0.5 (5.0–6.0)
O	7.4 ± 0.8 (6.4–9.0)	-	8.8 ± 0.8 (8.2–8.7)
Lip width	4.7 ± 0.2 (4.5–5.0)	3.5	4.7 ± 0.6 (4.0–5.0)
Median bulb length	26.7 ± 2.4 (24.0–31.0)	-	18.7 ± 0.6 (18.0–19.0)
Median bulb width	10.3 ± 1.2 (9.0–13.0)	-	8.8 ± 0.3 (8.5–9.0)
Anterior end to center median bulb	92.9 ± 6.4 (83.0–102.0)	-	72.0 ± 1.0 (71.0–73.0)
MB	69.3 ± 3.2 (61.5–72.2)	-	66.1 ± 0.3 (65.8–66.4)
Nerve ring to anterior end	109.7 ± 10.0 (91.0–123.0)	84	85.7 ± 1.5 (84.0–87.0)
Excretory pore to anterior end	94.1 ± 9.0 (82.0–109.0)	91	73.0 ± 2.0 (71.0–75.0)
Pharynx length	133.1 ± 9.4 (115.0–143.0)	98	109.0 ± 2.0 (107.0–111.0)
Maximum body diam.	16.0 ± 1.1 (14.5–18.0)	12.5	14.7 ± 0.6 (14.0–15.0)
Tail length	42.0 ± 4.4 (33.5–49.0)	32	22.3 ± 0.6 (22.0–23.0)
Anal body diam.	10.0 ± 1.1 (8.5–12.0)	11	8.2 ± 0.3 (8.0–8.5)
Spicules	-	19.5	-
Gubernaculum	-	5.5	-

* *Abbreviations: a = body length/greatest body diameter; b = body length/distance from anterior end to pharyngo-intestinal junction; DGO = distance between stylet base and orifice of dorsal pharyngeal gland; c = body length/tail length; c' = tail length/tail diameter at anus or cloaca; G1 = anterior genital branch length expressed as percentage (%) of the body length; L = overall body length; m = length of conus as percentage of total stylet length; MB = distance between anterior end of body and center of median pharyngeal bulb expressed as percentage (%) of the pharynx length; n = number of specimens on which measurements are based; O = DGO as percentage of stylet length; T = distance from cloacal aperture to anterior end of testis expressed as percentage (%) of the body length; V = distance from body anterior end to vulva expressed as percentage (%) of the body length.*

Molecular Characterization

Six D2-D3 sequences of 28S rRNA (MZ265108-MZ265113), five ITS (MZ265034-MZ265038), and six COI gene sequences (MZ262239-MZ262244) were generated for *P. macrodorus* without intraspecific sequence variations for ribosomal genes, and J4 and adult female sequences were identical, confirming the identity of these juvenile individuals as *P. macrodorus*. *Paratylenchus macrodorus* showed high molecular similarity with *P. pandatus* and *P. wuae*, being 99% similar for the D2-D3 of 28S rRNA (varying from 3 to 7 nucleotides and no indels). For the ITS region, similarity values found for *P. macrodorus* ranging from 96% (34 nucleotides and 13 indels) to 98% (13 nucleotides and 2 indels) to *P. wuae* (KM061783) and *P. pandatus* (MZ265041-MZ265042), respectively. Similarity values detected in the COI gene were lower than in the ribosomal genes, being 96% (14 nucleotides and no indels) to *P. wuae* and 94% (24 nucleotides and no indels) to *P. pandatus*. However, morphologi- cally and morphometrically *P. macrodorus*, *P. pandatus* and *P. wuae* can be clearly separated (see above).

2.1.7 Remarks on Paratylenchus pandatus (Raski, 1976) Siddiqi, 1986 (Figure 8, Table 8).

According to species grouping by Ghaderi et al. [2] this species belongs to group 10 characterized by stylet length more than 40 µm, four lateral lines and advulval flaps present. The Spanish population from Caravaca, Murcia province, was characterized by moderately long flexible stylet 57.0–68.5 µm, lip region rounded, continuous with body contour, with distinct submedian lobes, spermatheca elongate and filled with sperm, which indicates that males are required for reproduction but were not detected, female tail tapering gradually to rounded terminus. J4 was similar to female, except for absent stylet (stages confirmed belonging to this species by molecular markers). Morphometrics of the Spanish populations agree well with the original description, as well as Vietnam population with small differences in stylet length (57.0–68.5 µm vs. 63.0–70.0 µm), V ratio (74.5–77.7 vs. 70.0–76.0), and shape of tail terminus (finely rounded in Spanish and Vietnam populations while almost acute in original description), which may be due to geographical intraspecific variability [2]. This species was described from grapefruit in Nigeria [37] and has been reported from Vietnam [41] and Ethiopia [42], and this study comprises the first report from Spain. This species is closely related molecularly to *P. macrodorus*, but they have important morphological differences such as the presence vs. absence of advulval flaps and J4 without stylet vs. J4 with stylet.

Molecular Characterization

Two identical D2-D3 of 28S rRNA (MZ265116-MZ265117), two identical ITS sequences (MZ265041-MZ265042) and five identical COI gene sequences (MZ262247-MZ262251) were obtained from *P. pandatus* in the present study. Sequences obtained from J4 and females for all genes were identical, confirming that are the same species. *Paratylenchus pandatus* showed high molecular similarity with *P. macrodorus* and *P. wuae*, being 99% similar for the D2-D3 of 28S rRNA (varying from 7 to 8 nucleotides and no indels). For the ITS region, the similarity values were from 97% to 98% (differing by 13–15 nucleotides and from 2 to 6 indels) with *P. macrodorus* and *P. wuae*, respectively. Similarity values detected in the COI gene were lower than in the ribosomal genes, being 93% (23 nucleotides and no indels) to *P. wuae* and 94% (24 nucleotides and no indels) to *P. macrodorus*.

2.1.8 Remarks on Paratylenchus recisus Siddiqi, 1996 (Figure 9, Table 9).

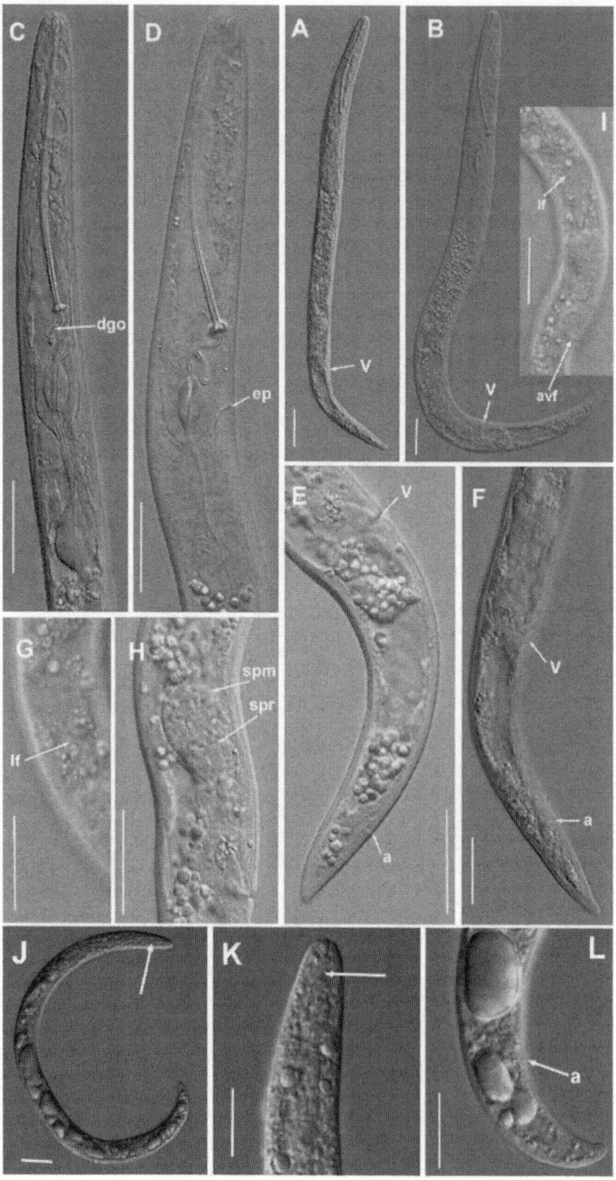

Figure 8. Light photomicrographs of *Paratylenchus pandatus* (Raski, 1976) Siddiqi, 1986. (A,B) entire females with vulva arrowed; (C,D) female pharyngeal region; (E,F) female posterior region showing vulva and anus (arrowed); (G) detail of lateral field (arrowed) at mid-body; (H) detail of spermatheca and sperm (arrowed); (I) detail of lateral field and advulval flap (arrowed); (J) entire fourth-stage juvenile, stylet absence arrowed; (K) fourth-stage juvenile lip region showing stylet absence (arrowed); (L) fourth-stage juvenile tail. Scale bars (A–L = 20 μm). (Abbreviations: a = anus; avf = advulval flap; dgo = pharyngeal dorsal gland orifice; ep = excretory pore; lf = laterl field; spm = spermatheca; spr = sperm; V = vulva).

Table 8. Morphometrics of Paratylenchus pandatus (Raski, 1976) Siddiqi, 1986 from Caravaca, Murcia, Spain. All measure- ments are in μm and in the form: mean ± s.d. (range).

Life Stage	Females	Fourth-Stage Juveniles	Females
Sample Code	PIN_AR	PIN_AR	Type
Locality, Province	Caravaca, Murcia	Caravaca, Murcia	Population [37]
n	12	3	10
L	317.2 ± 15.9 (290–339)	277.3 ± 22.7 (252–296)	330–420
a *	18.7 ± 1.8 (15.7–21.5)	18.1 ± 1.5 (16.8–19.7)	23–32
b	2.7 ± 0.1 (2.5–2.9)	3.8 ± 0.3 (3.5–4.0)	2.8–3.2
c	12.2 ± 2.2 (9.2–16.6)	13.2 ± 1.6 (11.5–14.2)	9–12
c′	2.5 ± 0.2 (2.2–3.0)	2.2 ± 0.1 (2.1–2.3)	4.7
V	75.8 ± 0.9 (74.5–77.7)	-	70–76
G1	32.0 ± 2.9 (28.6–38.6)	-	-
Stylet length	61.3 ± 3.8 (57.0–68.5)	-	63–70
(Stylet length/body length) × 100	19.3 ± 0.8 (17.8–21.1)	-	-
Conus length	53.2 ± 3.1 (49.0–59.0)	-	-
m	86.9 ± 1.1 (85.7–89.4)	-	-
DGO	5.5 ± 0.5 (5.0–6.5)	-	-
O	9.1 ± 0.9 (7.9–10.5)	-	-
Lip width	5.9 ± 0.7 (5.0–7.0)	4.7 ± 0.3 (4.5–5.0)	-
Median bulb length	23.7 ± 2.1 (19.0–26.5)	-	-
Median bulb width	11.0 ± 1.1 (10.0–14.0)	-	-
Anterior end to center median bulb	79.5 ± 6.7 (64.0–88.0)	-	-
MB	67.0 ± 2.5 (62.8–71.3)	-	-
Nerve ring to anterior end	97.7 ± 6.3 (82.0–106.0)	55.0 ± 1.0 (54.0–56.0)	-
Excretory pore to anterior end	84.9 ± 8.9 (71.0–111.0)	63.0 ± 1.0 (62.0–64.0)	94–119
Pharynx length	118.6 ± 7.6 (102.0–128.0)	73.0 ± 1.0 (72.0–74.0)	-
Maximum body diam.	17.1 ± 1.4 (15.0–19.5)	15.3 ± 0.6 (15.0–16.0)	17
Tail length	26.7 ± 4.6 (18.0–32.0)	21.0 ± 1.0 (20.0–22.0)	38
Anal body diam.	10.6 ± 1.5 (8.0–13.0)	9.3 ± 0.3 (9.0–9.5)	-

* *Abbreviations: a = body length/greatest body diameter; b = body length/distance from anterior end to pharyngo-intestinal junction; DGO = distance between stylet base and orifice of dorsal pharyngeal gland; c = body length/tail length; c′ = tail length/tail diameter at anus or cloaca; G1 = anterior genital branch length expressed as percentage (%) of the body length; L = overall body length; m = length of conus as percentage of total stylet length; MB = distance between anterior end of body and center of median pharyngeal bulb expressed as percentage (%) of the pharynx length; n = number of specimens on which measurements are based; O = DGO as percentage of stylet length; V = distance from body anterior end to vulva expressed as percentage (%) of the body length.*

Table 9. Morphometrics of Paratylenchus recisus Siddiqi, 1996 from Arroyo Frío, Jaén, Spain. All measurements are in μm and in the form: mean ± s.d. (range).

Life Stage	Females	Fourth-Stage Juveniles	Females
Sample Code	CAZ_06	CAZ_06	Type
Locality, Province	Arroyo Frío, Jaén	Arroyo Frío, Jaén	Population [43]
n	4	3	20
L	397.0 ± 36.5 (363–448)	353.0 ± 32.5 (329–390)	270–390
a *	18.4 ± 1.3 (17.2–20.2)	16.2 ± 1.0 (15.5–17.3)	16–27
b	4.6 ± 0.7 (3.8–5.4)	4.9 ± 0.3 (4.7–5.3)	3.8–5.0
c	12.6 ± 1.0 (11.2–13.6)	14.7 ± 0.8 (14.2–15.6)	13–16
c'	3.3 ± 0.4 (2.8–3.5)	2.4 ± 0.1 (2.4–2.6)	2.7–3.3
V	81.2 ± 0.7 (80.4–82.0)	-	78–83
G1	44.3 ± 1.6 (42.1–45.8)	-	-
Stylet length	15.1 ± 0.6 (14.5–16.0)	-	15–17
(Stylet length/body length) × 100	3.8 ± 0.4 (3.3–4.2)	-	-
Conus length	9.8 ± 0.5 (9.0–10.0)	-	-
m	64.5 ± 2.5 (62.1–66.7)	-	-
DGO	5.3 ± 0.5 (5.0–6.0)	-	-
O	34.8 ± 4.5 (31.3–41.4)	-	-
Lip width	5.6 ± 0.8 (5.0–6.5)	4.5 ± 0.5 (4.0–5.0)	-
Median bulb length	23.0 ± 2.2 (20.0–25.0)	14.0 ± 0.5 (13.5–14.5)	-
Median bulb width	10.4 ± 1.5 (9.0–12.5)	7.2 ± 0.3 (7.0–7.5)	-
Anterior end to center median bulb	47.5 ± 7.4 (39.0–57.0)	33.7 ± 1.5 (32.0–35.0)	-
MB	54.0 ± 4.9 (47.0–58.0)	46.7 ± 0.9 (45.7–47.3)	-
Nerve ring to anterior end	66.3 ± 7.1 (59.0–76.0)	62.0 ± 2.0 (60.0–64.0)	-
Excretory pore to anterior end	77.5 ± 9.7 (71.0–92.0)	68.7 ± 1.5 (67.0–70.0)	58–70
Pharynx length	87.8 ± 9.1 (81.0–101.0)	72.0 ± 2.0 (70.0–74.0)	-
Maximum body diam.	21.8 ± 3.3 (18.0–26.0)	21.8 ± 0.8 (21.0–22.5)	12–16
Tail length	31.5 ± 2.6 (28.0–34.0)	24.0 ± 1.0 (23.0–25.0)	18–29
Anal body diam.	9.8 ± 1.7 (8.0–12.0)	9.8 ± 0.8 (9.0–10.5)	-

* *Abbreviations: a = body length/greatest body diameter; b = body length/distance from anterior end to pharyngo-intestinal junction; DGO = distance between stylet base and orifice of dorsal pharyngeal gland; c = body length/tail length; c' = tail length/tail diameter at anus or cloaca; G1 = anterior genital branch length expressed as percentage (%) of the body length; L = overall body length; m = length of conus as percentage of total stylet length; MB = distance between anterior end of body and center of median pharyngeal bulb expressed as percentage (%) of the pharynx length; n = number of specimens on which measurements are based; O = DGO as percentage of stylet length; V = distance from body anterior end to vulva expressed as percentage (%) of the body length.*

According to species grouping by Ghaderi et al. [2] this species belongs to group 3 characterized by stylet length less than 40 μm, four lateral lines and advulval flaps present.

The Spanish population from Arroyo Frío, Jaén province, was characterized by a short stylet 14.5–16.0 μm with rounded basal knobs, lip region rounded to truncate, continuous with body contour, indistinct submedian lobes, spermatheca rounded and filled with sperm, which indicates that males are required for reproduction but were not detected, female tail ventrally arcuate, tapering gradually to rounded terminus. J4 was similar to female, except for absent stylet (stage confirmed belonging to this species by molecular markers). Morphometrics of the Spanish population agree well with original description from Colombia [43], with small differences in stylet length (14.5–16.0 μm vs. 15.0–17.0 μm), c' ratio (2.8–3.5 vs. 2.7–3.3), vulva–anus distance (1.5–1.8 times tail length), and tail length (28.0–34.0 μm vs. 18.0–29.0 μm), which may be due to geographical intraspecific variability. This species was described from Llanos Oriental in Colombia [43] and this study comprises the first report from Spain. This species is morphologically close to *P. microdorus*, from which can be differentiated by vulva–anus distance with regard to tail length and tail terminus, and probably has been misidentified in some previous records with *P. microdorus*, therefore additional studies need to clarify the real biodiversity in the *P. microdorus*-species complex in Spain by applying integrative taxonomy.

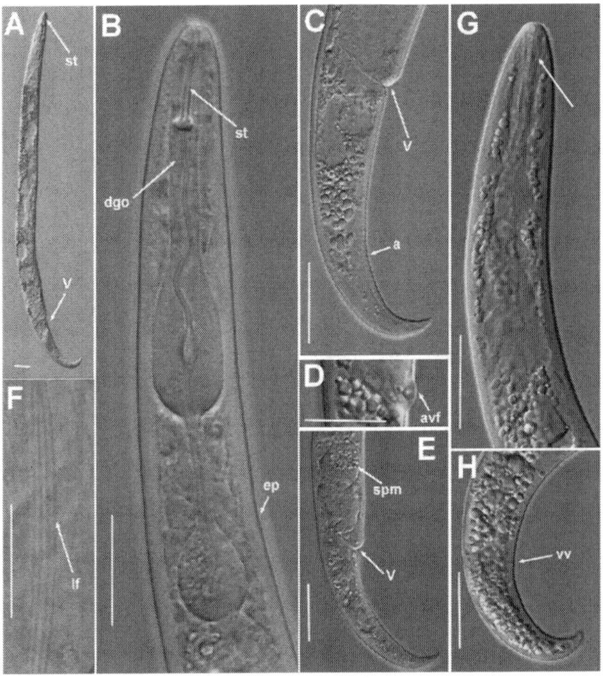

Figure 9. Light photomicrographs of *Paratylenchus recisus* Wu, 1974. (A) Entire female with stylet and vulva arrowed; (B) female pharyngeal region; (C–E) female posterior region with vulva, ad- vulval flap, spermatheca and anus (arrowed); (F) detail of lateral field at mid-body (arrowed); fourth-stage juvenile pharyngeal region showing absence of stylet (arrowed) and undeveloped pharynx; (H) fourth-stage juvenile posterior region showing vaginal vestigium (arrowed). Scale bars (A–H = 20 μm). (Abbreviations: a = anus; avf = advulval flap; dgo = pharyngeal dorsal gland orifice; ep = excretory pore; lf = lateral field; spm = spermatheca; st = stylet; vv = vaginal vestigium; V = vulva).

Molecular Characterization

Two D2-D3 of 28S rRNA (MZ265119-MZ265120), one ITS (MZ265043), and one COI gene sequence (MZ262252) were generated herein without intraspecific sequence variations. The closest *Paratylenchus* sequences to *P. recisus* were those of *P. microdorus* with 97, 93 and 91% similarity for the D2-D3 of 28S rRNA, ITS region and COI gene (MW421666-MW421667), respectively.

2.1.9 Remarks on *Paratylenchus sheri* (Raski, 1973) Siddiqi, 1986 (Figures 10 and 11, Table 10).

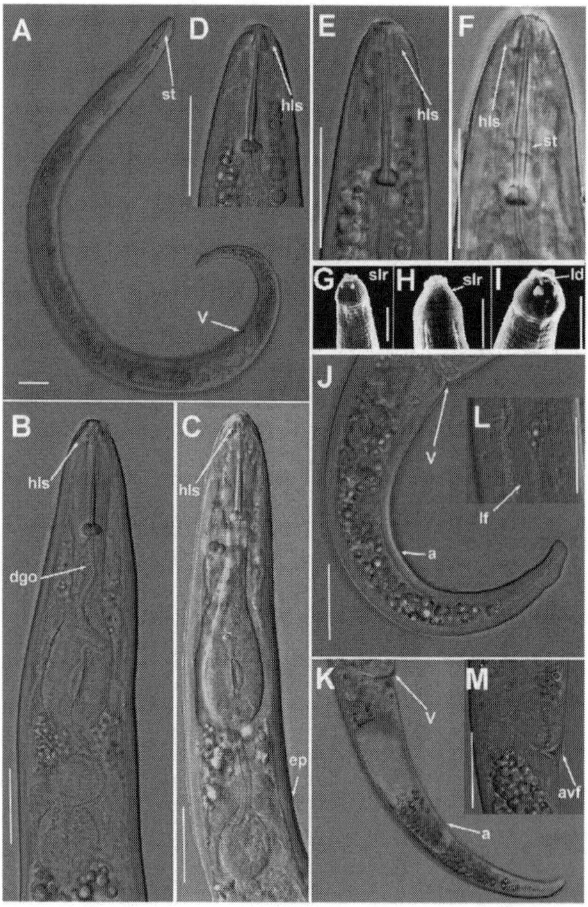

Figure 10. Light and SEM photomicrographs of *Paratylenchus sheri* (Raski, 1973) Siddiqi, 1986. (A) En- tire female with stylet and vulva arrowed; (B,C) female pharyngeal region showing heavy lip sclero- tization (arrowed); (D–F) female lip region showing heavy lip sclerotization (arrowed); (G–I) detail of lip region at SEM showing smooth lip region and labial disc (arrowed); (J,K) female posterior region with vulva and anus (arrowed); (L) detail of lateral field at mid body (arrowed); (M) detail of vulva showing advulval flap (arrowed). Scale bars (A–F = 20 μm; G–I =5 μm; J–M = 20 μm). (Abbreviations: a = anus; avf = advulval flap; dgo = pharyngeal dorsal

gland orifice; ep = excretory pore; hls = heavy lip sclerotization; ld = labial disc; slr = smooth lip region; st = stylet; V = vulva).

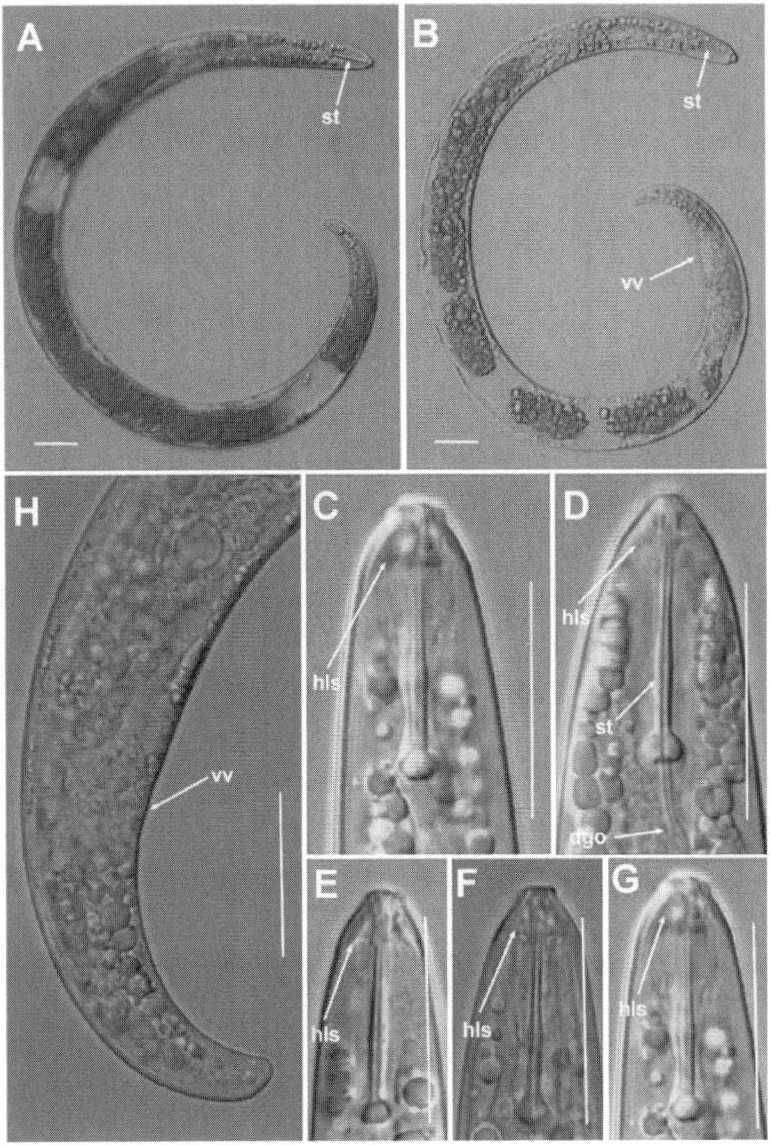

Figure 11. Light photomicrographs of *Paratylenchus sheri* (Raski, 1973) Siddiqi, 1986. (**A**,**B**) En- tire fourth-stage juvenile with stylet and vaginal vestigium arrowed; (**C**–**G**) fourth-stage juvenile lip region showing heavy lip sclerotization (arrowed); (**H**) fourth-stage juvenile with vaginal ves- tigium arrowed. Scale bars (A–H = 20 μm). (Abbreviations: dgo = pharyngeal dorsal gland orifice; hls = heavy lip sclerotization; st = stylet; vv = vaginal vestigium).

Table 10. Morphometrics of Paratylenchus sheri (Raski, 1973) Siddiqi, 1986 from Arroyo Frío and Coto Ríos, Jaén, Spain. All measurements are in μm and in the form: mean ± s.d. (range).

Life Stage	Females	Fourth-Stage Juveniles	Females
Sample Code	CAZ_04	CAZ_04	CAZ_07
Locality, Province	Arroyo Frío, Jaén	Arroyo Frío, Jaén	Coto Ríos, Jaén
n	14	5	4
L	548.6 ± 53.2 (459–626)	523.6 ± 11.6 (514–543)	540.0 ± 45.4 (492–595)
a *	20.5 ± 2.5 (15.8–24.6)	23.6 ± 1.3 (21.5–24.8)	19.9 ± 2.3 (16.6–22.0)
b	4.8 ± 0.5 (3.8–5.4)	4.9 ± 0.2 (4.6–5.1)	4.8 ± 0.3 (4.6–5.2)
c	11.1 ± 1.6 (9.2–14.6)	14.3 ± 2.9 (11.1–17.9)	11.5 ± 1.5 (10.3–13.5)
c′	3.8 ± 0.3 (3.2–4.2)	3.0 ± 0.2 (2.8–3.3)	3.6 ± 0.3 (3.2–3.9)
V	78.9 ± 1.7 (75.8–81.7)	-	79.5 ± 1.7 (77.0–80.9)
G1	49.7 ± 5.6 (40.7–56.7)	-	48.5 ± 7.0 (41.3–55.0)
Stylet length	23.8 ± 0.7 (22.5–25.0)	20.0 ± 0.4 (19.5–20.5)	23.6 ± 0.5 (23.0–24.0)
(Stylet length/body length) × 100	4.4 ± 0.4 (4.0–5.5)	3.9 ± 0.1 (3.8–4.0)	4.4 ± 0.4 (4.0–4.9)
Conus length	15.4 ± 0.8 (14.0–17.0)	12.3 ± 0.3 (12.0–12.5)	14.8 ± 0.5 (14.0–15.0)
m	64.5 ± 2.5 (62.0–70.8)	61.5 ± 1.1 (60.0–62.5)	62.4 ± 1.2 (60.9–63.8)
DGO	6.0 ± 0.5 (5.0–7.0)	4.2 ± 0.4 (4.0–5.0)	6.3 ± 0.5 (6.0–7.0)
O	25.3 ± 2.1 (20.8–30.4)	21.0 ± 1.9 (20.0–24.4)	26.5 ± 2.6 (25.0–30.4)
Lip width	5.1 ± 0.6 (4.5–6.0)	3.9 ± 0.4 (3.5–4.5)	5.0 ± 0.7 (4.5–6.0)
Median bulb length	27.2 ± 1.9 (24.0–31.0)	-	25.8 ± 1.3 (24.0–27.0)
Median bulb width	14.3 ± 1.9 (12.0–18.5)	-	13.3 ± 1.2 (12.0–14.0)
Anterior end to center median bulb	63.0 ± 3.8 (56.0–70.0)	-	61.8 ± 3.9 (57.0–65.0)
MB	55.6 ± 1.5 (52.5–58.1)	-	55.2 ± 1.0 (54.3–56.5)
Nerve ring to anterior end	86.0 ± 6.4 (76.0–99.0)	82.4 ± 1.7 (81.0–85.0)	83.3 ± 6.8 (76.0–90.0)
Excretory pore to anterior end	101.9 ± 6.4 (93.0–116.0)	99.2 ± 1.9 (97.0–102.0)	100.8 ± 6.9 (95.0–110.0)
Pharynx length	113.4 ± 6.3 (104.0–126.0)	107.0 ± 5.0 (102.0–115.0)	111.8 ± 6.4 (105.0–119.0)
Maximum body diam.	27.1 ± 4.0 (21.0–33.5)	22.2 ± 1.3 (21.0–24.0)	27.3 ± 2.7 (24.5–31.0)
Tail length	50.1 ± 7.0 (39.0–62.0)	38.0 ± 8.3 (29.0–49.0)	47.5 ± 5.5 (42.0–54.0)
Anal body diam.	13.3 ± 1.2 (11.5–15.5)	12.8 ± 2.1 (10.5–15.0)	13.3 ± 1.0 (12.0–14.0)

* *Abbreviations: a = body length/greatest body diameter; b = body length/distance from anterior end to pharyngo-intestinal junction; DGO = distance between stylet base and orifice of dorsal pharyngeal gland; c = body length/tail length; c′ = tail length/tail diameter at anus or cloaca; G1 = anterior genital branch length expressed as percentage (%) of the body length; L = overall body length; m = length of conus as percentage of total stylet length; MB = distance between anterior end of body and center of median pharyngeal bulb expressed as percentage (%) of the pharynx length; n = number of specimens on which measurements are based; O = DGO as percentage of stylet length; V = distance from body anterior end to vulva expressed as percentage (%) of the body length.*

According to species grouping by Ghaderi et al. [2] this species belongs to group 3 characterized by stylet length less than 40 μm, four lateral lines and advulval flaps present. The Spanish population from Arroyo Frío, Jaén province, was characterized by a conoid-truncate lip region, with an unstriated depression from body contour (4–5.5.0 μm wide) and strong sclerotization. Small projecting oral lips present. SEM face view (Figure 10) shows an unstriated, dorso-ventrally flattened lip region. Stylet robust, occupying 19 (15–23) annuli and 22.5–25.0 μm long. Stylet knobs weakly backwardly directed, 4.8 (4.0–5.5) μm across. Lateral field with four incisures with smooth margins (central two are very faint), 3.2 (2.5–4.0) μm wide. Orifice of dorsal pharyngeal gland 5.9 (4.5–6.5) μm from stylet base. Metacorpus with well-developed valvular apparatus 5.6 (4.5–7.5) μm long, its posterior margin situated at 70 (63–80) μm from anterior end. Excretory pore located near anterior end of basal bulb, immediately posterior to hemizonid. Cardia well developed, 2.5–3.5 μm wide. Distinct cuticular vulval flap, 6 (5.0–7.0) μm long. Large round spermatheca 13.5 (11.5–15.5) μm wide, filled with sperms 1–2 μm wide, which indicates that males are required for reproduction but their numbers are lower than females. Tail almost straight to slight ventrally curved, with rounded terminus, 0.8 (0.6–1.3) times vulva–anus distance or 3.8 (3.1–4.7) times anal body diameter. J4 with similar morphology to that of adult females, except sexual characters and shorter body length and stylet.

This species was described from Digne, France [44], and has been reported in Spain [17,18] and Italy [28]. This population was from the same locality as that reported by Gomez-Barcina et al. [17], which was confirmed by Prof. Raski [17]. The species was recently synonymized with *P. israelensis* by Ghaderi et al. [2] based on similar morphology, including strong labial sclerotization. However, the present results together with the recent integrative taxonomical diagnosis of *P. israelensis* [4] demonstrated that both species are closely related morphologically and molecularly (see below) and need to be considered as nominal valid species. This species has also been reported in Iran [45]; however, the single D2-D3 sequence provided for this Iranian population was 99.7% similar to *P. tateae* from Spain and Canada (see below) and needs to be revised by the authors.

Molecular Characterization

Six D2-D3 of 28S rRNA (MZ265121-MZ265126) with an intraspecific sequence variation of 0.5% (differing from 0 to 2 nucleotides), seven ITS (MZ265044-MZ265050) (99% similarity; nine nucleotides and no indels), and finally, nine COI gene sequences (MZ262253-MZ262261), with an intraspecific sequence variation of 5% (differing from 0 to 23 nu- cleotides), were generated. Two J4 from the Arroyo Frío population were sequenced, includ- ing D2–D3 of 28S (MZ265121-MZ265122), ITS region (MZ265044-MZ265045) and COI gene (MZ262253-MZ262254) being identical to the adult female sequences from this population. The D2-D3 of 28S rRNA sequences (MZ265121-MZ265126) showed high similarity with accession from *P. israelensis* (MW798301-MX798305) and *P. neoamblycephalus* (MW413660- MW413663) being 99% similar between them (differing from 2 to 7 nucleotides). For the ITS (MZ265044-MZ265050), the similarity detected was 98% (differing by 17–23 nucleotides and 6–8 indels) with *P. israelensis* (MW798343) and 95% (differing by 44–50 nucleotides and 20 indels) with *P. neoamblycephalus* (MW413607). Finally, for the COI gene sequences (MZ262253-MZ262261), *P. sheri* showed similarity values of 92–94% (differing from 21 to 29 nucleotides) with *P. israelensis* (MW797019-MW797020) and 89–91% (differing from 34 to 38 nucleotides) with *P. neoamblycephalus* (MW421677-MW421682). D2-D3 of 28S sequences from *P. sheri* obtained herein showed similarity values of 91% with the accession MN088374 of *P. sheri* from Iran, thus reinforcing the idea that this sequence belongs to *P. tateae* instead of *P. sheri,* as already suggested by Munawar et al. [8].

2.1.10. Remarks on *Paratylenchus variabilis* Raski, 1975 (Figure 12, Table 11).

According to species grouping by Ghaderi et al. [2] this species belongs to group 3 characterized by stylet length less than 40 μm, four lateral lines and advulval flaps present.

The Spanish population from Córdoba, Córdoba province, was characterized by a rounded lip region with indistinct submedian lobes, continuous with the rest of the body, short stylet 14.0–16.0 μm long, spermatheca oval and filled with sperm, which indicates that males are required for reproduction but were not found, and female tail narrows gradually to a bluntly rounded terminus. J4 with similar morphology to that of adult females, except sexual characters and shorter body length and stylet. This species was

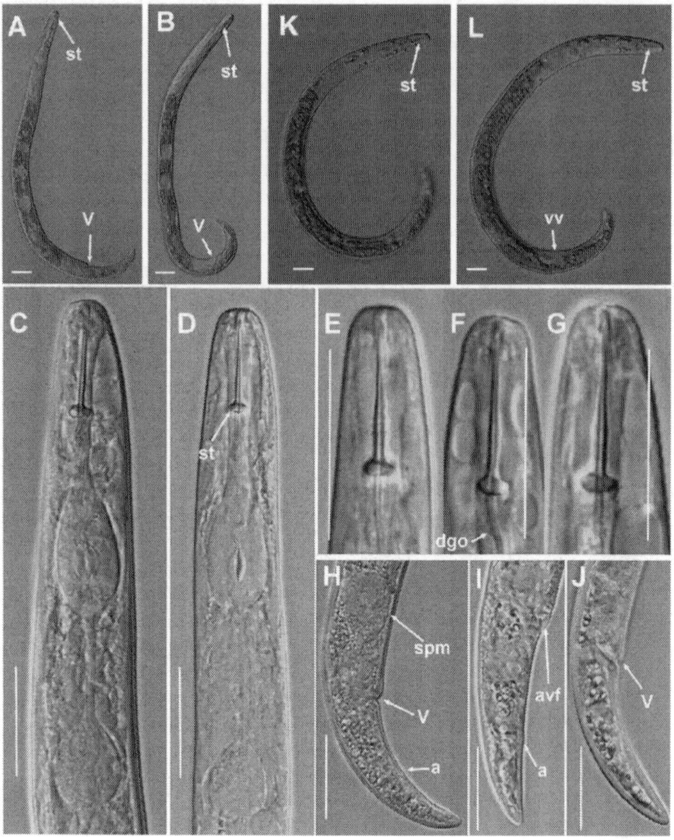

Figure 12. Light photomicrographs of *Paratylenchus variabilis* Raski, 1975. (A,B) Entire female with stylet and vulva arrowed; (C,D) female pharyngeal region; (E–G) female lip region; (H–J) female posterior region with vulva, spermatheca, advulval flap and anus (arrowed); (K–L) fourth-stage juvenile with stylet and vaginal vestigium arrowed. Scale bars (A–L = 20 μm). (Abbreviations: a = anus; avf = advulval flap; dgo = pharyngeal dorsal gland orifice; spm = spermatheca; st = stylet; vv=vaginalvestigium;V=vulva).

described from California and Utah [46] and has been reported in Israel and Iran [30], and this study comprises the first report from Spain. This species is morphologically close to P. microdorus, from which can be differentiated by the shape of female tail terminus, and probably has been misidentified in previous records with P. microdorus, therefore, additional studies need to clarify the real biodiversity in the P. microdorus-species complex in Spain by applying integrative taxonomy. In addition, P. variabilis is morphologically and morphometrically almost indistinguishable from P. zurgenerus [4], from which it can be separated by molecular markers (see below), and both can be considered as cryptic species.

Table 11. Morphometrics of Paratylenchus variabilis Raski, 1975 from Córdoba, Córdoba, Spain. All measurements are in μm and in the form: mean ± s.d

Life Stage	Females	Fourth-Stage Juveniles	Females
Sample Code	EP_ACA	EP_ACA	Type
Locality, Province	Córdoba, Córdoba	Córdoba, Córdoba	Population [46]
n	10	4	27
L	337.3 ± 30.5 (302–407)	312.3 ± 38.2 (282–367)	250–340
a *	21.7 ± 1.1 (20.1–23.5)	21.2 ± 2.3 (19.4–24.5)	19–25
b	3.9 ± 0.6 (2.8–4.8)	3.9 ± 0.2 (3.7–4.2)	3.6–4.7
c	15.3 ± 1.2 (13.1–17.2)	14.7 ± 1.3 (13.1–16.3)	12–18
c′	2.6 ± 0.2 (2.3–2.9)	2.2 ± 0.2 (1.9–2.4)	2.7
V	84.5 ± 1.0 (83.1–86.5)	-	82–85
G1	39.4 ± 6.6 (28.6–52.8)	-	-
Stylet length	14.9 ± 0.6 (14.0–16.0)	11.6 ± 0.8 (11.0–12.5)	13–16
(Stylet length/body length) × 100	4.5 ± 0.4 (3.4–4.9)	3.7 ± 0.3 (3.3–4.0)	-
Conus length	10.5 ± 0.5 (10.0–11.0)	7.5 ± 0.4 (7.0–8.0)	-
m	70.5 ± 3.7 (64.5–75.9)	64.6 ± 2.5 (62.5–68.2)	-
DGO	3.5 ± 0.6 (2.5–4.5)	2.6 ± 0.5 (2.0–3.0)	-
O	23.5 ± 3.9 (17.9–30.0)	22.6 ± 4.2 (18.2–27.3)	-
Lip width	5.6 ± 0.6 (5.0–6.5)	4.3 ± 0.3 (4.0–4.5)	-
Median bulb length	18.9 ± 1.7 (17.0–23.0)	18.0 ± 1.4 (17.0–20.0)	-
Median bulb width	8.9 ± 1.4 (7.5–12.0)	7.5 ± 0.6 (7.0–8.0)	-
Anterior end to center median bulb	46.0 ± 3.6 (40.0–52.0)	41.0 ± 0.8 (40.0–42.0)	-
MB	53.1 ± 7.4 (39.5–63.6)	51.9 ± 3.6 (46.6–54.5)	-
Nerve ring to anterior end	62.6 ± 3.0 (58.0–68.0)	57.8 ± 5.0 (51.0–63.0)	-
Excretory pore to anterior end	74.5 ± 7.8 (67.0–95.0)	67.0 ± 6.2 (60.0–75.0)	59–71
Pharynx length	87.9 ± 12.4 (77.0–119.0)	79.3 ± 5.9 (76.0–88.0)	-
Maximum body diam.	15.6 ± 1.6 (14.0–19.0)	14.8 ± 0.6 (14.0–15.5)	13
Tail length	22.2 ± 3.5 (20.0–31.0)	21.4 ± 2.6 (19.0–25.0)	21
Anal body diam.	8.8 ± 1.8 (7.0–13.0)	9.8 ± 1.0 (9.0–11.0)	-

* *Abbreviations: a = body length/greatest body diameter; b = body length/distance from anterior end to pharyngo-intestinal junction; DGO = distance between stylet base and orifice of dorsal pharyngeal gland; c = body length/tail length; c′ = tail length/tail diameter at anus or cloaca; G1 = anterior genital branch length*

expressed as percentage (%) of the body length; L = overall body length; m = length of conus as percentage of total stylet length; MB = distance between anterior end of body and center of median pharyngeal bulb expressed as percentage (%) of the pharynx length; n = number of specimens on which measurements are based; O = DGO as percentage of stylet length; V = distance from body anterior end to vulva expressed as percentage (%) of the body length.

Molecular Characterization

Three D2-D3 of 28S rRNA (MZ265127-MZ265129), three ITS (MZ265051-MZ265053) and three COI gene sequences (MZ262265-MZ262267) were generated in this study from two adult females and one J4 specimen without intraspecific sequence variations. *Paratylenchus variabilis* was closely related with *P. nanus*, showing similarity values of 96% (differ- ing by 31 nucleotides and 1 indel) for the D2-D3 region with several accessions of *P. nanus* (MW413657-MW413659, MW234449-MW234450). However, for the ITS region the similar- ity was lower, with values about 87% with accessions belonging to several *Paratylenchus* spp., such as, *P. veruculatus*, *P. goodeyi* and *P. nanus*. Finally, the closest species for the COI gene sequences was *P. goodeyi* (MW421648-MW421649), being 95% similar between them (19 nucleotides and no indels). Finally, *P. variabilis* can also be clearly separated molecularly from *P. zurgenerus* by D2-D3 and ITS, 88.9%, 78.3% similarity (differing in 79 bp, 166 and 20, 66 indels), respectively; low similarity was detected among COI sequences of both species.

2.1.11. Remarks on *Paratylenchus verus* (Brzeski, 1995) Brzeski, 1998 (Figure 13, Table 12).

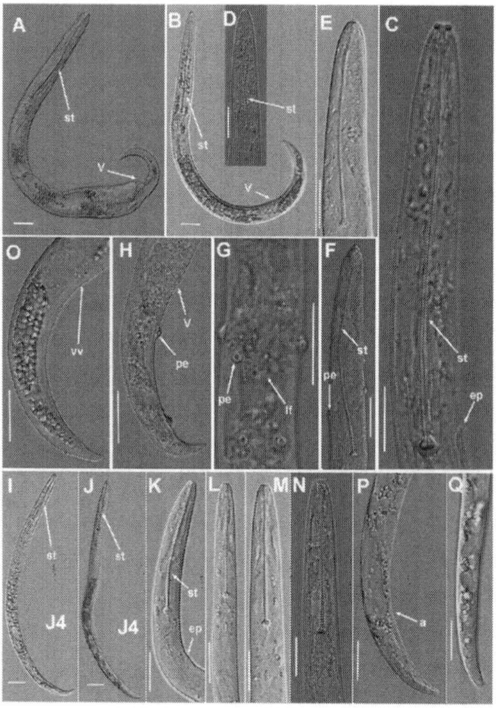

Figure 13. Light photomicrographs of *Paratylenchus verus* (Brzeski, 1995) Brzeski, 1998. (A,B) Entire female with stylet and vulva arrowed; (C–F) female lip region with stylet, excretory pore and *Pasteuria* endospore arrowed; (G) detail of lateral

field at mid-body and *Pasteuria* endospore arrowed; female posterior region with vulva and *Pasteuria* endospore arrowed; (I,J) entire fourth-stage juvenile with stylet arrowed; (K–N) fourth-stage juvenile lip region with stylet and excretory pore arrowed; (O) fourth-stage juvenile posterior region with vaginal vestigium arrowed; (P,Q) fourth- stage juvenile posterior region with anus arrowed. Scale bars (A–Q = 20 μm). (Abbreviations: a = anus; ep = excretory pore; lf = lateral field; pe = *Pasteuria* endospore; st = stylet; vv = vaginal vestigium; V = vulva).

Table 12. Morphometrics of Paratylenchus verus (Brzeski, 1995) Brzeski, 1998 from Santa María de Trasierra, Córdoba, Spain. All measurements are in μm and in the form: mean ± s.d. (range).

	Females	Fourth-Stage Juveniles	Females
Sample Code	AR_097	AR_097	[28]
Locality	Sta. Mª Trasierra, Córdoba	Sta. Mª Trasierra, Córdoba	Type Population
n	14	7	11
L	324.4 ± 25.0 (265–355)	276.9 ± 22.2 (253–319)	270–320
a *	19.4 ± 3.7 (12.2–24.7)	21.8 ± 2.3 (18.8–24.2)	20–24
b	2.4 ± 0.2 (2.1–2.7)	3.0 ± 0.5 (2.5–3.8)	2.2–2.5
c	12.1 ± 2.0 (9.1–16.9)	11.1 ± 1.1 (9.4–12.5)	13–19
c'	3.3 ± 0.3 (2.7–3.6)	3.0 ± 0.3 (2.8–3.3)	1.8–2.9
V	77.0 ± 1.8 (73.6–80.3)	-	76–79
G1	29.3 ± 4.1 (24.5–40.4)	-	-
Stylet length	89.1 ± 5.8 (79.0–97.0)	49.1 ± 1.2 (48.0–51.0)	66–86
(Stylet length/body length) × 100	27.6 ± 1.9 (24.8–30.2)	17.8 ± 1.4 (15.8–20.2)	-
Conus length	77.8 ± 5.3 (70.0–86.5)	41.9 ± 1.2 (40.0–43.0)	-
m	87.3 ± 1.9 (84.3–89.9)	85.2 ± 1.1 (83.3–86.9)	-
DGO	4.6 ± 0.7 (3.5–6.0)	4.2 ± 0.3 (4.0–4.5)	-
O	5.2 ± 0.8 (4.1–6.7)	8.6 ± 0.6 (7.8–9.4)	-
Lip width	4.4 ± 0.4 (4.0–5.0)	3.7 ± 0.3 (3.5–4.0)	-
Median bulb length	25.5 ± 3.3 (22.0–34.0)	15.2 ± 1.2 (14.0–17.0)	-
Median bulb width	10.6 ± 0.7 (9.5–12.0)	8.3 ± 0.4 (8.0–9.0)	-
Anterior end to center median bulb	101.7 ± 7.9 (87.0–115.0)	69.3 ± 3.4 (66.0–74.0)	-
MB	76.1 ± 4.0 (70.2–83.0)	73.7 ± 5.3 (68.3–83.5)	-
Nerve ring to anterior end	112.6 ± 11.3 (94.0–133.0)	77.9 ± 0.9 (77.0–79.0)	-
Excretory pore to anterior end	89.0 ± 5.2 (81.0–95.0)	82.0 ± 2.5 (78.0–85.0)	68–92
Pharynx length	133.9 ± 12.1 (112.0–161.0)	94.4 ± 7.8 (85.0–105.0)	-
Maximum body diam.	17.4 ± 4.2 (13.0–27.0)	12.9 ± 2.3 (11.0–17.0)	14
Tail length	27.1 ± 2.6 (21.0–30.0)	25.3 ± 3.9 (23.0–34.0)	18
Anal body diam.	8.1 ± 0.5 (7.0–9.0)	8.4 ± 1.4 (7.5–11.5)	-

* *Abbreviations: a = body length/greatest body diameter; b = body length/distance from anterior end to pharyngo-intestinal junction; DGO = distance between stylet base and orifice of dorsal pharyngeal gland; c = body length/tail length; c' = tail length/tail diameter at anus or cloaca; G1 = anterior genital branch length expressed as percentage (%) of the body length; L = overall body length; m = length of conus as percentage of total stylet length; MB = distance between anterior end of body and center of median pharyngeal bulb*

expressed as percentage (%) of the pharynx length; n = number of specimens on which measurements are based; O = DGO as percentage of stylet length; V = distance from body anterior end to vulva expressed as percentage (%) of the body length.

According to species grouping by Ghaderi et al. [2] this species belongs to group 10 characterized by stylet length more than 40 μm, four lateral lines and advulval flaps present. The Spanish population from Sta. Maria de Trasierra, Córdoba province, was characterized by a rounded lip region with distinct submedian lobes, continuous with the rest of the body, long flexible stylet 79.0–97.0 μm long, excretory pore opposite to median bulb, spermatheca oval and filled with sperm, which indicates that males are required for reproduction but were not found, and female tail narrows gradually to a rounded terminus. J4 with similar morphology to that of adult females, except sexual characters and shorter body length and stylet. This species was described from Texcoco, Mexico [28], and this study comprises the first report from Spain. Several females of the Spanish population had conspicuous infections of *Pasteuria* sp. on cuticle, especially on anterior and posterior ends (Figure 13).

Molecular Characterization

Four D2-D3 of 28S rRNA (MZ265130-MZ265133), five ITS (MZ265054-MZ265058) and four COI (MZ262268-MZ262271) gene sequences were generated for the first time from this species, including J4 and adult females, without intraspecific sequence variations, except for the ITS sequences with 98–100% similarity (differing from 2 to 11 nucleotides and 0 to 2 indels). The closest *Paratylenchus* spp. was *P. idalimus* being 96% similar (22 nucleotides and no indels) for the D2-D3 of 28S rRNA, 90% similar for the ITS region (differing by 69–75 nucleotides and from 21 to 23 indels) and, finally, 93% for COI sequences (MW411839) (differing by 26 nucleotides and no indels).

2.1.12. Remarks on *Paratylenchus vitecus* (Pramodini et al., 2006) Ghaderi et al., 2014 (Figure 14, Table 13).

According to species grouping by Ghaderi et al. [2] this species belongs to group 11 characterized by stylet length more than 40 μm, four lateral lines in and advulval flaps absent. The Spanish population from Córdoba, Córdoba province, was characterized by a conoid-rounded lip region with distinct submedian lobes, continuous with the rest of the body, long flexible stylet 62.0–70.0 μm long, spermatheca elongate and filled with sperm, which indicates that males are required for reproduction but not found, and female tail finely rounded. J4 with similar morphology to that of adult females, except sexual characters and shorter body length and stylet. Morphometrics of the Spanish population agree well with original description with small differences in stylet length (62.0–70.0 μm vs. 42.0–65.0 μm), V ratio (68.1–75.4 vs. 72.0–77.0), which may be due to geographical intraspecific variability [2]. Molecularly, *P. vitecus* is close to *P. teres* (see below), however, it can be morphologically separated by clear differences in stylet length (42.0–65.0 μm, 62.0–70.0 μm vs. 69.0–83.0 μm, 67.0–96.0 μm) and c' ratio (2.9, 2.7–3.5 vs. 4.2, 3.1–3.9). This species was described from Manipur, India [47], and this study comprises the first report from Spain.

Molecular Characterization

Six D2-D3 of 28S rRNA (MZ265136-MZ265141) and four ITS (MZ265059-MZ265062) with one and two variable positions, respectively, and three identical COI gene sequences (MZ262272-MZ262274) were generated for this species, including sequences from J4 and adult females. The closest *Paratylenchus* spp. was *P. teres* with 97% similarity for the D2-D3 of 28S rRNA (differing by 25 nucleotides) to MN088376. Unfortunately, no data for ITS or COI from *P. teres* are available in the GenBank.

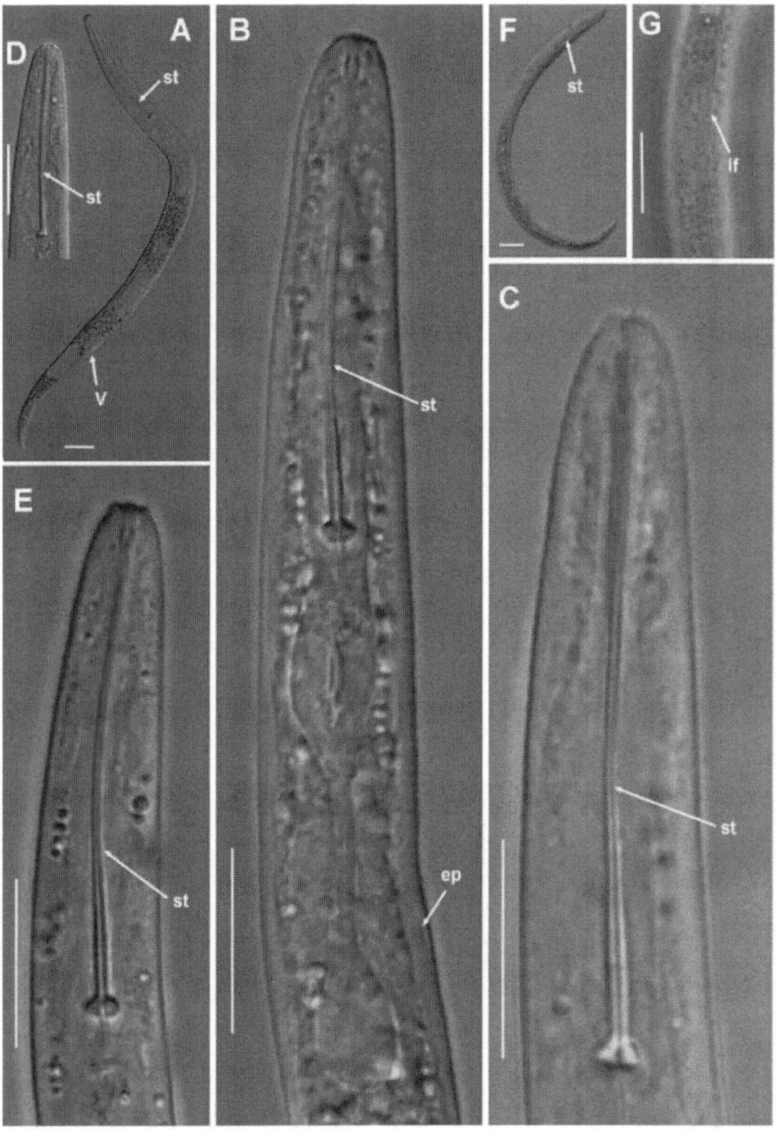

Figure 14. Light photomicrographs of *Paratylenchus vitecus* (Pramodini et al., 2006) Ghaderi et al., 2014. (A) Entire female with stylet and vulva arrowed; (B) female pharyngeal region with stylet and excretory pore arrowed; (C–E) female

lip region; (F) entire fourth-stage juvenile with stylet arrowed; (G) detail of lateral field at mid-body (arrowed). Scale bars (A–G = 20 μm). (Abbreviations: ep = excretory pore; lf = lateral field; st = stylet; V = vulva).

Table 13. Morphometrics of Paratylenchus vitecus (Pramodini et al., 2006) Ghaderi et al., 2014 from Córdoba, Córdoba, Spain. All measurements are in μm and in the form: mean ± s.d. (range).

	Females	Fourth-Stage Juveniles	Females
Sample Code	EP_ACA	EP_ACA	Type Population
Locality	Córdoba, Córdoba	Córdoba, Córdoba	[47]
n	3	7	7
L	341.2 ± 22.1 (323–366)	271.6 ± 12.0 (259–288)	220–350
a *	20.3 ± 1.7 (18.3–21.5)	19.5 ± 1.4 (18.8–24.2)	20–26
b	2.7 ± 0.1 (2.5–2.8)	2.9 ± 0.2 (2.6–3.1)	2.7–2.9
c	11.6 ± 1.6 (9.8–12.9)	10.6 ± 0.6 (9.6–11.5)	9–16
c′	3.1 ± 0.4 (2.7–3.5)	2.9 ± 0.2 (2.7–3.1)	2.9
V	72.8.0 ± 4.1 (68.1–75.4)	-	72–77
G1	28.3 ± 3.8 (24.5–32.1)	-	-
Stylet length	66.0 ± 4.0 (62.0–70.0)	42.5 ± 1.6 (40.0–44.0)	42–65
(Stylet length/body length) × 100	19.4 ± 2.1 (16.9–20.8)	15.7 ± 1.1 (13.9–16.7)	-
Conus length	59.3 ± 1.5 (58.0–61.0)	35.6 ± 1.0 (34.0–37.0)	-
m	90.0 ± 3.2 (87.1–93.5)	83.7 ± 1.5 (81.8–85.7)	-
DGO	4.3 ± 0.6 (4.0–5.0)	3.1 ± 0.4 (2.5–4.0)	-
O	6.1 ± 0.8 (5.4–7.0)	7.2 ± 1.1 (6.3–9.5)	-
Lip width	4.5 ± 0.5 (4.0–5.0)	4.5 ± 0.5 (4.0–5.0)	-
Median bulb length	24.0 ± 2.0 (22.0–26.0)	17.6 ± 3.8 (15.0–26.0)	-
Median bulb width	9.3 ± 2.3 (8.0–12.0)	8.1 ± 0.6 (7.5–9.0)	-
Anterior end to center median bulb	88.3 ± 4.9 (85.0–94.0)	57.1 ± 3.1 (53.0–61.0)	-
MB	68.9 ± 1.8 (67.2–70.8)	59.9 ± 0.9 (58.2–61.1)	-
Nerve ring to anterior end	105.7 ± 7.2 (101.0–114.0)	70.0 ± 3.7 (63.0–74.0)	-
Excretory pore to anterior end	84.3 ± 6.8 (79.0–92.0)	78.4 ± 6.6 (72.0–89.0)	85–94
Pharynx length	128.3 ± 8.5 (120.0–137.0)	95.4 ± 4.4 (91.0–102.0)	-
Maximum body diam.	17.0 ± 2.6 (15.0–20.0)	14.0 ± 1.6 (12.5–16.5)	10–15
Tail length	29.8 ± 3.5 (26.0–33.0)	25.6 ± 1.5 (24.0–27.5)	14–39
Anal body diam.	9.8 ± 1.5 (8.5–11.5)	8.8 ± 0.6 (8.0–9.5)	-

* *Abbreviations: a = body length/greatest body diameter; b = body length/distance from anterior end to pharyngo-intestinal junction; DGO = distance between stylet base and orifice of dorsal pharyngeal gland; c = body length/tail length; c′ = tail length/tail diameter at anus or cloaca; G1 = anterior genital branch length expressed as percentage (%) of the body length; L = overall body length; m = length of conus as percentage of total stylet length; MB = distance between anterior end of body and center of median pharyngeal bulb expressed as percentage (%) of the pharynx length; n = number of specimens on which measurements are based;*

O = DGO as percentage of stylet length; V = distance from body anterior end to vulva expressed as percentage (%) of the body length.

2.2 Distribution of Paratylenchus spp. in Spain

In the exhaustive review of the geographical distribution of Paratylenchus species in cultivated and natural environments in Spain, we detected that pin nematodes exhibited a wide distribution across an extensive variety of herbaceous and woody hosts, including 39 species (Figure 15). It should be noted that the highest diversity seems to be associated with southern Spain (Andalucia), with 35 out of 39 species in the country (Figure 15). Although the data suggest that the nematode survey efforts were higher in southern than in central and northern parts of the country, the biodiversity of Paratylenchus in Andalucia is really remarkable (Figure 15). In any case, the Paratylenchus species distribution observed herein revealed that this genus is adapted to a wide variety of host plants and heterogeneous environmental conditions (climatic, edaphic) from all over the country (ca. 1000 km across north–south, and ca. 600 km across east–west).

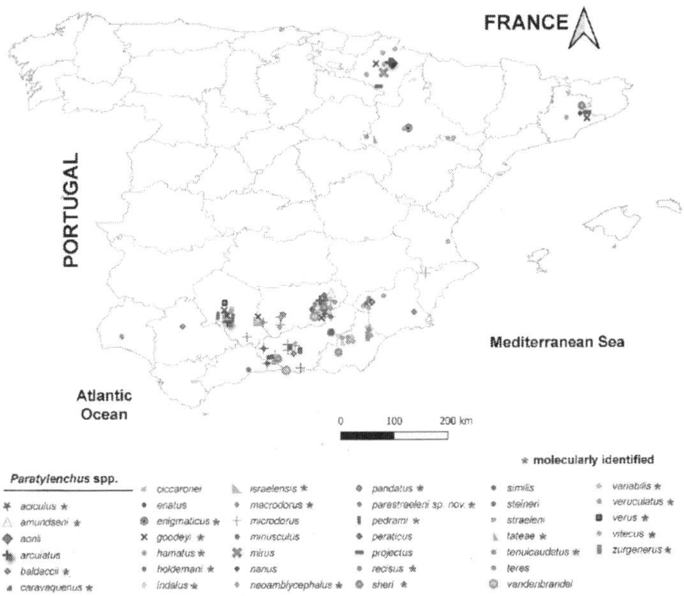

Figure 15. Spain map distribution of *Paratylenchus* species across all of the country. Species list with asterisk (*) indicated species identified by integrative taxonomy and including molecular analyses confirmation.

2.3 Phylogenetic Analyses of Paratylenchus spp.

The D2-D3 domains of the 28S rRNA gene alignment (702 bp long) included 148 sequences of 64 *Paratylenchus* species and three outgroup species (*Basiria gracillis* (DQ328717), *Aglenchus agricola* (AY780979), and *Coslenchus costatus* (DQ328719)). Seventy-eight new sequences were included in this analysis. The Bayesian 50% majority rule consensus tree inferred from the D2-D3 alignment is given in Figure 16. The tree contained two moderately supported clades (PP = 0.94, PP = 0.84). These clades are mainly coincident with other recent studies on *Paratylenchus* spp. [3,4]. The new species, *P. parastraeleni* sp.

nov., clustered with several accessions of *P. straeleni* from Belgium, Iran, South Africa, and Turkey, but clearly separated into two different subclades (PP = 1.00) (Figure 16). Newly sequenced species clustered in separated clusters and subclusters, *viz. P. variabilis, P. amundseni, P. recisus, P. verus, P. macrodorus, P. pandatus, P. vitecus* and *P. aciculus*, but with mixed stylet patterns (long and flexible stylet > 40 µm with conus representing about more than 70% of the total stylet and short and rigid stylet < 40 µm with conus about 50% of the total stylet) within the main clus- ters, except for a basal clade moderately supported (PP = 0.84) comprising 14 species with stylet > 40 µm, including the four species newly sequenced herein (*P. aciculus, P. macrodorus, P. pandatus*, and *P. vitecus*) (Figure 16).

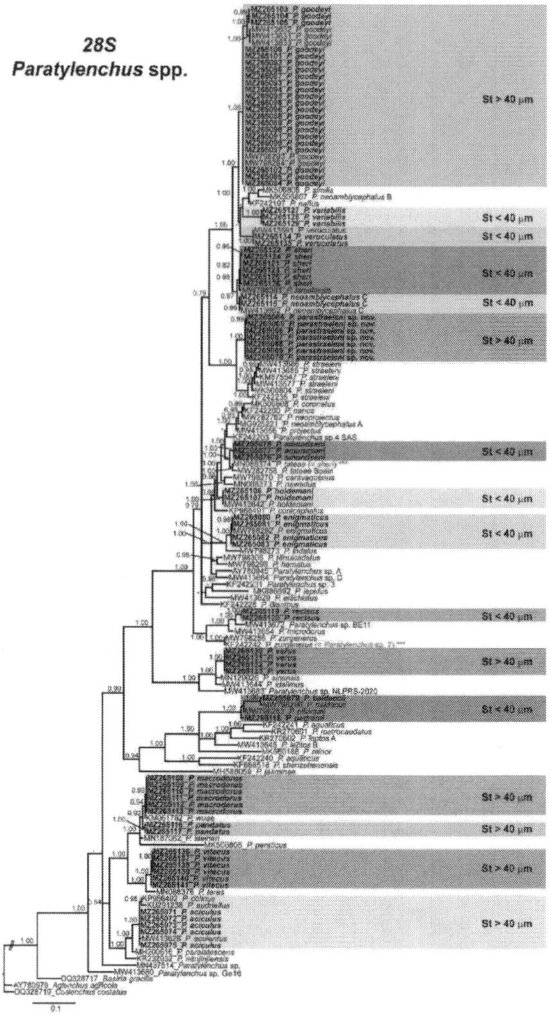

Figure 16. **Phylogenetic relationships within the genus** ***Paratylenchus*****. Bayesian 50% majority rule consensus tree as inferred from D2-D3 expansion domains of the 28S rRNA sequence alignment under the general time-reversible model of sequence evolution with correction for invariable sites and a gamma-shaped distribution (GTR + I + G). Posterior probabilities of more than 0.70 are given for appropriate clades.**

Newly obtained sequences in this study are shown in bold. The scale bar indicates expected changes per site. * Red font names refer to the previous consideration in NCBI.**

The ITS rRNA gene alignment (836 bp long) included 117 sequences of 55 *Paratylenchus* species and three outgroup species (*Hemicycliophora lutosa* (GQ406237), *H. wyei* (KC329575) and *H. poranga* (KF430598)). Fifty-nine new sequences were included in this analysis. The Bayesian 50% majority rule consensus tree inferred from the ITS alignment is given in Figure 17. The tree contained two highly supported major clades I and II (PP = 0.99 and PP = 1.00, respectively) and several subclades (Figure 17). Clade I includes mostly species with short stylet (<40 μm), but also species with long stylet (>40 μm), including all isolates of *P. goodeyi*, the new species *P. parastraeleni* sp. nov., *P. straeleni*, *P. verus* and *P. idalimus* (Figure 17). Clade II mostly includes species with long stylet (>40 μm), but also species with short stylet (<40 μm), including *P. baldaccii*, *P. pedrami*, *P. jasminae*, *P. minor*, and *P. rostrocaudatus* (Figure 17). These clades were partially coincident with previous studies with, in some cases, similar or different clade support [3,4].

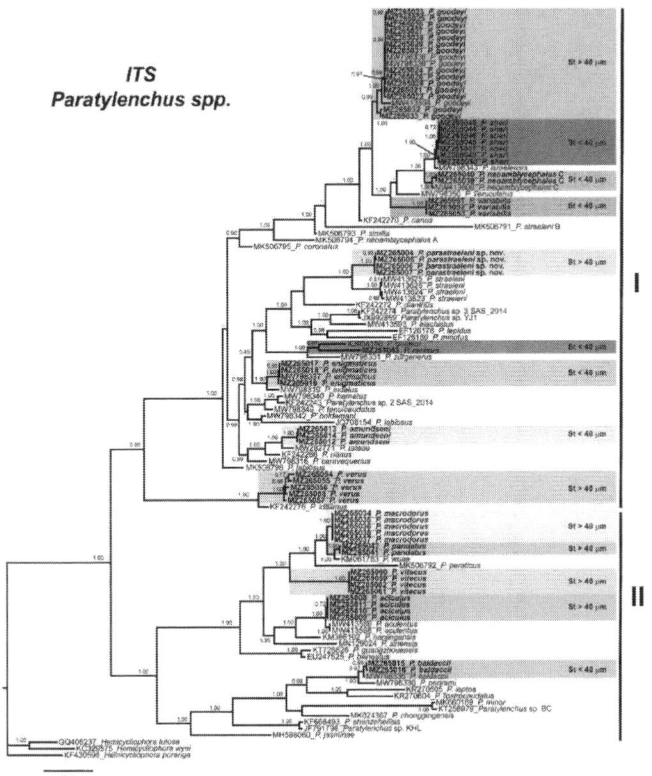

Figure 17. Phylogenetic relationships within the genus *Paratylenchus*. Bayesian 50% majority rule consensus tree as inferred from ITS rRNA sequence alignment under the general time-reversible model of sequence evolution with correction for invariable sites and a gamma-shaped distribution (GTR + I + G). Posterior probabilities of more than 0.70 are given for appropriate clades. Newly obtained sequences in this study are shown in bold. The scale bar indicates expected changes per site.

The COI gene alignment (384 bp long) included 245 sequences of 51 *Paratylenchus* species and three outgroup species (Hemicriconemoides californianus (KM516192), *Hemicycliophora floridensis* (MG019867) and *H. poranga* (MG019892)). Sixty-seven new sequences were included in this analysis. The Bayesian 50% majority rule consensus tree inferred from the COI sequence alignment is given in Figure 18. The tree contained four major clades, but only one basal clade (IV) was well supported (PP = 1.00), including two uniden- tified *Paratylenchus* species and *P. verus* and *P. idalimus*, and all others (clades I, II, and III) low supported (PP < 0.70 to 0.89). The *P. straeleni*-complex clustered in a well-

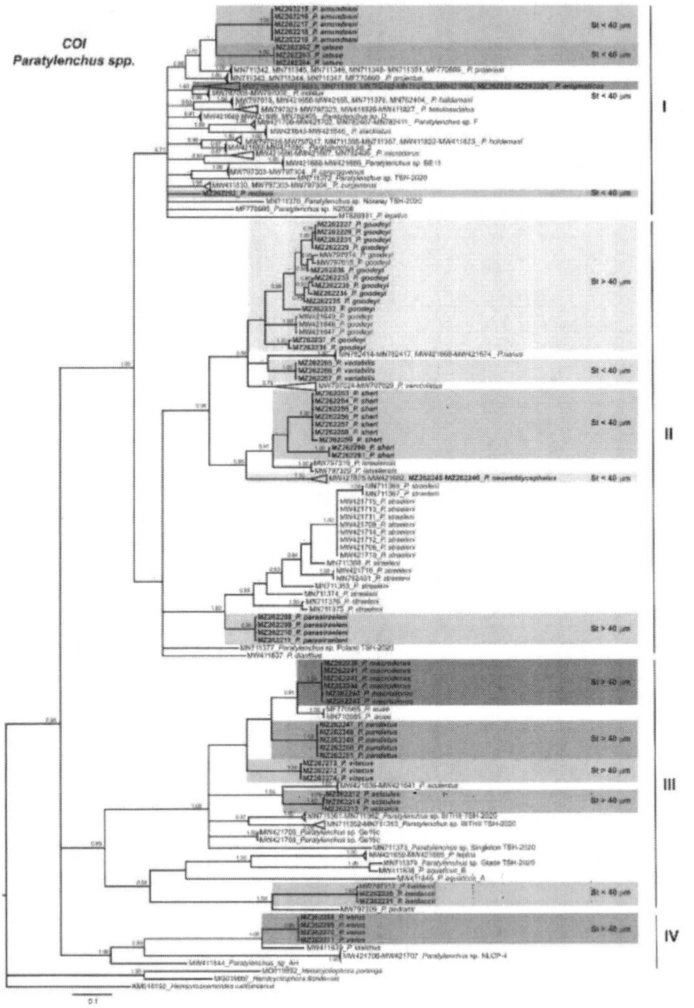

Figure 18. Phylogenetic relationships within the genus *Paratylenchus*. Bayesian 50% majority rule consensus tree as inferred from cytochrome c oxidase subunit 1 (COI) sequence alignment under the general time-reversible model of sequence evolution with a gamma-shaped distribution (GTR + G). Posterior probabilities of more than 0.70 are given for appropriate clades. Newly obtained sequences in this study are shown in bold. The scale bar indicates expected changes per site.

supported subclade (PP = 1.00) within clade II, and the new species, *P. parastraeleni* sp. nov., was clearly separated from all other isolates of *P. straeleni* from Belgium, Canada, Ireland, and USA (Figure 18). Similar as in ribosomal markers, stylet length patterns (> or <40 μm) were mixed in clusters II and III, whereas cluster I comprises species with short stylets and clade IV species with long stylets (Figure 18). These clades were partially coincident with other studies with, in some cases, similar or different clade support [3,4].

3. Discussion

This research comprises the second part focused on the integrative taxonomical identification of pin nematodes of the genus *Paratylenchus* in Spain. These results increase the number of species with morphological and molecular data for their unequivocal identifica- tion, as well as confirming the huge biodiversity of this group including the description of a new species *viz. P. parastraeleni* sp. nov., within the *P. straeleni*-complex.

Eighteen *Paratylenchus* spp. from nine different localities, including almond and natu- ral environment soil samples, were identified. All of them except one, were already known (*P. amundseni, P. aciculus, P. baldaccii, P. enigmaticus, P. goodeyi, P. holdemani, P. macrodorus, P. neoamblycephalus, P. pandatus, P. pedrami, P. recisus, P. sheri, P. tateae, P. variabilis, P. verucula- tus, P. verus*, and *P. vitecus*), and eight considered as first reports for Spain in this work (*viz. P. amundseni, P. aciculus, P. neoamblycephalus, P. pandatus, P. recisus, P. variabilis, P. verus* and *P. vitecus*). Finally, one of the 18 species detected was identified as a new species, *P. parastrae- leni* sp. nov., which confirmed the cryptic diversity within the *P. straeleni*-species complex group by applying integrative taxonomical approaches verifying an outstanding example of the cryptic diversity. Overall, the results of this and previous studies reported a total of 39 species of *Paratylenchus* in Spain, widespread in cultivated and natural ecosystems.

In *Paratylenchus* spp. with longer stylet (>40 μm) most juveniles bear elongate flexible stylet (formerly belonging to the genus *Gracilacus*), but some species are found to have what appears to be fourth-stage juveniles with very length reduced and rigid stylets, a characteristic most frequently found in species of *Paratylenchus sensu stricto* with female stylets of 40 μm or less [36]. Since many soil samples from natural environments comprise mixed species (even four different species), it is very difficult to associate specimens of one developmental stage with the appropriate adult state [3,4]. However, applying integrative taxonomical approaches (molecular barcoding of juvenile and adult individuals) we can accurately study juvenile and adult forms in each soil sample. For the first time, morphological and molecular data (D2-D3, ITS and COI for the same individual) of J4 for the majority of the species detected in this study were provided herein, allowing the first report for authenticating a clear example of stylet and lip region metamorphosis between J4 and adult female. Within several isolates of *P. goodeyi* studied here, we verified that short rigid stylet and conoid-truncate lip region with strong labial sclerotization in J4 moved to a long and slender flexible stylet and a conoid-rounded lip region without labial sclerotization in adult females. Apart from the unequivocal identification of juvenile stages of each species, the integrative taxonomical identification of J4 allows to document some important biological aspects for some species, as well as a useful tool for the species identification in periods when the resting-stage accumulates predominantly in soil under adverse environmental conditions (*viz.* drought conditions) [3,4].

Although we are aware that nematological efforts on *Paratylenchus* species in Southern Spain have been higher than that carried out in central and northern parts of the country, the present distribution of the genus in Spain, with about 90% of species (35 out of 39 species, and 24 of them confirmed by integrative taxonomy) only reported in Southern Spain, suggest that this part of the country can be considered as a potential hotspot of biodiversity. Nevertheless, further research is needed to definitely confirm this hypothesis. This study also ratifies the previous proposed hypothesis [4] that we have only deciphered just a small part of the species diversity of pin nematodes reported in Spain, indicating that the biodiversity of this group is far from being adequately explored all over the world [3,4]. The present data also suggest that species richness was higher in natural environments than in cultivated areas, since the number of *Paratylenchus* species detected within the same sample in natural environments included four different species (*viz. P. holdemani, P. macrodorus, P. pedrami*, and *P. veruculatus* in wild olive sample code as AR_102), and more than 60% of soil samples from natural environments exhibited at least two *Paratylenchus* species, whereas in the majority of samples from cultivated areas only one or maximum two mixed species were frequently detected in the same sample [3,4,8]. Nevertheless, this hypothesis needs to be contrasted with further investigations.

Paratylenchus microdorus has been extensively reported in Spain in cultivated and natural environments [15–18]. The present results, comprising integrative studies on some geographical areas with previous records of *P. microdorus*, suggest that probably this species was misidentified in previous records. In this case, *P. recisus, P. variabilis, P. veruculatus*, and *P. zurgenerus*, are close morphologically to *P. microdorus*, but molecularly well separated. This study suggests that previous records of *P. microdorus* could be misidentified, since only detailed traits can separate these species (short differences in stylet length, shape of tail terminus, vulva anus distance with regard to tail length), and therefore additional studies are needed to clarify the real biodiversity in the *P. microdorus*-species complex in Spain by applying integrative taxonomy. Probably, this potential misidentification can also be referred to the numerous records of *P. microdorus* in other countries such as Bulgaria, Germany, Hungary, Poland, and Romania [2], which need further investigations. Molecularly, *P. microdorus*-species complex was separated in two subgroups, one comprising *P. microdorus, P. recisus* and *P. zurgenerus*, and another very separate subclade including *P. variabilis* and *P. veruculatus*, being consistent for ribosomal and mitochondrial genes. The genus *Paratylenchoides* was proposed by Raski [44] to accommodate two *Paraty-lenchus* species populations from France and Israel with heavy sclerotization in the lip region and narrow lip region dorso-ventrally. This action was partially followed by Sid- diqi [48] proposing a subgeneric rank within the genus *Paratylenchus*. However, Raski and Luc [49] considered that differences between *Paratylenchoides* and *Paratylenchus* were minor and cannot be considered important to separate both taxa, synonymizing *Paratylenchoides* with *Paratylenchus*. The present results confirm that, molecularly, *P. sheri* and *P. israelensis* (formerly *Paratylenchoides* species) clustered together in two separate subclades in D2-D3, ITS and COI trees, but always together with other *Paratylenchus* species with long and short stylets, such as *P. neoamblycephalus, P. veruculatus* or *P. parastraeleni* sp. nov. and *P. goodeyi* and cannot be considered a separate genus as it is the case for *Gracilacus* already discussed [3,4,6].

The results obtained in the present study, reinforce the idea that for accurate identification of *Paratylenchus* spp. it is essential to carry out an integrative identification, including morphological, morphometrical and molecular analysis, the latter of which should be based on multilocus approaches (D2-D3 region of 28S rRNA and COI) [3,4,6]. In our case several species demonstrate low differences in ribosomal markers (98–99%) among species, but clear differences on COI and are also clearly different morphologically. This situation has been observed among *P. sheri–P. israelensis–P. neoamblycephalus, P. macrodorus–P. pandatus–P. wuae* or between *P. aciculus–P. aculentus.* This is because mitochondrial DNA display a high mutation rate and maternal inheritance, which also enables better discrimination of closely related species [50,51]. On the other hand, several species showed some molecular intraspe- cific variability in the three regions studied herein (0.5–4%) but with identical morphology and morphometry, such as *P. goodeyi, P. enigmaticus.*

Phylogenetic analyses based on D2-D3, ITS, and COI gene using BI mostly agree with the clustering obtained by other authors [3,4,6]. Ribosomal and mitochondrial phylogenies did not separate the long stylet length (>40 μm) with the short stylet length (>40 μm) supporting the synonymy of *Gracilacus* and suggesting that stylet length in *Paratylenchus* has evolved independently several times during the evolution of this genus [3,4,6].

4. Materials and Methods

4.1 Nematode Sampling and Morphological Identification

Fifteen soil samples were collected mainly from the rhizosphere of herbaceous and woody plants including 5 samples from almond with different rootstocks (*Prunus* spp.), 2 samples from Portuguese oak (*Quercus faginea* Lam.), 3 samples from Aleppo pine (*Pinus halepensis* Mill.), 3 samples from wild olive (*Olea europaea* sbsp. *silvestris* (Mill.) Lehr), and 2 samples from grasses, in 10 localities in Spain (Table 1). Samples were collected using a shovel and considering the upper 5–40 cm depth of soil. Nematodes were extracted from a 500-cm^3 subsample of soil by centrifugal flotation [52].

A total of 232 individuals including 160 females, 5 males and 67 juveniles were used for morphological and morphometrical analyses. Specimens for study using light microscopy (LM) and morphometrical studies were killed and fixed in an aqueous solution of 4% formaldehyde + 1% glycerol, dehydrated using alcohol-saturated chamber and processed to pure glycerine using Seinhorst's method [53] as modified by De Grisse [54]. The developmental stage of the juveniles was determined according to the body length and the degree of development of genital cells [26]. Light micrographs were taken using fresh nematodes and measurements of each nematode population, including important diagnostic characteristics (i.e., de Man indices, body length, stylet length, lip region, tail shape) [55], were performed using a Leica DM6 compound microscope with a Leica DFC7000 T digital camera using fixed and embedded nematodes in glycerin. Nematodes were identified at the species level using an integrative approach combining molecular and morphological techniques to achieve efficient and accurate identification [3,4,8]. For each nematode population, key diagnostic characters were determined, including body length, stylet length, a ratio (body length/maximum body width), b ratio (body length/ total pharynx length), c ratio (body length/tail length), c' ratio (tail length/body width at anus), V ratio ((distance from anterior end to vulva/body length) 100), and o ratio ((distance from stylet base to dorsal pharyngeal opening/stylet length) 100) [3,4,8], and the sequencing of specific DNA fragments (described below) confirmed the identity of the nematode species for each population. Specimens for SEM observations were processed using Wergin's method [56], coated with gold and observed with a JEOL 50A scanning electron microscope at 10 kV of accelerating voltage.

Nematode populations of *Paratylenchus* species already described were analyzed morphologically and molecularly in this study and proposed as standard and reference populations for each species given until topotype material becomes available and molecularly characterized. Voucher specimens of these described species have been deposited in the nematode collection of Institute for Sustainable Agriculture, IAS-CSIC, Córdoba, Spain.

4.2 Nematode Molecular Characterization

For molecular analyses, and in order to avoid mistakes in case of mixed populations in the same sample (being common in several soil samples), single specimens from the sample were temporarily mounted in a drop of 1 M NaCl containing glass beads (to avoid nematode crushing/damaging specimens) to ensure specimens conformed with the unidentified population. All necessary morphological and morphometrical data by taking pictures and measurements using the above camera-equipped microscope were recorded. Then DNA extraction from single individuals was performed as described by Palomares- Rius et al. [57], and more importantly, for all the 24 studied isolates, all the three molecular markers of each *Paratylenchus* isolate belong to the same single extracted individual in each PCR tube without any exception. In addition, male and juveniles conspecificity was proven by single DNA extraction of male or juveniles for each species.

The D2 and D3 expansion domains of the 28S rRNA were amplified using the D2A (5′-ACAAGTACCGTGAGGGAAAGTTG-3′) and D3B (5′-TCGGAAGG-AACCA-GCTACTA- 3′) primers [58]. The Internal Transcribed Spacer region (ITS) was amplified by using forward primer TW81 (5′- GTTTCCGTAGGTGAACCTGC -3′) and reverse primer AB28 (5′- ATATGCTT-AAGTTCAGCGGGT -3′) [59]. The COI gene was amplified using the primers JB3 (5′-TTTTTTGGGCATCCTGAGGTTTAT-3′) and JB5 (5′-AGCACCTAAACTTAAAACAT AATGAAAATG-3′) [60]. The PCR cycling conditions for the 28S rRNA and ITS regions were as follows: 95 °C for 15 min, followed by 35 cycles of 94 °C for 30 s, an annealing temperature of 55 °C for 45 s, and 72 °C for 1 min, and one final cycle of 72 °C for 10 min. The PCR cycling for COI primers was as follows: 95 °C for 15 min, 39 cycles at 94 °C for 30 s, 53 °C for 30 s, and 68 °C for 1 min, followed by a final extension at 72 °C for 7 min. PCR volumes were adapted to 25 μL for each reaction, and primer concentrations were as described in De Ley et al. [58], Subbotin et al., [59] and Bowles et al. [60]. We used 5 HOT FIREpol Blend Master Mix (Solis Biodyne, Tartu, Estonia) in all PCR reactions. The PCR products were purified using ExoSAP-IT (Affimetrix, USB products, Kandel, Germany) and used for direct sequencing in both directions with the corresponding primers. The resulting products were run in a DNA multicapillary sequencer (Model 3130XL Genetic Analyzer; Applied Biosystems, Foster City, CA, USA), using the BigDye Terminator Sequencing Kit v.3.1 (Applied Bio-systems) at the Stab Vida sequencing facility (Caparica, Portugal). The sequence chromatograms of the 3 markers (ITS, COI and D2-D3 expansion segments of 28S rRNA) were analyzed using DNASTAR LASERGENE SeqMan v. 7.1.0. Basic local alignment search tool (BLAST) at the National Center for Biotechnology Information (NCBI) was used to confirm the species identity of the DNA sequences obtained in this study [61]. The newly obtained sequences were deposited in the GenBank database under accession numbers indicated on the phylogenetic trees and in Table 1.

4.3 Phylogenetic Analyses

D2-D3 expansion segments of 28S rRNA, ITS rRNA, and COI mtDNA sequences of the 24 *Paratylenchus* isolates were obtained in this study. These sequences and other sequences from species of *Paratylenchus* from GenBank were used for phylogenetic analyses. Selection of outgroup taxa for each dataset were based on previously published studies [3,4,7,62]. Multiple sequence alignments of the different genes were completed using the FFT-NS-2 algorithm of MAFFT V.7.450 [63]. BioEdit program V. 7.2.5 [64] was used for sequence alignments visualization and edited by Gblocks ver. 0.91b [65] in Castresana Laboratory server (http://molevol.cmima.csic.es/castresana/Gblocks_server.html accessed on 13 May 2021) using options for a less stringent selection (minimum number of sequences for a conserved or a flanking position: 50% of the number of sequences +1; maximum number of contiguous non-conserved positions: 8; minimum length of a block: 5; allowed gap positions: with half). Phylogenetic analyses of the sequence datasets were based on Bayesian inference (BI) using MrBayes 3.1.2 [66]. The best-fit model of DNA evolution was achieved using JModelTest V.2.1.7 [67] with the Akaike information criterion (AIC). The best-fit model, the base frequency, the proportion of invariable sites, and the gamma distribution shape parameters and substitution rates in the AIC were then used in MrBayes for the phylogenetic analyses. The general time-reversible model with invariable sites and a gamma-shaped distribution (GTR +I+ G) for the D2-D3 segments of 28S rRNA and the partial ITS rRNA and the general time-reversible model with a gamma-shaped distribution (GTR + G) for COI gene, were run with four chains for 4, 4, and 10 10^6 generations, respectively. A combined analysis of the three ribosomal genes was not undertaken due to some sequences not being available for all species. The sampling for Markov chains was carried out at intervals of 100 generations. For each analysis, two runs were conducted. After discarding burn-in samples of 30% and evaluating convergence, the remaining samples were retained for more in-depth analyses. The topologies were used to generate a 50% majority-rule consensus tree. On each appropriate clade, posterior probabilities (PP) were given. FigTree software version v.1.42 [68] was used for visualizing trees from all analyses.

5. Conclusions

This study reveals the existence of a huge cryptic biodiversity within the genus *Paratylenchus*, increasing and expanding the diversity of this group in Spain. For the first time, morphological and molecular data (D2-D3, ITS and COI for the same individual) of J4 allowed to authenticate an example of stylet and lip region metamorphosis between J4 and adult females in *P. goodeyi* (from short rigid stylet and conoid-truncate lip region with strong labial sclerotization in J4 to a long and slender flexible stylet and a conoid-rounded lip region without labial sclerotization in adult females). This study also ratifies the previous proposed hypothesis that we have only deciphered just a small part of the species diversity within pin nematodes reported in Spain and most probably all over the world. Our data also suggest that *P. microdorus* comprise a complex of species morphologically very close, but molecularly well separated, and therefore additional studies are needed to clarify the real biodiversity within the *P. microdorus*-species complex in Spain and all over the world by applying integrative taxonomy.

References

1. Micoletzky, H. Die freilebenden Erd-Nematoden. *Arch. Für Nat.* **1922**, *87*, 1–650.
2. Ghaderi, R.; Geraert, E.; Karegar, A. *The Tylenchulidae of the World, Identification of the Family Tylenchulidae (Nematoda: Tylenchida)*; Academia Press: Ghent, Belgium, 2016.
3. Singh, R.; Karssen, G.; Coureur, M.; Subbotin, S.; Bert, W. Integrative taxonomy and molecular phylogeny of the plant-parasitic nematode genus *Paratylenchus* (Nematoda: Paratylenchinae): Linking species with molecular barcodes. *Plants* **2021**, *10*, 408. [CrossRef] [PubMed]
4. Clavero-Camacho, I.; Cantalapiedra-Navarrete, C.; Archidona-Yuste, A.; Castillo, P.; Palomares-Rius, J.E. Remarkable cryptic diversity of *Paratylenchus* spp. (Nematoda: Tylenchulidae) in Spain. *Animals* **2021**, *11*, 1161. [CrossRef] [PubMed]
5. Raski, D.J. Paratylenchidae n.fam. with descriptions of five new species of *Gracilacus* n.g. and an emendation of *Cacopaurus* Thorne, 1943, *Paratylenchus* Micoletzky, 1922 and Criconematidae Thorne, 1943. *Proc. Helminthol. Soc. Wash.* **1962**, *29*, 189–207.
6. Van den Berg, E.; Tiedt, L.R.; Subbotin, S.A. Morphological and molecular characterisation of several *Paratylenchus* Micoletzky, 1922 (Tylenchida: Paratylenchidae) species from South Africa and USA, together with some taxonomic notes. *Nematology* **2014**, *16*, 323–358. [CrossRef]
7. Munawar, M.; Miao, W.; Castillo, P.; Zheng, J.-W. A new pin nematode, *Paratylenchus sinensis* n. sp. (Nematoda: Paratylenchinae) in the rhizosphere of white mulberry from Zhejiang Province, China. *Eur. J. Plant Pathol.* **2020**, *156*, 1023–1029.
8. Munawar, M.; Yevtushenko, D.P.; Palomares-Rius, J.E.; Castillo, P. Species diversity of pin nematodes (*Paratylenchus* spp.) from potato growing regions of southern Alberta, Canada. *Plants* **2021**, *10*, 188. [CrossRef]
9. Castillo, P.; Gómez-Barcina, A. Some species of Tylenchida from natural habitats in southeastern Spain. *Nematol. Medit.* **1988**, *16*, 75–86.
10. Akyazi, F.; Felek, A.F.; Cˇ ermák, V.; Cˇ udejková, M.; Foit, J.; Yildiz, S.; Háneˇl, L. Description of *Paratylenchus* (*Gracilacus*) *straeleni* (De Coninck, 1931) Oostenbrink, 1960 (Nematoda: Criconematoidea, Tylenchulidae) from hazelnut in Turkey and its comparison with other world populations. *Helminthologia* **2015**, *52*, 270–279. [CrossRef]
11. Powers, T.O.; Harris, T.S.; Higgins, R.S.; Mullin, P.G.; Powers, K.S. Nematode biodiversity assessments need vouchered databases: A BOLD reference library for plant-parasitic nematodes in the superfamily Criconematoidea. *Genome* **2020**, *64*, 232–241. [CrossRef]

12. Hernandez, M.; Mateo, M.D.; Jordana, R. Estudio comparativo entre grupos tróficos de Nematodos del suelo de cinco bosques de Navarra (tres naturales y dos de repoblación). *Actas II Congr. Mund. Vasco. Sec. Biol. Ambient.* **1988**, *2*, 323–335.

13. Brzeski, M.; Hanel, L.; Nico, A.; Castillo, P. Paratylenchinae: Redescription of *Paratylenchus arculatus* Luc & de Guiran, 1962, a new senior synonym of *P. nainianus* Edward & Misra, 1963 (Nematoda: Tylenchulidae). *Nematology* **1999**, *1*, 375–380.

14. Nico, A.I.; Rapoport, H.F.; Jiménez-Díaz, R.M.; Castillo, P. Incidence and population density of plant-parasitic nematodes associated with olive planting stocks at nurseries in southern Spain. *Plant Dis.* **2002**, *86*, 1075–1079. [CrossRef] [PubMed]

15. Peña-Santiago, R. Plant-parasitic nematodes associated with olive (*Olea europea* L.) in the province of Jaén, Spain. *Rev. Némato1.* **1990**, *13*, 113–115.

16. Archidona-Yuste, A.; Wiegand, T.; Castillo, P.; Navas-Cortés, J.A. Dataset on the diversity of plant-parasitic nematodes in cultivated olive trees in southern Spain. *Data Brief* **2019**, *27*, 104658. [CrossRef]

17. Gomez-Barcina, A.; Castillo, P.; Pais, M.A.G. Four species of the genus *Paratylenchus* Micoletzky from southeasthern Spain. *Nematol. Medit.* **1990**, *18*, 169–177.

18. Talavera, M.; Navas, A. Incidence of plant-parasitic nematodes in natural and semi-natural mountain grassland and the host status of some common grass species. *Nematology* **2002**, *4*, 541–552.

19. Escuer, M. Els Nematodes. In *El Patrimoni Biològic del Montseny. Catàleg de Flora i Fauna*; Servei de Parcs Naturals, Diputació de Barcelona: Barcelona, Spain, 1995; Volume 2, pp. 17–22.

20. Castillo, P.; González-País, M.A.; Gómez-Barcina, A. El género *Gracilacus* Raski, 1962 en España (Paratylenchinae: Tylenchida). *Rev. Ibér. Parasitol.* **1989**, *49*, 321–328.

21. Gomez-Barcina, A.; Castillo, P.; González-País, M.A. Nematodos fitoparásitos de la subfamilia Criconematinae Taylor, 1936 en la Sierra de Cazorla. *Rev. Ibér. Parasitol.* **1989**, *49*, 241–255.

22. Talavera, M.; Tobar Jimenez, A. Plant parasitic nematodes from unirrigated fields in Alhama, southeastern Spain. *Nematol. Medit.* **1997**, *25*, 73–81.

23. Imaz, A.; Hernández, M.A.; Ariño, A.H.; Armendáriz, I.; Jordana, R. Diversity of soil nematodes across a Mediterranean ecotone. *Appl. Soil Ecol.* **2002**, *20*, 191–198. [CrossRef]

24. Peña Santiago, R.; Geraert, E. New data on *Aorolaimus perscitus* (Doucet, 1980) and *Gracilacus teres* Raski, 1976 (Nematoda: Tylenchida) associated with olive (*Olea europea* L.) in the province of Jaén, Spain. *Nematologica* **1990**, *36*, 408–416.

25. Brzeski, M.W. Seasonal dynamics of *Paratylenchus bukowinensis* Micol. and some other nematodes. *Rocz. Nauk Rol. Ser. E* **1977**, *7*, 67–74.

26. Brzeski, M.W.; Hanel, L. Paratylenchinae: Postembryonic developmental stages of *Paratylenchus straeleni* (De Coninck, 1931) and *P. steineri* Golden, 1961 (Nematoda: Tylenchulidae). *Nematology* **1999**, *1*, 673–680. [CrossRef]

27. De Coninck, L.A.P. Sur trois espèces nouvelles de nématodes libres trouvés en Belgique. *Bull. Musée R. d'Hist. Nat. Belg.* **1931**, *7*, 1–15.

28. Brzeski, M.W. Paratylenchinae: Morphology of some known species and descriptions of *Gracilacus Bilineata* sp. n. and *G. Vera* sp. n. (Nematoda: Tylenchulidae). *Nematologica* **1995**, *41*, 535–565. [CrossRef]

29. Brzeski, M.W. *Nematodes of Tylenchina in Poland and Temperate Europe*; Muzeum i Instytutu Zoologii, PAN: Warsaw, Poland, 1998; p. 397.

30. Ghaderi, R.; Karegar, A. Some species of *Paratylenchus* (Nematoda: Tylenchulidae) from Iran. *Iran. J. Plant Pathol.* **2013**, *49*, 137–156.

31. Brown, G.L. Three new species of the genus *Paratylenchus* from Canada (Nematoda: Criconematidae). *Proc. Helminthol. Soc. Wash.* **1959**, *26*, 1–8.

32. Bernard, E.C. Criconematina (nematoda: Tylenchida) from the aleutian islands. *J. Nematol.* **1982**, *14*, 323–331.

33. Castillo, P.; Gomez-Barcina, A. Plant-parasitic nematodes associated with tropical and subtropical crops in southern Spain. *Nematol. Medit.* **1993**, *21*, 45–47.

34. Escuer, M.; Cano, A.; Bello, A. Nematodos fitoparásitos de la Región de Murcia y alternativas de control. In *Desinfección de Suelos en Invernaderos de Pimientos*; Serie Jornadas y Congresos, 16; Consejería de Agricultura, Agua y Medioambiente: Murcia, Spain, 2004; pp. 27–57.

35. Geraert, E. The Genus *Paratylenchus*. *Nematologica* **1965**, *11*, 301–334. [CrossRef]

36. Oostenbrink, M. A note on *Paratylenchus* in the Netherlands with the description of *P. goodeyi* n. sp. (Nematoda, Criconematidae). *Tijdschr. Plantenziekten* **1953**, *59*, 207–216. [CrossRef]

37. Raski, D.J. Revision of the genus *Paratylenchus* Micoletzky, 1922 and descriptions of new species. Part III of three parts—*Gracilacus*. *J. Nematol.* **1976**, *8*, 97–115. [PubMed]

38. Yu, Q.; Ye, W.; Powers, T. Morphological and molecular characterization of *Gracilacus wuae* n. sp. (Nematoda: Criconematoidea) associated with cow parsnip (*Heracleum maximum*) in Ontario, Canada. *J. Nematol.* **2016**, *48*, 203–213. [CrossRef] [PubMed]

39. Brzeski, M. Paratylenchus macrodorus n. sp. (Nematoda, Paratylenchidae), a new plant parasitic nematode from Poland. *Bull. L'Acad. Pol. Sci.* **1963**, *11*, 277–280.

40. Van den Berg, E.; Quénéhervé, P.; Tiedt, L. Two *Paratylenchus* species (Nemata: Tylenchulidae) from Martinique and New Caledonia. *J. Nematode Morphol. Syst.* **2006**, *9*, 1–10.

41. Nguyen, C.N.; Baldwin, J.G.; Choi, Y.E. New records of *Paratylenchus* Micoletzky, 1922 (Nematoda: Paratylenchinae) from Viet Nam with description of *Paratylenchus lapcaiensis* sp. n. *J. Nematode Morphol. Syst.* **2004**, *7*, 51–75.

42. Van Den Berg, E.; Mekete, T.; Tiedt, L.R. New records of Criconematidae (Nemata) from Ethiopia. *J. Nematode Morphol. Syst.* **2004**, *6*, 161–174.

43. Siddiqi, M. *Paratylenchus recisus* sp.n. and *P. perminimus* sp.n. (Criconematina: Paratylenchidae). *Afro-Asian J. Nematol.* **1996**, *6*, 55–58.

44. Raski, D.J. *Paratylenchoides* gen. n. and two new species (Nematoda: Paratyleiichidae). *Proc. Helminthol. Soc. Wash.* **1973**, *40*, 230–233.

45. Mirbabaei, H.; Eskandari, A.; Ghaderi, R.; Karegar, A. On the synonymy of *Trophotylenchulus asoensis* and *T. okamotoi* with *T. arenarius*, and intra-generic structure of *Paratylenchus* (Nematoda: Tylenchulidae). *J. Nematol.* **2019**, *51*, 1–22. [CrossRef]

46. Raski, D.J. Revision of the genus *Paratylenchus* Micoletzky, 1922 and descriptions of new species. Part I of three parts. *J. Nematol.* **1975**, *7*, 15–34. [PubMed]

47. Pramodini, M.; Mohilal, N.; Dhanachand, C. *Gracilacus vitecus* sp.n. and record of *G. raskii* Phukan & Sanwal from Manipur, India. *Ind. J. Nematol.* **2006**, *36*, 272–276.

48. Siddiqi, M.R. *Tylenchida: Parasites of Plants and Insects*; Commonwealth Institute of Parasitology: Slough, UK, 1986; p. 645.

49. Raski, D.; Luc, M. A reappraisal of Tylenchina (Nemata): 10. The superfamily Criconematoidea Taylor, 1936. *Rev. Nématol.* **1987**, *10*, 409–444.

50. Derycke, S.; Backeljau, T.; Moens, T. Dispersal and gene flow in free-living marine nematodes. *Front. Zool.* **2013**, *10*, 109–152. [CrossRef]

51. Jex, A.R.; Littlewood, D.T.; Gasser, R.B. Toward next-generation sequencing of mitochondrial genomes-focus on parasitic worms of animals and biotechnological implications. *Biotechnol. Adv.* **2010**, *28*, 151–159. [CrossRef]

52. Coolen, W.A. Methods for extraction of *Meloidogyne* spp. and other nematodes from roots and soil. In *Root-Knot Nematodes (Meloidogyne Species): Systematics, Biology and Control*; Lamberti, F., Taylor, C.E., Eds.; Academic Press: New York, NY, USA, 1979; pp. 317–329.

53. Seinhorst, J.W. Killing nematodes for taxonomic study with hot F.A. 4:1. *Nematologica* **1966**, *12*, 178. [CrossRef]

54. De Grisse, A.T. Redescription ou modifications de quelques techniques utilisées dans l'étude de nématodes phytoparasitaires. *Meded. Rijksfac. Landbouwwet. Gent* **1969**, *34*, 315–359.

55. Hunt, D.J.; Palomares-Rius, J.E. General morphology and morphometries of plant-parasitic nematodes. In *Practical Plant Nematology*; Biblioteca Basica de Agricultura: Texcoco, Mexico, 2012; pp. 25–64.

56. Wergin, W.P. Scanning electron microscopic techniques and applications for use in nematology. In *Plant Parasitic Nematodes*; Zuckerman, B.M., Rohde, R.A., Eds.; Academic Press: New York, NY, USA; London, UK, 1981; Volume 3, pp. 175–204.

57. Palomares-Rius, J.E.; Clavero-Camacho, I.; Archidona-Yuste, A.; Cantalapiedra-Navarrete, C.; León-Ropero, G.; Braun Miyara, S.; Karssen, G.; Castillo, P. Global distribution of the reniform nematode genus *Rotylenchulus* with the synonymy of *Rotylenchulus macrosoma* with *Rotylenchulus borealis*. *Plants* **2021**, *10*, 7. [CrossRef] [PubMed]

58. De Ley, P.; Felix, M.A.; Frisse, L.; Nadler, S.; Sternberg, P.; Thomas, W.K. Molecular and morphological characterisation of two reproductively isolated species with mirror-image anatomy (Nematoda: Cephalobidae). *Nematology* **1999**, *1*, 591–612. [CrossRef]

59. Subbotin, S.A.; Vierstraete, A.; De Ley, P.; Rowe, J.; Waeyenberge, L.; Moens, M.; Vanfleteren, J.R. Phylogenetic relationships within the cyst-forming nematodes (Nematoda, Heteroderidae) based on analysis of sequences from the ITS regions of ribosomal DNA. *Mol. Phylogenet. Evol.* **2001**, *21*, 1–16. [CrossRef] [PubMed]

60. Bowles, J.; Blair, D.; McManus, D.P. Genetic variants within the genus *Echinococcus* identified by mitochondrial DNA sequencing. *Mol. Biochem. Parasitol.* **1992**, *54*, 165–173. [CrossRef]

61. Altschul, S.F.; Gish, W.; Miller, W.; Myers, E.W.; Lipman, D.J. Basic local alignment search tool. *J. Mol. Biol.* **1990**, *215*, 403–410. [CrossRef]

62. Subbotin, S.A.; Yan, G.; Kantor, M.; Handoo, Z. On the molecular identity of *Paratylenchus nanus* Cobb, 1923 (Nematoda: Tylenchida). *J. Nematol.* **2020**, *52*, 1–7. [CrossRef] [PubMed]

63. Katoh, K.; Rozewicki, J.; Yamada, K.D. MAFFT online service: Multiple sequence alignment, interactive sequence choice and visualization. *Brief. Bioinform.* **2019**, *20*, 1160–1166. [CrossRef]

64. Hall, T.A. BioEdit: A user-friendly biological sequence alignment editor and analysis program for Windows 95/98/NT. *Nucleic Acids Symp. Ser.* **1999**, *41*, 95–98.

65. Castresana, J. Selection of conserved blocks from multiple alignments for their use in phylogenetic analysis. *Mol. Biol. Evol.* **2000**, *17*, 540–552. [CrossRef]

66. Ronquist, F.; Huelsenbeck, J.P. MrBayes 3: Bayesian phylogenetic inference under mixed models. *Bioinformatics* **2003**, *19*, 1572–1574. [CrossRef]

67. Darriba, D.; Taboada, G.L.; Doallo, R.; Posada, D. jModelTest 2: More models, new heuristics and parallel computing. Nat. Methods 2012, 9, 772. [CrossRef]

68. Rambaut, A. FigTree v1.4.2, A Graphical Viewer of Phylogenetic Trees. Available online: http://tree.bio.ed.ac.uk/software/ figtree/ (accessed on 11 June 2021).

Chapter-7

Current Insights into Migratory Endoparasitism: Deciphering the Biology, Parasitism Mechanisms, and Management Strategies of Key Migratory Endoparasitic Phytonematodes

1. Introduction

Of approximately 27,000 described nematode species, roughly 4100 utilize higher terrestrial plants as a predominant source of nutrition [1]. These plant-parasitic nematodes (PPNs), cause ~\$80–\$157 billion crop losses annually worldwide [2,3]. The earliest evidence of a nematode identified within a plant is the fossilized eggs, juveniles, and adults of *Palaeonema phyticum* (Poinar, Kerp, and Hass, 2008) in the stem of the terrestrial plant *Aglaophyton major* (Kidson and Lang, 1920) in the Devonian era [4]. This discovery of *P. phyticum* uncovered an ancient and a pivotal point in the timeline of transition of nematodes from free-living to parasites of land plants. Since microbes were found in the stomatal spaces invaded by the nematode, *P. phyticum* was tentatively categorized as a facultative plant-parasite, belonging to clade 1 of the phylum Nematoda (due to the morphological similarities to clade 1 nematodes) [4–6]. Feeding on fungi or bacteria or other microbes is also considered to be one of the most important determinants for the evolution of facultative or obligate plant-parasitism in nematodes. This evolution is thought to have occurred independently four times within Nematoda [4,5,7,8]. In the phylogenetic tree outlined by Van Megen et al., these events place PPNs in four of the twelve clades: 1 (Triplonchida), 2 (Dorylaimida), 10 (Aphelenchoididae), and 12 (Tylenchida). A prominent morphologically distinctive feature of all PPNs is a protrusible needle-like apparatus known as the stylet. The stylet, an artifact of convergent evolution (within the above-mentioned four clades) is utilized by PPNs for two main reasons: (1) puncturing the plant cell wall to extract cell nutrients, and (2) in certain nematode species, delivering secretory molecules into the apoplast and/or cytoplasm to manipulate host cells to develop a permanent feeding site [9,10]. Feeding mechanisms differ among PPNs; they are, therefore, a useful tool to group PPN species. Broadly speaking, PPNs are divided into two main categories based on feeding mechanism: endoparasitic and ectoparasitic. Ectoparasitic nematodes occupy clades 1 (Trichodoridae) and 2 (Longidoridae) in the Nematoda lineage; they feed on cortical root cells from outside the root. Clade 2 ectoparasitic nematodes of the genera *Xiphinema* and *Longidorus* utilize an odontostyle (as opposed to the stomatostylet, a feature common to clades 10 and 12 or onchiostyle in clade 1 PPN) to induce cell enlargement. Although the feeding apparatus used by *Xiphinema* and *Longidorus* is the same, the terminal plant cell feeding sites formed after this intimate biotrophic association are slightly different, with those formed by *Xiphinema* inducing plant cell karyokinesis and those formed by *Longidorus* remaining mononucleate [5]. In contrast to ectoparasitic nematodes, endoparasitic nematodes penetrate and feed within plant roots and have been a tremendous burden on global agricultural production, especially in developing areas like Sub-Saharan Africa, where the resources to diagnose and combat them are limited [11]. Endoparasitic nematodes can be further divided into two sub-categories: migratory and sedentary. Sedentary endoparasitic nematodes, such as the root-knot and the cyst nematodes, form specialized feeding sites, which act as nutritional sinks for the developing nematode. Migratory endoparasitic nematodes such as *Scutellonema bradys* (Steiner and LeHew, 1933) Andrassy, 1958, are a tremendous agricultural and economic burden on yam (*Dioscorea* spp.), a crop that has been a major source of income and an important part of the diet in the western regions of Africa [12,13]. A massive amount of genomic and transcriptomic information has been obtained about many migratory endoparasitic nematodes primarily through application of next-generation sequencing (NGS) technologies [14–20]. Analysis of this information has opened a novel gateway for researchers worldwide to formulate evidence-based conclusions regarding the biology, phylogenetic relationships, and parasitism mechanisms of these nematodes. In this review, we focus only on some of the economically important migratory endoparasitic PPNs from clades 10 and 12. For in-depth analysis of the effectors and processes targeted by sedentary endoparasites of clade 12, the authors suggest several other reviews [10,21,22].

2. The Biology of Migratory Endoparasitic Nematodes

A survey of PPN researchers worldwide ranked the top 10 economically significant and scientifically relevant PPNs [23]. Not surprisingly, migratory endoparasitic nematodes, such as root lesion nematodes (*Pratylenchus* spp.), burrowing nematode (*Radopholus similis* (Cobb, 1893) Thorne, 1949), stem or stem and bulb nematode (*Ditylenchus dipsaci* (Kuhn, 1857) Filipjev, 1936) and pine wood nematode (*Bursaphelenchus xylophilus* (Steiner and Buhrer, 1934) Nickle, 1970) occupy positions 3 to 6 in this list, following sedentary root-knot (*Meloidogyne* spp.) and cyst (*Heterodera and Globodera* spp.) nematode species [23]. With the exception of *B. xylophilus*, in clade 10, both migratory and sedentary endoparasitic nematodes are clustered together in clade 12, indicating the possibility that the evolution of *B. xylophilus* into a plant parasite is a recent and convergent one [23]. *B. xylophilus* is also an exception when compared to the other migratory nematodes in the top ten list, in that, it is vectored by insects—specifically adult *Monochamus* (Cerambycidae) beetles (principal vectors), ovipositing on host pine trees [24]. *B. xylophilus* is a facultative plant-parasite, as it feeds on fungi (fungal mycelial mats) as well as xylem parenchyma of live trees. This feature is unique to *Bursaphelenchus* when compared to species of *Pratylenchus, Ditylenchus*, and *Radopholus* and most plant-parasitic nematodes, as they are obligate plant-parasites. However, in yet another exception, a newly discovered species of *Bursaphelenchus* named *B. sycophilus*, does not grow on fungal mycelial mats of *Botrytis cinerea* Pers or possess morphological characteristics like the fungal feeding species in this genus, making this nematode an obligate plant-parasite [25].

In the migratory endoparasitic nematodes, the genus *Pratylenchus*, which comprises of ~70 species, ranks third after sedentary root-knot and cyst nematodes, as an economically devastating pest of numerous agriculturally important crops and fruits [26]. An important feature that aids this pathogen in colonizing such a broad host range is the ability for different species to thrive in tropical and temperate climactic conditions. On the contrary, the other three migratory endoparasitic nematode genera, namely *R. similis* (distributed in tropical and temperate greenhouse conditions), *D. dipsaci* (worldwide specifically temperate regions), and *B. xylophilus* (the northern hemisphere, which is home to its vector, *Monochamus* beetles) have a comparatively narrower geographic distribution [27]. Different survival strategies are employed by these nematodes to survive or overcome harsh climactic conditions. For instance, *D. dipsaci* has been shown to survive for more than 20 years by entering into long-term anhydrobiosis [28]. Another migratory PPN, *B. xylophilus*, has been shown to overwinter in both living and dead tissues of coniferous trees, allowing it to endure long, harsh winters [29]. However, the molecular mechanisms underpinning cold tolerance in PPNs are multipartite and poorly understood at present [30]. Homologs of dauer genes have also been found in the genomic and transcriptome analysis of the burrowing nematode *R. similis*; however, since *R. similis* has not been shown to form dauers, the roles of these genes remain unclear [14,19]. There exist six distinct stages: eggs, four juvenile stages (J1–J4), and adults (female and rarely male) in the typical lifecycle of a PPN. The first molt, J1, occurs in the egg. Following this, hatching of the infective juvenile stage, J2, takes place. In most migratory endoparasitic nematodes, all stages (J2–J4 and adult females) possess the ability to infect host cells. Reproductive strategies employed by the different migratory endoparasitic nematode species described above are also oddly similar. Most of the species are dioecious; however, in the absence of males, alternate strategies are pursued, such as hermaphroditism in case of *R. similis* and parthenogenesis in case of certain lesion nematodes [23].

3. Feeding Strategies Employed by Migratory Endoparasitic Nematodes

Sedentary endoparasitic PPNs form multinucleate hypertrophied, permanent feeding sites, referred to as syncytium for cyst nematodes and giant cells for root-knot nematodes, which serve as a source of nutrition for the nematode. In contrast, for the migratory endoparasitic nematodes, the feeding mechanisms are fairly straightforward and comparatively less complex. Of these, the penetration and feeding mechanisms of the root-lesion nematode *P. penetrans* (Cobb, 1917) Filipjev and Schuurmans-Stekhoven, 1941, have been extensively studied [31–35]. *Pratylenchus penetrans* has been shown to feed ectoparasitically and endoparasitically [36]. Its endoparasitic feeding behavior can be classified into three distinct stages, namely root surface probing, root penetration, and infection. In the first stage, nematodes migrate primarily to the root hair region and sometimes closer the root tip, around the zone of elongation [33]. Once there, they ectoparasitically probe local epidermal cells and initiate stylet thrusting at an intensity probably proportional to the cell wall structure and thickness [35]. In the second stage, penetration, the regions next to the root intercellular walls are punctured with the stylet accompanied by a slight pressure from the labial region to allow the nematode to enter and create a gateway for subsequent nematodes to enter the root. Intense stylet thrusting continues during this stage as well. Following penetration and entry, the third stage, infection, is initiated. This stage is divided into two sub-stages, brief and extensive feeding. At the beginning of brief feeding, a small salivation zone is present surrounding the stylet tip when inserted into a host cell. The period of brief feeding differs between different nematode developmental stages with juvenile stages consuming less time (approximately 5 minutes) and the adult stages consuming more (approximately 10 minutes). Cortical cell response to the nematode's brief feeding period includes cytoplasmic streaming and rarely cell death, but nematode migration following these brief feeding periods induces cell death along the migration path. Additionally, preferential feeding behavior is also seen, with J2 and J3 life stages feeding on the root hair and the higher developmental stages (J4 and adults) feeding on the cortical cells [37]. During extended feeding periods, a relatively prominent salivation zone is formed, following which several cell-wall modifying enzymes (CWME) and effectors are secreted into the host cell prior to ingestions.

Over many years, numerous CWMEs, such as cellulases, pectate lyases, xylanases and arabinases, have been discovered in migratory and sedentary endoparasitic phytonematodes [38–43]. Of these, cellulases have been of particular interest to understand the underlying mechanism of host-parasitism and the role of horizontal gene transfer in the origin of parasitism in phytonematodes. Cellulases breakdown cellulose, the major structural component of plant cell walls. Cellulases of root-knot and cyst nematodes have been extensively studied since they are involved in the preliminary interaction with host tissues. The cellulases secreted by sedentary endoparasitic nematodes during the mechanical root puncturing and migration activity have been shown to be homologous to bacterial cellulases, suggesting a possible ancient horizontal gene transfer (HGT) event between bacteria and nematodes with plant parasitism. Molecular characterization studies have confirmed the presence of four cellulases belonging to the GHF5 family (mainly bacterial cellulases) in the burrowing nematode, *R. similis* [39]. Of these, three showed relatively lower expression in males, most likely since males are non-feeding. In another migratory phytonematode, *B. xylophilus*, three cellulases belonging to the GHF45 family have been discovered [41]. GHF45 cellulases are secreted by many microbes including bacteria and fungi as well as some animals; however, the *B. xylophilus* cellulases have been shown to be of fungal origin as opposed to bacterial origin (and *B. xylophilus* feeds upon and lives in close association with fungi). In-situ hybridization studies have confirmed the expression of nematode cellulases specifically within the esophageal gland secretory cells of these parasites, corroborating the hypothesis that these proteins are secreted during the initial phases of parasitism (penetration and migration). However, parasitism proteins involved in disparate stages of host interaction can also be released by other tissues such as the amphids and hypodermis [10]. Cellulases have also been characterized in the molecular and transcriptomic studies of the lesion nematode *P. penetrans*, *P.*

thornei (Sher and Allen), *P. vulnus* (Allen and Jensen, 1951), and *P. zeae* Graham, 1951 [15,16,37,44,45]. The *P. penetrans* GHF5 cellulases have shown similarity to those in cyst and root-knot nematodes, enforcing a previous finding that some of the early members of the Pratylenchidae family could be gene donors to the root-knot and cyst nematodes [16,43]. Additionally, due to their expression in all migratory endoparasitic nematode life stages, cellulases, specifically β-1,4-endoglucanases, have been utilized to design diagnostic PCR markers for classification of different *Pratylenchus* species from soil and root samples [46].

In addition to cellulases, another secretory protein group extensively studied in PPNs are pectate lyases. Pectate lyases are involved in the breakdown of pectin, an essential component of the plant cell-wall involved in supporting the cellulose and hemicellulose fiber molecules within and between plant cell-walls [47]. Pectate lyases, specifically PL3, have been cloned and reported in numerous sedentary endoparasitic nematodes such as *Meloidogyne, Globodera*, and *Heterodera* [48–52]. In migratory phytonematodes, PL3 has been discovered in the genomes and transcriptomes of *R. similis* [14], *P. coffeae* (Campos and Villain, 2005), *P. penetrans* [16,17], *D. destructor* Thorne, 1945 [53], and *B. xylophilus* [40]. These PL3s are involved in softening the plant cell wall, thereby allowing the nematode to migrate through the intercellular spaces. In sedentary endoparasitic nematodes, close homologs of these pectate lyases secreted from the subventral esophageal glands have been shown to be of a bacterial and fungal origin [54]. Furthermore, RNAi knockdown studies of these pectate lyases in the sedentary endoparasitic phytonematode, *H. schachtii* Schmidt, 1871, culminated in fewer nematodes per root tissue sample of *Arabidopsis thaliana* (L.) Heynhold, 1976 plants, implicating the need for these enzymes in parasitism [55]. CWMEs such as arabinases, xylanases, and polygalactouranases have also been found during genomic and transcriptomic analyses of multiple parasitic phytonematodes [2,38,47,56]. However, extensive molecular characterization studies elucidating the interplay between different CWMEs during parasitism in migratory PPNs have not been performed, thereby leaving a significant gap in the knowledge necessary to understand the processes taking place during the course of infection by a migratory PPN.

4. Parasitism Gene Repertoire and Effectors of Migratory Endoparasitic Nematodes

Low-cost and resource-efficient NGS technologies have opened gateways for researchers to analyze the arsenal of putative effectors secreted by PPNs. The broad definition of effectors proposed by Hogenhout et al., "all pathogen proteins and small molecules that alter host-cell structure and function", encompasses the spectrum of effectors secreted by PPNs [57]. PPN effectors are generally expressed and secreted by esophageal gland cells, specifically one dorsal and two subventral esophageal gland cells, through the stylet. The role of these esophageal gland cells in sedentary endoparasitic PPN is generally regulated; the subventral gland cells are metabolically active during the pre-parasitic and early infective stages, and the dorsal gland cells are active during a more permanent association with the host tissues [10,37]. An overview of some of the key parasitism genes discovered in economically important migratory PPNs since 2013 is provided in Table 1.

Table 1. List of some of the significant parasitism genes in the four key migratory plant-parasitic nematode (PPN) genera published since 2013.

Gene Name/Cluster ID	Species	Function/Annotation	Reference
		Pratylenchus spp.	
Ppen12895_c0_seq1 (FAR)	*P. penetrans*	Fatty-acid metabolism	[45]
Ppen12103_c0_seq1 (SXP-RAL2)	*P. penetrans*	Function in parasitism: unclear	[45]
Vap-1	*P. zeae*	Host defense suppression	[15]
Sec-2	*P. zeae*	Overcoming host defense	[15]
		Radopholus similis	
Rs-scp-1	*R. similis*	Development, invasion, and pathogenesis in some PPN	[58]
Rs-cps	*R. similis*	Embryonic development invasion and pathogenesis	[59]
Rs-cb-1	*R. similis*	Reproduction and invasion	[60]
Rs-crt	*R. similis*	Reproduction and pathogenicity	[61]
Rs-far-1	*R. similis*	Development, reproduction, infection, and disruption of plant defense	[62]
		Ditylenchus spp.	
DD03093(VAP-1)	*D. destructor*	Host defense suppression	[63]
DDC03397(VAP-2)	*D. destructor*	Host defense suppression	[63]
DD03835 (Sec-2)	*D. destructor*	Overcome host defense	[63]
		Bursaphelenchus spp.	
BxSapB1	*B. xylophilus*	Contributes to virulence and cell death	[64]
1-3-endoglucanase	*B. xylophilus*	Cell-wall degrading enzymes	[65]
Expansin-like protein	*B. xylophilus*	Cell-wall degrading enzymes	[65]
Peroxiredoxin	*B. xylophilus*	Detoxifying enzyme	[65]
Cytochrome-P450	*B. xylophilus*	Detoxifying enzyme	[66]
Glutathione-S-transferase	*B. xylophilus*	Detoxifying enzyme	[66]

4.1 Pratylenchus spp.

In *P. penetrans*, numerous effectors have been identified and their putative role in parasitism has been deciphered [16,45]. Some of the significant effectors discovered in the *P. penetrans* transcriptome include catalases (with N-terminal signal peptide) and glutathione peroxidase that play role in shielding the nematode against host induced reactive oxygen species (ROS) molecules. Moreover, notable effectors identified in the transcriptome dataset include the venom allergen-like proteins (VAPs) that have also been identified in *B. xylophilus* and have been hypothesized to be involved in movement within the host plant [67]. Furthermore, effectors such as transthyretin-like proteins (TTLs) and fatty-acid and retinol-binding proteins (FARs) were also identified in the *P. penetrans* transcriptome dataset [45]. TTLs have been implicated to play an important role in the nervous system of *R. similis* and FARs from the cyst, and root-knot nematodes have been shown to bind precursor molecules involved in the defense related jasmonic acid signaling pathway [68–70]. During the comparative studies of root-lesion nematodes, several genes coding for PPN effectors involved in initiating a feeding site in host tissues were noted to be absent [15–17].

Transcriptome analysis of another member of this genus, *P. coffeae*, revealed several proteins homologous to effectors involved in parasitism [17]. Notable amongst them are the genes coding for chorismate mutase, glutathione peroxidase, glutathione-S-transferase peroxiredoxin, TTLs, and VAPs. A noteworthy finding of this study was the identification of proteins homologous to RBP-1, a cyst nematode secretory protein with an SP1a and ryanodine receptor domain (SPRYSEC). SPRYSECs have been shown to both suppress and elicit immune response in plants and within PPNs; these proteins have been identified only in the sedentary cyst nematodes [71]. The authors identified four putative SPRYSEC proteins in *P. coffeae* with a significantly high homology to the cyst nematode SPRYSECs [17]. In addition, the authors identified other SPRY domain containing proteins in *P. coffeae* as well. However, these do not show any similarity with the cyst nematode SPRYSEC proteins. A putative gene coding for an arabinogalactan galactosidase, a protein found only in the cyst nematodes, was also identified in *P. coffeae*. Additionally, many genes involved in cell-wall modification, such as β-1,4-endoglucanase, pectate lyase, xylanase, and GH16 (β-1,3-endoglucanases), were uncovered during the transcriptome analysis of *P. coffeae* [17]. However, no GH16s were identified in the transcriptome of *P. penetrans*, indicating the evolution of a distinct effector set in *P. penetrans* as a result of host/niche specialization [16]. Transcriptome analysis of another root-lesion nematode, *P. zeae*, an important pest of high-value crops such as sugarcane (*Saccharum officinarum* L.) and sorghum (*Sorghum bicolor* (L.) Moench), also revealed a similar trend in CWME occurrence [15]. *Pratylenchus zeae* possesses a similar suite of effectors when compared with close relatives, *P. penetrans* and *P. coffeae*. Notable in the transcriptome were the genes involved in combating host-derived oxidative stresses, such as glutathione S-transferase, peroxiredoxin, thioredoxin, glutathione peroxidase, and superoxide dismutase [15]. Additionally, genes identified in the transcriptome dataset of *P. coffeae* such as the cyst nematode secreted SPRYSEC proteins were also identified in *P. zeae*. Analysis of the putative secretome of *P. zeae* revealed several sequences involved in a variety of functions such as stress response, energy metabolism, protein digestion, host defense evasion, and plant cell wall modification [15]. Notably, the authors found expression of a gene with a SPRY domain localized within the nematode esophageal gland cells, implicating it in the parasitism process. The authors also noted that several genes involved in feeding site formation induced by sedentary nematodes such as C-terminally encoded peptides (CEP), CLE-like peptides, 16D10, and 7E12 were missing from the transcriptome analysis of *P. zeae*. Another important absence is of a gene coding for a putative chorismate mutase in *P. zeae*, which is present in the transcriptomes of *P. coffeae* and *P. thornei*.

4.2 Radopholus Similis

Of the 30 described species of *Radopholus*, *R. similis* is the only major burrowing nematode species considered significant worldwide [72,73]. The transcriptome of *R. similis* has been sequenced twice [14,18]. Genes identified in *R. similis* include those involved in plant cell wall modification such as β-1,4-endoglucanase, pectate lyase, endoxylanase, arabinase, and expansins and in parasitism, genes were identified in few root-lesion nematodes such as SXP/RAL-2, FAR and chorismate mutase. Proteinase-coding genes such as serine carboxypeptidase, calreticulin, and cathepsin that have been linked to different aspects of plant parasitism were also identified in *R. similis* [58,59,61,69]. Similar to few root-lesion nematodes, homologues of sequences involved in sedentary endoparasitism such as CLE, CEP, 16D10, and 7E12 were absent in *R. similis*. Several gene sequences coding for SPRY domains were identified in the genome analysis. However, no N-terminal signal peptides were found in these proteins. Another notable finding in the *R. similis* genome, as well as transcriptome analysis, is the presence of several genes involved in the dauer pathway. A study by Chabrier et al. demonstrated after 180 days under no host conditions, *R. similis* males increased in number (21.7%) compared to females (9.8%), indicating a possibility of higher expression of dauer genes in males than females (or lower survival of females during unfavorable conditions, and hence a lower expression of dauer genes) [74].

The recent genome analysis of *R. similis* has also shed more light into the relationship of this migratory nematode with the sedentary cyst nematodes. A recent robust phylogenetic analysis of the small subunit ribosomal DNA (SSU rDNA) by Holterman et al. revealed five lineages leading to sedentary endoparasitism. Among these, the subfamilies Pratylenchinae and Hirschmanniellinae were hypothesized as ancestors of the sedentary root-knot nematodes. Additionally, a common ancestral link between the semi-endoparasitic nematode, *Rotylenchulus reniformis* Linford and Oliveira, 1940 and the endoparasitic cyst nematodes has also been suggested by rDNA analysis as well as the effector analysis [75,76]. In the same phylogenetic analysis, it can be noted that *R. similis* and the members of the genus *Heterodera*, *Globodera*, and *Rotylenchulus* share a common ancestor. However, although it is relatively phylogenetically close, *R. similis* shares no overlap of key effectors such as CLE and SPRYSEC with the sedentary cyst nematodes. Holterman et al. also demonstrated the close association between members of the *Radopholus* genus, specifically *R. bridgei* and *R. similis* with *Hoplolaimus femina*. *Hoplolaimus* is comprised of phytonematodes with a wide range of lifestyle (ecto, endo, and semi-endoparasites) [77]. However, limited information is available regarding the molecular interactions governing parasitism for this nematode (*Hoplolaimus* genus) or the similarity/differences of effector repertoire between the other two *Radopholus* species (*R. bridgei* Siddiqi and Hahn, 1995 and *R. similis*).

4.3 Ditylenchus spp.

Expressed sequence tag (EST) analyses has been used previously to identify several genes involved in parasitism in the root-knot and cyst forming nematodes [78,79]. EST analysis of the potato rot nematode, *D. destructor*, revealed homologs of several important effectors such as VAP and calreticulin, which play vital roles in host defense induction. Moreover, effectors involved in circumventing host defense in cyst nematodes such as SEC-2 proteins and numerous cell wall modifying enzymes, such as pectate lyase, cellulase, and expansin, were also identified in *D. destructor* [63,70]. Additionally, 14-3-3b a secretory protein identified in the gland cell of *M. incognita* (Kofoid and White, 1919) Chitwood, 1949 and implicated in playing an essential role in signal transduction pathways, has also been identified in the EST dataset of *D. destructor* [63,80]. Since *D. destructor* feeds on fungi as well, it is unsurprising that proteins involved in fungal cell wall degradation, such as chitinases and GH16 (1-3(4))-beta-glucanase genes, were identified in the EST dataset of this nematode.

EST analysis of the peanut pod nematode, *D. africanus* Wendt, Swart, Vrain, and Webster, 1995, identified a similar suite of expressed parasitism genes [81]. Homologs of key genes participating in anhydrobiosis, such as the late embryogenesis abundant protein (LEA), were also identified in *D. africanus*. This is on par with certain members of the genus *Ditylenchus*, such as *D. dipsaci*, which have been shown to be capable of anhydrobiosis [82]. Genes involved in providing structural integrity to the nematode cell membrane, such as fatty acid desaturase and stomatin, were also identified in the EST dataset [81].

4.4 *Bursaphelenchus* spp.

Analysis of over 13,000 and 3000 ESTs from *B. xylophilus* and *B. mucronatus* (Mamiya and Enda, 1979), respectively, revealed several genes involved in parasitism [83]. In addition to the conventional parasitism genes, genes such as chitinase, expressed in the esophageal gland cells of the cyst nematode, have also been found in *B. xylophilus* [83,84]. Presence of this and other chitin degrading enzymes, like GH16, are necessary because *Bursaphelenchus* feeds on fungi (cell wall made of chitin) and is vectored by *Monochamus* insects (insect cuticle is made of chitin). Additionally, with *D. africanus*, the authors found several dauer genes such as LEA homologs as well as 18 *Caenorhabditis elegans* (Maupas, 1899) dauer-formation (*daf*) homologs. Identification and characterization of parasitism genes has led to the discovery of a multilayered enzymatic detoxification strategy in *B. xylophilus*, with detoxification enzymes such as glutathione S-transferase being secreted as the first line of response, and other parasitism effectors such as VAP being secreted later [66]. Detoxification enzymes such as glutathione S-transferase have been shown to play significant roles towards plant parasitism in root-knot nematodes [66,85]. Recently, the effector gene, *BxSapB1*, has been identified in *B. xylophilus* [66]. *BxSapB1* possesses a functional signal peptide and has been demonstrated to contribute towards host cell death and increased virulence of *B. xylophilus* [64].

5. Management of Migratory PPNs

A variety of management strategies has been employed for different types of PPN due to their ability to parasitize a wide range of hosts. Migratory PPNs such as those belonging to the genera *Radopholus*, *Ditylenchus*, and *Pratylenchus* can be easily spread by contaminated vegetative plant parts and timber (in the case of *B. xylophilus*) [86]. Integrating multiple management strategies is essential; however, the specific strategy should depend upon accurate diagnosis of the PPN species, growing conditions, available resources, and economic feasibility, which vary among cropping systems and in developing vs. developed countries.

5.1 Cultural Practices

Cultural practices such as crop rotation with a non-host crop or instituting a period of fallow (including elimination of weeds), can reduce nematode populations. Crop rotation is less effective as a control strategy, however, due to the polyphagous nature of some PPNs. For instance, rotation with crops such as cassava (*Manihot esculenta*) and sweetpotato (*Ipomoea batatas*) has been useful in bringing down *R. similis* populations [87], but sweet potato is a susceptible host for other types of PPN such as *M. enterolobii* Yang and Eisenback, 1981 and *Rotylenchulus reniformis*. However, in the case of *P. zeae*, effective control has been achieved by rotating rice with leguminous beans such as black gram (*Vigna Mungo* (L.) Hepper) and mung bean (*Vigna radiata* (L.) R. Wilczek) [88]. Control of another lesion nematode, *P. thornei* in wheat (*Triticum aestivum*), was also achieved in Mexico by rotating with crops such as cotton (*Gossypium* spp.), corn (*Zea mays*), and soybean (*Glycine max*) for two successive years [89]. The use of certified nematode-free starting material such as seed potato (*Solanum tuberosum*) or banana (*Musa* spp.) corm has been helpful in reducing nematode populations of *D. dipsaci* and *R. similis*, respectively [90–92]. Use of clean starting material has also been effective in reducing populations of the yam nematode *S. bradys*, the causative agent of dry rot disease in yams. Utilizing hot water treatment for disinfecting starting plant material has been useful for managing populations of *D. dipsaci*, *P. vulnus*, and *R. similis* [90–92].

Cover crops are an indispensable unit in an agricultural ecosystem since they provides a host of benefits to the soil such as increased nutrients, reduced pest populations, as well as an increase in beneficial soil microbes. Use of marigold (*Tagetes* spp.), which has been shown to produce thiophene α terthienyl (a nematicidal compound), as a cover crop has provided effective control against *P. penetrans* [93–96]. Sunn hemp (*Crotalaria juncea* L.), a leguminous cover crop, has also shown significant potential in reducing populations of several sedentary nematodes of the *Meloidogyne* and *Heterodera* genera as well as migratory nematodes such as the sting nematode (*Belonolaimus longicaudatus* Rau, 1958) and stubby root nematodes (*Paratrichodorus* and *Trichodorus* spp.) [97,98]. Moreover, recent studies with sunn hemp, pigeon pea (*Cajanus cajan*), and *Gliricidia sepium* have delivered promising results in managing the endoparasitic yam nematode, *S. bradys* [99]. However, a recent trial with sunn hemp and pigeon pea showed no nematode suppression activity on the migratory nematode *R. similis* [100]. In addition, cover crops such as mustard (*Brassica* and *Sinapis* spp.) contain a wide variety of PPN antagonistic compounds such as isothiocyanates and degradation products of glucosinolates, which have provided moderate nematode suppression effects [101]. For a detailed review of phytochemicals for nematode control, the authors recommend Chitwood [102].

5.2 Resistance as a Tool for Nematode Control

Due to the growing environmental and human health concerns caused by the use of toxic nematicides and the associated regulations surrounding their use, incorporating natural resistance (R) genes into desirable crop cultivars has been a successful strategy to provide crop protection against PPNs [103]. The first cloned naturally occurring R gene that provided resistance against a PPN was the $Hs1^{pro1}$ gene from sugar beet (*Beta vulgaris*) [104,105]. This gene confers resistance to the sedentary cyst-forming nematode, *H. schachtii* [104]. Since then, many nematode R genes have been cloned and identified including *Mi-1*, *Hero A*, *Gpa2*, *Gro1-4*, *Rhg1*, and *Rhg4*; however, these genes confer resistance only to sedentary PPNs [106–112].

With regards to migratory nematodes, Atkinson et al. (2004) [113] demonstrated the efficacy of the first transgenic bananas expressing rice cystatins which provided significant resistance (around 70% ± 10%) against *R. similis*. Transgenic *Nicotiana benthamiana* plants expressing the cysteine proteinase cathepsin S also demonstrated enhanced resistance to

R. similis [114]. Musa varieties such as the Yangambi Km5 and Pisang Jari Buaya also provide some degree of resistance against *R. similis*, primarily through mechanisms involving phenol accumulation and lignification, respectively, at the nematode infection site [115,116]. Furthermore, evidence of a mutualistic Fusarium endophyte (*Fusarium oxysporum* isolate A1 and *Fusarium* cf. *diversisporum*) inducing systemic resistance against *R. similis* in bananas has also been demonstrated [117]. Another instance of resistance against a migratory PPN is the bread wheat line Gatcher selection 50a, which provides partial resistance against the root lesion nematode *P. thornei* [118]. Partial resistance to *P. thornei* has also been identified in other varieties of wheat such as "CPI133872", grown in multiple regions in Australia, where *P. thornei* is an extremely damaging pathogen [119–121]. Although, several trials on potato varieties have been conducted to identify sources of resistance or tolerance ('tolerance" as defined by Mwaura et al.) against nematodes in the *Ditylenchus* genus, few varieties are commercially available [122]. In a recent study based on the relative susceptibility score, the potato variety "Spunta" was classified as resistant against *D. destructor* and *D. dipsaci* under greenhouse conditions [122]. However, these results have not been confirmed under field conditions.

5.3 Nematicides

Due to the deleterious effects of certain nematicides/pesticides on the environment and non-target organisms, severe regulatory restrictions including the ban of important nematicides have been established. One such broad spectrum pesticide, methyl bromide, which is also a popular fumigant nematicide, was phased out of agricultural use in the United States in 2005 due to its atmospheric ozone depletion properties [123–125]. Currently there exists a huge gap in the number of effective and economical nematicide products that are available to growers. Some of the important fumigant nematicides currently approved for use on high-value crops and ornamentals in the United States include metam sodium (Vapam®), metam potassium (Metam CLR™), chloropicrin (Metapicrin), and 1,3-dichloropropene (Telone®) [125]. An alternative to methyl bromide with regards to the lowest per-unit cost is the registered fumigant chloropicrin, which has been used in combination with the fumigant nematicide 1,3-D [126]. This combination treatment has provided effective control against soilborne pathogens in a broad variety of crops such as almonds (*Prunus dulcis,* syn. *Prunus amygdalus)*, sweetpotatoes, strawberries (*Fragaria anannasa*), grapes (*Vitis vinifera*), and carrots (*Daucus carota* subsp. *sativus*), but limited control of weeds and high input costs concern many growers. Additionally, application of 1,3-D has been restricted within California townships, with a ban of 1,3-D in the month of December due to air quality concerns [127,128]. Another example is the non-fumigant nematicide, fenamiphos (Nemacur®). Fenamiphos has been utilized by *Anthurium* growers in Hawaii as a post-plant application for managing *R. similis* and other PPNs [129]. However, with a phase-out of this chemical in the U.S., growers are now integrating cleaner management practices such as using micropropogated *Anthurium* plantlets and disinfested starting materials [130]. An alternative microbial nematicide, DiTera® (nematicide synthesized from the fungus *Myrothecium verrucaria*) has also been recommended as an effective low-risk nematicide for reducing *R. similis* populations in *Anthurium* [130]. In a recent study by Zouhar et al., treatment of garlic (*Allium sativum)* cloves with the fumigant hydrogen cyanide at a concentration of 20g/m^3 caused significant increase in *D. dipsaci* mortality [131]. Post-plant application of the systemic nonfumigant nematicide oxamyl (Vydate®) has provided effective control against the root-lesion nematode, *P. penetrans*, in raspberries (*Rubus idaeus)* during field trials in Washington [132]. Another nonfumigant nematicide that has been registered in several countries for crop protection against multiple nematodes, predominantly the root-lesion, potato rot, and pine wood nematode, is the contact nematicide fosthiazate (Nemathorin®) [133]. Migratory PPNs such as root-lesion nematodes generally occur as a complex with other plant-pathogens such as *Fusarium* spp. [134]. Trials with a combination of abamectin–fungicide coated seeds reduced root infection by *P. penetrans* in maize [135]. Seed treatments are also an important form of control for the stem and bulb nematode *D. dipsaci* [131]. Novel seed treatment strategies such as electrospinning of agrichemicals has also recently shown promise to manage plant pathogens under laboratory conditions [136]. Essential oils and volatiles derived from several families of Portuguese aromatic flora have also shown effective nematicidal properties against the pine-wilt nematode *B. xylophilus* [137]. Although, several significant nematicides have been phased out and severe regulations have been imposed on some of the remaining ones, a management strategy combining multiple control practices should be practiced for effective nematode control.

6. Conclusions

Migratory PPNs are distributed across multiple clades in the phylum Nematoda. Although there exist similarities in the biology and life cycle of several migratory PPNs, huge contrasts and divergences are seen with respect to the anatomical adjustments made to survive in the absence of a viable host. Parallel evolution of feeding enzymes that allow successful feeding in the host cells and on other microbes such as fungi is a trait that has undergone divergence several times in the course of evolution of migratory PPNs in the Nematoda lineage. By coupling spatio-temporal NGS approaches with molecular/functional characterization studies, insights into some of the key effectors and their targets can be gained, which shed light on the multifaceted interaction of a PPN with its host. Migratory PPNs, like their sedentary counterparts, have been distributed across the globe, probably as a result of increased international trade. A multipronged control strategy that depends less on chemical means and integrates factors such as the geographical location, nematode species and host range, plant resistance, as well as sound agricultural practices should be considered during the decision-making process for managing any plant-parasitic nematode.

References

1. Decraemer, W.; Hunt, D. Plant nematology. In *Structure and Classification*; Perry, R.N., Moens, M., Eds.; CABI: Cambridge, MA, USA, 2006; pp. 3–32.

2. Abad, P.; Gouzy, J.; Aury, J.-M.; Castagnone-Sereno, P.; Danchin, E.G.; Deleury, E.; Perfus-Barbeoch, L.; Anthouard, V.; Artiguenave, F.; Blok, V.C. Genome sequence of the metazoan plant-parasitic nematode Meloidogyne incognita. *Nat. Biotechnol.* **2008**, *26*, 909. [CrossRef] [PubMed]

3. Nicol, J.M.; Turner, S.J.; Coyne, D.L.; Den Nijs, L.; Hockland, S.; Maafi, Z.T.; Jones, J.T.; Gheysen, G.; Fenoll, C. Genomics and molecular genetics of plant-nematode interactions. *Curr. Nematode Threats Word Agric. Cap* **2011**, *2*, 21–26.

4. Hass, H.; Kerp, H.; Poinar, G. *Palaeonema phyticum gen. n.*, sp. n.(Nematoda: Palaeonematidae fam. n.), a Devonian nematode associated with early land plants. *Nematology* **2008**, *10*, 9–14. [CrossRef]

5. Smant, G.; Helder, J.; Goverse, A. Parallel adaptations and common host cell responses enabling feeding of obligate and facultative plant parasitic nematodes. *Plant J.* **2018**, *93*, 686–702. [CrossRef]

6. Holterman, M.; van der Wurff, A.; van den Elsen, S.; van Megen, H.; Bongers, T.; Holovachov, O.; Bakker, J.; Helder, J. Phylum-wide analysis of SSU rDNA reveals deep phylogenetic relationships among nematodes and accelerated evolution toward crown clades. *Mol. Biol. Evol.* **2006**, *23*, 1792–1800. [CrossRef]

7. Giblin-Davis, R.M.; Davies, K.A.; Morris, K.; Thomas, W.K. Evolution of parasitism in insect-transmitted plant nematodes. *J. Nematol.* **2003**, *35*, 133.

8. Kikuchi, T.; Eves-van den Akker, S.; Jones, J.T. Genome evolution of plant-parasitic nematodes. *Annu. Rev. Phytopathol.* **2017**, *55*, 333–354. [CrossRef]

9. Bird, D.M.; Jones, J.T.; Opperman, C.H.; Kikuchi, T.; Danchin, E.G. Signatures of adaptation to plant parasitism in nematode genomes. *Parasitology* **2015**, *142*, S71–S84. [CrossRef]

10. Mitchum, M.G.; Hussey, R.S.; Baum, T.J.; Wang, X.; Elling, A.A.; Wubben, M.; Davis, E.L. Nematode effector proteins: An emerging paradigm of parasitism. *New Phytol.* **2013**, *199*, 879–894. [CrossRef]

11. Coyne, D.L.; Cortada, L.; Dalzell, J.J.; Claudius-Cole, A.O.; Haukeland, S.; Luambano, N.; Talwana, H. Plant-Parasitic Nematodes and Food Security in Sub-Saharan Africa. *Annu. Rev. Phytopathol.* **2018**, *56*, 381–403. [CrossRef]

12. FAOSTAT. Available online: http://www.fao.org/faostat/en/#rankings/countries_by_commodity (accessed on 20 September 2019).

13. Yusuf, A.; Claudius-Cole, A.O. Evaluation of Culturing Methods for the Yam Nematode Scutellonema bradys. *J. Exp. Agric. Int.* **2018**, 1–8. [CrossRef]

14. Huang, X.; Xu, C.-L.; Yang, S.-H.; Li, J.-Y.; Wang, H.-L.; Zhang, Z.-X.; Chen, C.; Xie, H. Life-stage specific transcriptomes of a migratory endoparasitic plant nematode, *Radopholus similis* elucidate a different parasitic and life strategy of plant parasitic nematodes. *Sci. Rep.* **2019**, *9*, 6277. [CrossRef] [PubMed]

15. Fosu-Nyarko, J.; Tan, J.-A.C.; Gill, R.; Agrez, V.G.; Rao, U.; Jones, M.G. D e novo analysis of the transcriptome of *Pratylenchus zeae* to identify transcripts for proteins required for structural integrity, sensation, locomotion and parasitism. *Mol. Plant Pathol.* **2016**, *17*, 532–552. [CrossRef] [PubMed]

16. Vieira, P.; Eves-Van Den Akker, S.; Verma, R.; Wantoch, S.; Eisenback, J.D.; Kamo, K. The *Pratylenchus penetrans* transcriptome as a source for the development of alternative control strategies: Mining for putative genes involved in parasitism and evaluation of in planta RNAi. *PLoS ONE* **2015**, *10*, e0144674. [CrossRef] [PubMed]

17. Haegeman, A.; Joseph, S.; Gheysen, G. Analysis of the transcriptome of the root lesion nematode *Pratylenchus coffeae* generated by 454 sequencing technology. *Mol. Biochem. Parasitol.* **2011**, *178*, 7–14. [CrossRef] [PubMed]

18. Jacob, J.; Mitreva, M.; Vanholme, B.; Gheysen, G. Exploring the transcriptome of the burrowing nematode Radopholus similis. *Mol. Genet. Genom.* **2008**, *280*, 1–17. [CrossRef] [PubMed]

19. Mathew, R.; Opperman, C.H. The genome of the migratory nematode, *Radopholus similis*, reveals signatures of close association to the sedentary cyst nematodes. *PLoS ONE* **2019**, *14*, e0224391. [CrossRef] [PubMed]

20. Mathew, R.; Burke, M.; Opperman, C.H. A Draft Genome Sequence of the Burrowing Nematode *Radopholus similis. J. Nematol.* **2019**, *51*, e2019-51. [CrossRef]

21. Juvale, P.S.; Baum, T.J. "Cyst-ained" research into Heterodera parasitism. *PLoS Pathog.* **2018**, *14*, e1006791. [CrossRef] [PubMed]

22. Vieira, P.; Gleason, C. Plant-parasitic nematode effectors—Insights into their diversity and new tools for their identification. *Curr. Opin. Plant Biol.* **2019**, *50*, 37–43. [CrossRef]

23. Jones, J.T.; Haegeman, A.; Danchin, E.G.; Gaur, H.S.; Helder, J.; Jones, M.G.; Kikuchi, T.; Manzanilla-López, R.; Palomares-Rius, J.E.; Wesemael, W.M. Top 10 plant-parasitic nematodes in molecular plant pathology. *Mol. Plant Pathol.* **2013**, *14*, 946–961. [CrossRef] [PubMed]

24. Kikuchi, T.; Cotton, J.A.; Dalzell, J.J.; Hasegawa, K.; Kanzaki, N.; McVeigh, P.; Takanashi, T.; Tsai, I.J.; Assefa, S.A.; Cock, P.J. Genomic insights into the origin of parasitism in the emerging plant pathogen *Bursaphelenchus xylophilus. PLoS Pathog.* **2011**, *7*, e1002219. [CrossRef] [PubMed]

25. Kanzaki, N.; Tanaka, R.; Giblin-Davis, R.M.; Davies, K.A. New plant-parasitic nematode from the mostly mycophagous genus Bursaphelenchus discovered inside figs in Japan. *PLoS ONE* **2014**, *9*, e99241. [CrossRef]

26. Castillo, P.; Vovlas, N. *Pratylenchus (Nematoda: Pratylenchidae): Diagnosis, Biology, Pathogenicity and Management*; Brill: Leiden-Boston, MA, USA, 2007; ISBN 90-474-2401-8.

27. CABI. In: EPPO Bulletin. 2019. Available online: www.cabi.org/isc (accessed on 1 November 2019).

28. Perry, R.N.; Wharton, D.A. *Molecular and Physiological Basis of Nematode Survival*; CABI: Cambridge, MA, USA, 2011; ISBN 1-84593-711-2.

29. Braasch, H. Bursaphelenchus species in conifers in Europe: Distribution and morphological relationships. *EPPO Bull.* **2001**, *31*, 127–142. [CrossRef]

30. Liu, Z.; Li, Y.; Pan, L.; Meng, F.; Zhang, X. Cold adaptive potential of pine wood nematodes overwintering in plant hosts. *Biol. Open* **2019**, *8*, bio041616. [CrossRef]

31. Han, Z.; Boas, S.; Schroeder, N.E. Serotonin regulates the feeding and reproductive behaviors of *Pratylenchus penetrans. Phytopathology* **2017**, *107*, 872–877. [CrossRef]

32. Karakas,, M. Penetration and Feeding Behavior of *Pratylenchus penetrans* (Nematoda: Pratylenchidae) In Red Radish Roots. *Kafkas Üniversitesi Fen Bilimleri Enstitüsü Dergisi* **2009**, *2*, 31–36.

33. Kurppa, S.; Vrain, T.C. Penetration and feeding behavior of *Pratylenchus penetrans* in strawberry roots. *Rev. Nématologie* **1985**, *8*, 273–276.

34. Rebois, R.V.; Huettel, R.N. Population dynamics, root penetration, and feeding behavior of *Pratylenchus agilis* in monoxenic root cultures of corn, tomato, and soybean. *J. Nematol.* **1986**, *18*, 392. [PubMed]

35. Zunke, U. Observations on the invasion and endoparasitic behavior of the root lesion nematode Pratylenchus penetrans. *J. Nematol.* **1990**, *22*, 309. [PubMed]

36. Zunke, U. Ectoparasitic feeding behaviour of the root lesion nematode, *Pratylenchus penetrans*, on root hairs of different host plants. *Rev. Nématologie* **1990**, *13*, 331–337.

37. Fosu-Nyarko, J.; Jones, M.G. Advances in understanding the molecular mechanisms of root lesion nematode host interactions. *Annu. Rev. Phytopathol.* **2016**, *54*, 253–278. [CrossRef] [PubMed]

38. Haegeman, A.; Vanholme, B.; Gheysen, G. Characterization of a putative endoxylanase in the migratory plant-parasitic nematode *Radopholus similis*. *Mol. Plant Pathol.* **2009**, *10*, 389–401. [CrossRef] [PubMed]

39. Haegeman, A.; Jacob, J.; Vanholme, B.; Kyndt, T.; Gheysen, G. A family of GHF5 endo-1, 4-beta-glucanases in the migratory plant-parasitic nematode *Radopholus similis*. *Plant Pathol.* **2008**, *57*, 581–590. [CrossRef]

40. Kikuchi, T.; Shibuya, H.; Aikawa, T.; Jones, J.T. Cloning and characterization of pectate lyases expressed in the esophageal gland of the pine wood nematode *Bursaphelenchus xylophilus*. *Mol. Plant-Microbe Interact.* **2006**, *19*, 280–287. [CrossRef]

41. Kikuchi, T.; Jones, J.T.; Aikawa, T.; Kosaka, H.; Ogura, N. A family of glycosyl hydrolase family 45 cellulases from the pine wood nematode Bursaphelenchus xylophilus. *FEBS Lett.* **2004**, *572*, 201–205. [CrossRef]

42. Momota, Y.; Uehara, T.; Kushida, A. PCR-based cloning of two Ξ-1, 4-endoglucanases from the root-lesion nematode *Pratylenchus penetrans*. *Nematology* **2001**, *3*, 335–341. [CrossRef]

43. Rybarczyk-Mydłowska, K.; Maboreke, H.R.; van Megen, H.; van den Elsen, S.; Mooyman, P.; Smant, G.; Bakker, J.; Helder, J. Rather than by direct acquisition via lateral gene transfer, GHF5 cellulases were passed on from early Pratylenchidae to root-knot and cyst nematodes. *BMC Evol. Biol.* **2012**, *12*, 221. [CrossRef]

44. Fanelli, E.; Troccoli, A.; Picardi, E.; Pousis, C.; De Luca, F. Molecular characterization and functional analysis of four β-1, 4-endoglucanases from the root-lesion nematode *Pratylenchus vulnus*. *Plant Pathol.* **2014**, *63*, 1436–1445. [CrossRef]

45. Vieira, P.; Maier, T.R.; Eves-van den Akker, S.; Howe, D.K.; Zasada, I.; Baum, T.J.; Eisenback, J.D.; Kamo, K. Identification of candidate effector genes of *Pratylenchus penetrans*. *Mol. Plant Pathol.* **2018**, *19*, 1887–1907. [CrossRef]

46. Peetz, A.B.; Zasada, I.A. Species-specific diagnostics using a β-1, 4-endoglucanase gene for *Pratylenchus* spp. occurring in the Pacific Northwest of North America. *Nematology* **2016**, *18*, 1219–1229. [CrossRef]

47. Davis, E.L.; Haegeman, A.; Kikuchi, T. Degradation of the Plant Cell Wall by Nematodes. In *Genomics and Molecular Genetics of Plant-Nematode Interactions*; Jones, J., Gheysen, G., Fenoll, C., Eds.; Springer: Dordrecht, The Netherlands, 2011; pp. 255–272. ISBN 978-94-007-0434-3. [CrossRef]

48. De Boer, J.M.; Davis, E.L.; Hussey, R.S.; Popeijus, H.; Smant, G.; Baum, T.J. Cloning of a putative pectate lyase gene expressed in the subventral esophageal glands of Heterodera glycines. *J. Nematol.* **2002**, *34*, 9. [PubMed]

49. Doyle, E.A.; Lambert, K.N. Cloning and characterization of an esophageal-gland-specific pectate lyase from the root-knot nematode Meloidogyne javanica. *Mol. Plant-Microbe Interact.* **2002**, *15*, 549–556. [CrossRef] [PubMed]

50. Huang, G.; Dong, R.; Allen, R.; Davis, E.L.; Baum, T.J.; Hussey, R.S. Developmental expression and molecular analysis of two Meloidogyne incognita pectate lyase genes. *Int. J. Parasitol.* **2005**, *35*, 685–692. [CrossRef] [PubMed]

51. Kudla, U.; MILAC, A.-L.; Qin, L.; Overmars, H.; Roze, E.; Holterman, M.; PETRESCU, A.-J.; Goverse, A.; Bakker, J.; Helder, J. Structural and functional characterization of a novel, host penetration-related pectate lyase from the potato cyst nematode Globodera rostochiensis. *Mol. Plant Pathol.* **2007**, *8*, 293–305. [CrossRef] [PubMed]

52. Popeijus, H.; Overmars, H.; Jones, J.; Blok, V.; Goverse, A.; Helder, J.; Schots, A.; Bakker, J.; Smant, G. Enzymology: Degradation of plant cell walls by a nematode. *Nature* **2000**, *406*, 36. [CrossRef]

53. Zheng, J.; Peng, D.; Chen, L.; Liu, H.; Chen, F.; Xu, M.; Ju, S.; Ruan, L.; Sun, M. The Ditylenchus destructor genome provides new insights into the evolution of plant parasitic nematodes. *Proc. R. Soc. B* **2016**, *283*, 20160942. [CrossRef]

54. Williamson, V.M.; Gleason, C.A. Plant–nematode interactions. *Curr. Opin. Plant Biol.* **2003**, *6*, 327–333. [CrossRef]

55. Vanholme, B.; Van Thuyne, W.; Vanhouteghem, K.; De Meutter, J.A.N.; Cannoot, B.; Gheysen, G. Molecular characterization and functional importance of pectate lyase secreted by the cyst nematode *Heterodera schachtii*. *Mol. Plant Pathol.* **2007**, *8*, 267–278. [CrossRef]

56. Mitreva-Dautova, M.; Roze, E.; Overmars, H.; de Graaff, L.; Schots, A.; Helder, J.; Goverse, A.; Bakker, J.; Smant, G. A symbiont-independent endo-1, 4-β-xylanase from the plant-parasitic nematode Meloidogyne incognita. *Mol. Plant-Microbe Interact.* **2006**, *19*, 521–529. [CrossRef]

57. Hogenhout, S.A.; Van der Hoorn, R.A.; Terauchi, R.; Kamoun, S. Emerging concepts in effector biology of plant-associated organisms. *Mol. Plant-Microbe Interact.* **2009**, *22*, 115–122. [CrossRef] [PubMed]

58. Huang, X.; Xu, C.-L.; Chen, W.-Z.; Chen, C.; Xie, H. Cloning and characterization of the first serine carboxypeptidase from a plant parasitic nematode, *Radopholus similis*. *Sci. Rep.* **2017**, *7*, 4815. [CrossRef] [PubMed]

59. Wang, K.; Li, Y.; Huang, X.; Wang, D.; Xu, C.; Xie, H. The cathepsin S cysteine proteinase of the burrowing nematode Radopholus similis is essential for the reproduction and invasion. *Cell Biosci.* **2016**, *6*, 39. [CrossRef] [PubMed]

60. Li, Y.; Wang, K.; Xie, H.; Wang, D.-W.; Xu, C.-L.; Huang, X.; Wu, W.-J.; Li, D.-L. Cathepsin B cysteine proteinase is essential for the development and pathogenesis of the plant parasitic nematode *Radopholus similis. Int. J. Biol. Sci.* **2015**, *11*, 1073. [CrossRef] [PubMed]

61. Li, Y.; Wang, K.; Xie, H.; Wang, Y.-T.; Wang, D.-W.; Xu, C.-L.; Huang, X.; Wang, D.-S. A nematode calreticulin, Rs-CRT, is a key effector in reproduction and pathogenicity of *Radopholus similis. PLoS ONE* **2015**, *10*, e0129351. [CrossRef]

62. Zhang, C.; Xie, H.; Cheng, X.; Wang, D.-W.; Li, Y.; Xu, C.-L.; Huang, X. Molecular identification and functional characterization of the fatty acid-and retinoid-binding protein gene Rs-far-1 in the burrowing nematode *Radopholus similis* (Tylenchida: Pratylenchidae). *PLoS ONE* **2015**, *10*, e0118414. [CrossRef]

63. Peng, H.; Gao, B.; Kong, L.; Yu, Q.; Huang, W.; He, X.; Long, H.; Peng, D. Exploring the host parasitism of the migratory plant-parasitic nematode Ditylenchus destuctor by expressed sequence tags analysis. *PLoS ONE* **2013**, *8*, e69579. [CrossRef]

64. Hu, L.-J.; Wu, X.-Q.; Li, H.-Y.; Zhao, Q.; Wang, Y.-C.; Ye, J.-R. An Effector, BxSapB1, Induces Cell Death and Contributes to Virulence in the Pine Wood Nematode *Bursaphelenchus xylophilus. Mol. Plant-Microbe Interact.* **2019**, *32*, 452–463. [CrossRef]

65. Shinya, R.; Morisaka, H.; Kikuchi, T.; Takeuchi, Y.; Ueda, M.; Futai, K. Secretome analysis of the pine wood nematode *Bursaphelenchus xylophilus* reveals the tangled roots of parasitism and its potential for molecular mimicry. *PLoS ONE* **2013**, *8*, e67377. [CrossRef]

66. Espada, M.; Silva, A.C.; Eves van den Akker, S.; Cock, P.J.; Mota, M.; Jones, J.T. Identification and characterization of parasitism genes from the pinewood nematode *Bursaphelenchus xylophilus* reveals a multilayered detoxification strategy. *Mol. Plant Pathol.* **2016**, *17*, 286–295. [CrossRef]

67. Kang, J.S.; Koh, Y.H.; Moon, Y.S.; Lee, S.H. Molecular properties of a venom allergen-like protein suggest a parasitic function in the pinewood nematode *Bursaphelenchus xylophilus. Int. J. Parasitol.* **2012**, *42*, 63–70. [CrossRef] [PubMed]

68. Iberkleid, I.; Vieira, P.; de Almeida Engler, J.; Firester, K.; Spiegel, Y.; Horowitz, S.B. Fatty acid-and retinol-binding protein, Mj-FAR-1 induces tomato host susceptibility to root-knot nematodes. *PLoS ONE* **2013**, *8*, e64586. [CrossRef] [PubMed]

69. Jacob, J.; Vanholme, B.; Haegeman, A.; Gheysen, G. Four transthyretin-like genes of the migratory plant-parasitic nematode *Radopholus similis*: Members of an extensive nematode-specific family. *Gene* **2007**, *402*, 9–19. [CrossRef] [PubMed]

70. Prior, A.; Jones, J.T.; Beauchamp, J.; McDermott, L.; Cooper, A.; Kennedy, M.W. A surface-associated retinol-and fatty acid-binding protein (Gp-FAR-1) from the potato cyst nematode Globodera pallida: Lipid binding activities, structural analysis and expression pattern. *Biochem. J.* **2001**, *356*, 387–394. [CrossRef] [PubMed]

71. Moffett, P.; Ali, S.; Magne, M.; Chen, S.; Obradovic, N.; Jamshaid, L.; Wang, X.; Bélair, G. Analysis of Globodera rostochiensis effectors reveals conserved functions of SPRYSEC proteins in suppressing and eliciting plant immune responses. *Front. Plant Sci.* **2015**, *6*, 623.

72. De Waele, D.; Elsen, A. Challenges in tropical plant nematology. *Annu. Rev. Phytopathol.* **2007**, *45*, 457–485. [CrossRef]

73. Perry, R.N.; Moens, M. *Plant Nematology*; CABI: Cambridge, MA, USA, 2006; ISBN 1-84593-057-6.

74. Chabrier, C.; Tixier, P.; Duyck, P.-F.; Cabidoche, Y.-M.; Quénéhervé, P. Survival of the burrowing nematode Radopholus similis (Cobb) Thorne without food: Why do males survive so long? *Appl. Soil Ecol.* **2010**, *45*, 85–91. [CrossRef]

75. Holterman, M.; Karegar, A.; Mooijman, P.; van Megen, H.; van den Elsen, S.; Vervoort, M.T.; Quist, C.W.; Karssen, G.; Decraemer, W.; Opperman, C.H. Disparate gain and loss of parasitic abilities among nematode lineages. *PLoS ONE* **2017**, *12*, e0185445. [CrossRef]

76. Wubben, M.J.; Ganji, S.; Callahan, F.E. Identification and molecular characterization of a β-1, 4-endoglucanase gene (Rr-eng-1) from Rotylenchulus reniformis. *J. Nematol.* **2010**, *42*, 342.

77. Bae, C.; Szalanski, A.; Robbins, R. Molecular Analysis of the Lance Nematode, Hoplolaimus spp., Using the First Internal Transcribed Spacer and the D1-D3 Expansion Segments of 28S Ribosomal DNA1. *J. Nematol.* **2008**, *40*, 201–209.

78. Jones, J.T.; Furlanetto, C.; Bakker, E.; Banks, B.; Blok, V.; Chen, Q.; Phillips, M.; Prior, A. Characterization of a chorismate mutase from the potato cyst nematode Globodera pallida. *Mol. Plant Pathol.* **2003**, *4*, 43–50. [CrossRef] [PubMed]

79. Rosso, M.-N.; Favery, B.; Piotte, C.; Arthaud, L.; De Boer, J.M.; Hussey, R.S.; Bakker, J.; Baum, T.J.; Abad, P. Isolation of a cDNA encoding a β-1, 4-endoglucanase in the root-knot nematode Meloidogyne incognita and expression analysis during plant parasitism. *Mol. Plant-Microbe Interact.* **1999**, *12*, 585–591. [CrossRef] [PubMed]

80. Jaubert, S.; Laffaire, J.-B.; Ledger, T.N.; Escoubas, P.; Amri, E.-Z.; Abad, P.; Rosso, M.N. Comparative analysis of two 14-3-3 homologues and their expression pattern in the root-knot nematode *Meloidogyne incognita. Int. J. Parasitol.* **2004**, *34*, 873–880. [CrossRef] [PubMed]

81. Haegeman, A.; Jacob, J.; Vanholme, B.; Kyndt, T.; Mitreva, M.; Gheysen, G. Expressed sequence tags of the peanut pod nematode Ditylenchus africanus: The first transcriptome analysis of an Anguinid nematode. *Mol. Biochem. Parasitol.* **2009**, *167*, 32–40. [CrossRef] [PubMed]

82. Nema Wool—ZooTerms (Dictionary of Invertebrate Zoology). Available online: https://species-id.net/ zooterms/nema_wool (accessed on 23 March 2020).

83. Kikuchi, T.; Aikawa, T.; Kosaka, H.; Pritchard, L.; Ogura, N.; Jones, J.T. Expressed sequence tag (EST) analysis of the pine wood nematode *Bursaphelenchus xylophilus* and *B. mucronatus. Mol. Biochem. Parasitol.* **2007**, *155*, 9–17. [CrossRef] [PubMed]

84. Gao, B.; Allen, R.; Maier, T.; McDermott, J.P.; Davis, E.L.; Baum, T.J.; Hussey, R.S. Characterisation and developmental expression of a chitinase gene in *Heterodera glycines*. *Int. J. Parasitol.* **2002**, *32*, 1293–1300. [CrossRef]

85. Dubreuil, G.; Magliano, M.; Deleury, E.; Abad, P.; Rosso, M.N. Transcriptome analysis of root-knot nematode functions induced in the early stages of parasitism. *New Phytol.* **2007**, *176*, 426–436. [CrossRef]

86. Moens, M.; Perry, R.N. Migratory plant endoparasitic nematodes: A group rich in contrasts and divergence. *Annu. Rev. Phytopathol.* **2009**, *47*, 313–332. [CrossRef]

87. Price, N.S. Alternate cropping in the management of *Radopholus similis* and Cosmopolites sordidus two important pests of banana and plantain. *Int. J. Pest Manag.* **1994**, *40*, 237–244. [CrossRef]

88. Prasad, J.S.; Rao, Y.S. Influence of crop rotations on the population densities of the root lesion nematode, *Pratylenchus indicus* in rice and rice soils. *Annales de Zoologie Ecologie Animale* **1978**, *10*, 627–634.

89. Van Gundy, S.D. A pest management approach to the control of *Pratylenchus thornei* on wheat in Mexico. *J. Nematol.* **1974**, *6*, 107. [PubMed]

90. Qiu, J.; Westerdahl, B.B.; Giraud, D.; Anderson, C.A. Evaluation of hot water treatments for management of *Ditylenchus dipsaci* and fungi in daffodil bulbs. *J. Nematol.* **1993**, *25*, 686. [PubMed]

91. Towson, A.J.; Lear, B. Control of nematodes in rose plants by hot-water treatment preceded by heat-hardening. *Nematologica* **1982**, *28*, 339–353.

92. Tsang, M.M.C.; Hara, A.H.; Sipes, B. Hot-water treatments of potted palms to control the burrowing nematode, *Radopholus similis*. *Crop Prot.* **2003**, *22*, 589–593. [CrossRef]

93. Alexander, S.A.; Waldenmaier, C.M. Suppression of Pratylenchus penetrans populations in potato and tomato using African marigolds. *J. Nematol.* **2002**, *34*, 130.

94. Evenhuis, A.; Korthals, G.; Molendijk, L. Tagetes patula as an effective catch crop for long-term control of Pratylenchus penetrans. *Nematology* **2004**, *6*, 877–881. [CrossRef]

95. Kimpinski, J.; Arsenault, W.J.; Gallant, C.E.; Sanderson, J.B. The effect of marigolds (*Tagetes* spp.) and other cover crops on *Pratylenchus penetrans* and on following potato crops. *J. Nematol.* **2000**, *32*, 531.

96. Pudasaini, M.; Viaene, N.; Moens, M. Effect of marigold (*Tagetes patula*) on population dynamics of *Pratylenchus penetrans* in a field. *Nematology* **2006**, *8*, 477–484. [CrossRef]

97. Sipes, B.S.; Schmitt, D.P. Crotalaria as a cover crop for nematode management: A review. *Nematropica* **2002**, *32*, 35–58.

98. Wang, K.H.; McSorley, R. *Management of Nematodes and Soil Fertility with Sunn Hemp Cover Crop*; University of Florida IFAS Extension Publication No ENY-717; The Institute of Food and Agricultural Sciences (IFAS): Gainesville, FL, USA, 2004.

99. Moraes, A.d.M.; Muniz, M.; Lima, R.d.S.; Moura Filho, G.; Castro, J.d.C. *Organic-Matter Effects on Populations of Dry Rot of Yam Nematodes*; Embrapa Semiárido-Artigo em Periódico Indexado (ALICE); Empresa Brasileira de Pesquisa Agropecuária: Brasília, Brazil, 2016.

100. Henmi, V.H.; Marahatta, S.P. Impacts of sunnhemp and pigeon pea on plant-parasitic nematodes, Radopholus similis and Meloidogyne spp. and beneficial bacterivorous nematodes. *Int. J. Phytopathol.* **2015**, *4*, 29–33. [CrossRef]

101. Halbrendt, J.M. Allelopathy in the management of plant-parasitic nematodes. *J. Nematol.* **1996**, *28*, 8. [PubMed]

102. Chitwood, D.J. Phytochemical based strategies for nematode control. *Annu. Rev. Phytopathol.* **2002**, *40*, 221–249. [CrossRef] [PubMed]

103. Davies, L.J.; Elling, A.A. Resistance genes against plant-parasitic nematodes: A durable control strategy? *Nematology* **2015**, *17*, 249–263. [CrossRef]

104. Cai, D.; Kleine, M.; Kifle, S.; Harloff, H.-J.; Sandal, N.N.; Marcker, K.A.; Klein-Lankhorst, R.M.; Salentijn, E.M.; Lange, W.; Stiekema, W.J. Positional cloning of a gene for nematode resistance in sugar beet. *Science* **1997**, *275*, 832–834. [CrossRef] [PubMed]

105. Williamson, V.M.; Kumar, A. Nematode resistance in plants: The battle underground. *Trends Genet.* **2006**, *22*, 396–403. [CrossRef] [PubMed]

106. Ernst, K.; Kumar, A.; Kriseleit, D.; Kloos, D.-U.; Phillips, M.S.; Ganal, M.W. The broad-spectrum potato cyst nematode resistance gene (Hero) from tomato is the only member of a large gene family of NBS-LRR genes with an unusual amino acid repeat in the LRR region. *Plant J.* **2002**, *31*, 127–136. [CrossRef]

107. Hauge, B.M.; Wang, M.L.; Parsons, J.D.; Parnell, L.D. Nucleic Acid Molecules and Other Molecules Associated with Soybean Cyst Nematode Resistance. U.S. Patents US7154021B2, 26 December 2006.

108. Milligan, S.B.; Bodeau, J.; Yaghoobi, J.; Kaloshian, I.; Zabel, P.; Williamson, V.M. The root knot nematode resistance gene Mi from tomato is a member of the leucine zipper, nucleotide binding, leucine-rich repeat family of plant genes. *Plant Cell* **1998**, *10*, 1307–1319. [CrossRef]

109. Paal, J.; Henselewski, H.; Muth, J.; Meksem, K.; Menéndez, C.M.; Salamini, F.; Ballvora, A.; Gebhardt, C. Molecular cloning of the potato Gro1-4 gene conferring resistance to pathotype Ro1 of the root cyst nematode Globodera rostochiensis, based on a candidate gene approach. *Plant J.* **2004**, *38*, 285–297. [CrossRef]

110. Sobczak, M.; Avrova, A.; Jupowicz, J.; Phillips, M.S.; Ernst, K.; Kumar, A. Characterization of susceptibility and resistance responses to potato cyst nematode (*Globodera* spp.) infection of tomato lines in the absence and presence of the broad-spectrum nematode resistance Hero gene. *Mol. Plant-Microbe Interact.* **2005**, *18*, 158–168. [CrossRef]

111. Van Der Vossen, E.A.; Van Der Voort, J.N.R.; Kanyuka, K.; Bendahmane, A.; Sandbrink, H.; Baulcombe, D.C.; Bakker, J.; Stiekema, W.J.; Klein-Lankhorst, R.M.

Homologues of a single resistance-gene cluster in potato confer resistance to distinct pathogens: A virus and a nematode. *Plant J.* **2000**, *23*, 567–576. [CrossRef]

112. Vos, P.; Simons, G.; Jesse, T.; Wijbrandi, J.; Heinen, L.; Hogers, R.; Frijters, A.; Groenendijk, J.; Diergaarde, P.; Reijans, M. The tomato Mi-1 gene confers resistance to both root-knot nematodes and potato aphids. *Nat. Biotechnol.* **1998**, *16*, 1365. [CrossRef] [PubMed]

113. Atkinson, H.J.; Grimwood, S.; Johnston, K.; Green, J. Prototype demonstration of transgenic resistance to the nematode *Radopholus similis* conferred on banana by a cystatin. *Transgenic Res.* **2004**, *13*, 135–142. [CrossRef]

114. Li, Y.; Wang, K.; Lu, Q.; Du, J.; Wang, Z.; Wang, D.; Sun, B.; Li, H. Transgenic *Nicotiana benthamiana* plants expressing a hairpin RNAi construct of a nematode Rs-cps gene exhibit enhanced resistance to *Radopholus similis*. *Sci. Rep.* **2017**, *7*, 13126. [CrossRef]

115. Fogain, R.; Gowen, S.R. Investigations on possible mechanisms of resistance to nematodes in Musa. *Euphytica* **1995**, *92*, 375–381. [CrossRef]

116. Hölscher, D.; Dhakshinamoorthy, S.; Alexandrov, T.; Becker, M.; Bretschneider, T.; Buerkert, A.; Crecelius, A.C.; De Waele, D.; Elsen, A.; Heckel, D.G. Phenalenone-type phytoalexins mediate resistance of banana plants (Musa spp.) to the burrowing nematode *Radopholus similis*. *Proc. Natl. Acad. Sci. USA* **2014**, *111*, 105–110.

117. Vu, T.; Sikora, R.; Hauschild, R. *Fusarium oxysporum endophytes* induced systemic resistance against *Radopholus similis* on banana. *Nematology* **2006**, *8*, 847–852. [CrossRef]

118. Thompson, J.P.; Brennan, P.S.; Clewett, T.G.; Sheedy, J.G.; Seymour, N.P. Progress in breeding wheat for tolerance and resistance to root-lesion nematode (*Pratylenchus thornei*). *Australas. Plant Pathol.* **1999**, *28*, 45–52. [CrossRef]

119. Sheedy, J.G.; Thompson, J.P. Resistance to the root-lesion nematode *Pratylenchus thornei* of Iranian landrace wheat. *Australas. Plant Pathol.* **2009**, *38*, 478–489. [CrossRef]

120. Thompson, J.P.; Haak, M.I. Resistance to root-lesion nematode (*Pratylenchus thornei*) in *Aegilops tauschii Coss.*, the D-genome donor to wheat. *Aust. J. Agric. Res.* **1997**, *48*, 553–560. [CrossRef]

121. Zwart, R.S.; Thompson, J.P.; Godwin, I.D. Genetic analysis of resistance to root-lesion nematode (*Pratylenchus thornei*) in wheat. *Plant Breed.* **2004**, *123*, 209–212. [CrossRef]

122. Mwaura, P.; Niere, B.; Vidal, S. Resistance and tolerance of potato varieties to potato rot nematode (Ditylenchus destructor) and stem nematode (*Ditylenchus dipsaci*). *Ann. Appl. Biol.* **2015**, *166*, 257–270. [CrossRef]

123. Ristaino, J.B.; Thomas, W. Agriculture, methyl bromide, and the ozone hole: Can we fill the gaps? *Plant Dis.* **1997**, *81*, 964–977. [CrossRef] [PubMed]

124. Martin, F.N. Development of alternative strategies for management of soilborne

pathogens currently controlled with methyl bromide. *Annu. Rev. Phytopathol.* **2003**, *41*, 325–350. [CrossRef] [PubMed]

125. Zasada, I.A.; Halbrendt, J.M.; Kokalis-Burelle, N.; LaMondia, J.; McKenry, M.V.; Noling, J.W. Managing nematodes without methyl bromide. *Annu. Rev. Phytopathol.* **2010**, *48*, 311–328. [CrossRef]
126. Carpenter, J.; Gianessi, L.; Lynch, L. *The Economic Impact of the Scheduled Phase-out of Methyl Bromide*; National Center for Food and Agricultural Policy: Washington, DC, USA, 2000; 466p.
127. Carpenter, J.; Lynch, L.; Trout, T. Township limits on 1, 3-D will impact adjustment to methyl bromide phase-out. *Calif. Agric.* **2001**, *55*, 12–18. [CrossRef]
128. New Rules Governing Use of Fumigant Pesticide 1,3-D. Available online: https://www.cdpr.ca.gov/docs/ pressrls/2016/161006.htm (accessed on 30 September 2019).
129. Sipes, B.; Myers, R. Plant Parasitic Nematodes in Hawaiian Agriculture. In *Plant Parasitic Nematodes in Sustainable Agriculture of North America: Vol.1—Canada, Mexico and Western USA*; Subbotin, S.A., Chitambar, J.J., Eds.; Sustainability in Plant and Crop Protection; Springer International Publishing: Cham, Switzerland, 2018; pp. 193–209. ISBN 978-3-319-99585-4. [CrossRef]
130. Sipes, B.; Kuehnle, A.; Lichty, J.; Sewake, K.; Hara, A. Anthurium Decline: Options for Controlling Burrowing Nematodes. *Plant Dis.June* **2004**, 26.
131. Zouhar, M.; Douda, O.; Dlouhý, M.; Lišková, J.; Manˇasová, M.; Stejskal, V. Using of hydrogen cyanide against Ditylenchus dipsaci nematode present on garlic. *Plant Soil Environ.* **2016**, *62*, 184–188. [CrossRef]
132. Zasada, I.A.; Walters, T.W. Effect of application timing of oxamyl in nonbearing raspberry for *Pratylenchus penetrans* management. *J. Nematol.* **2016**, *48*, 177. [CrossRef] [PubMed]
133. Burelle, N. Review of the non-fumigant nematicides oxamyl and fosthiazate. In Proceedings of the International Conference on Methyl Bromide Alternatives and Emissions Reductions, San Diego, CA, USA, 1–3 November 2005; Volume 110, pp. 1–2.
134. Jordaan, E.M.; Loots, G.C.; Jooste, W.J.; De Waele, D. Effects of root-lesion nematodes (Pratylenchus brachyurus Godfrey and P. zeae Graham) and Fusarium moniliforme Sheldon alone or in combination, on maize. *Nematologica* **1987**, *33*, 213–219.
135. da Silva, M.P.; Tylka, G.L.; Munkvold, G.P. Seed treatment effects on maize seedlings coinfected with *Fusarium* spp. and *Pratylenchus penetrans. Plant Dis.* **2016**, *100*, 431–437. [CrossRef]
136. Farias, B.V.; Pirzada, T.; Mathew, R.; Sit, T.L.; Opperman, C.; Khan, S.A. Electrospun Polymer Nanofibers as Seed Coatings for Crop Protection. *ACS Sustain. Chem. Eng.* **2019**, *7*, 19848–19856. [CrossRef]
137. Barbosa, P.; Lima, A.S.; Vieira, P.; Dias, L.S.; Tinoco, M.T.; Barroso, J.G.; Pedro, L.G.; Figueiredo, A.C.; Mota, M. Nematicidal activity of essential oils and volatiles derived from Portuguese aromatic flora against the pinewood nematode, *Bursaphelenchus xylophilus. J. Nematol.* **2010**, *42*, 8. [PubMed]

Chapter-8

Nematicidal Activity of Essential Oils on a Psychrophilic *Panagrolaimus* sp. (Nematoda: Panagrolaimidae)

1. Introduction

Nematodes are mostly microscopic size invertebrates that inhabit terrestrial and aquatic areas. Beside their significant economic importance as human, animal and plant parasites, they can also be beneficial, free-living microbivorous organisms. It has been estimated that about 2.5 million tons of pesticides are used on crops each year [1]. Such a practice has resulted in the decline of many beneficial organisms, such as nitrogen-fixing soil bacteria [2], blue-green algae [3], mycorrhizal fungi [4], water fishes [5], aquatic mammals [6], and birds [7]. In addition, the pesticide residues in food and water are massive long-term threats for human health at a global level. According to the European legislation (Regulation (EC) no. 1107/2009), the application of non-chemical and natural alternatives should be the first choice in plant protection and integrated pest management. The Regulation requires that substances or products produced or placed on the market do not have any harmful effect on human or animal health or any unacceptable effects on the environment, such as an impact on non-target species and impact on biodiversity and the ecosystem.

The use of essential oils (EOs) is known from folk medicine centuries ago [8]. Nowadays, it seems to be a promising control strategy against different nematode plant and animal parasites (*Bursaphelenchus xylophilus*, *Cooperia* spp., *Ditylenchus dipsaci*, *Haemonchus contortus*, *Meloidogyne chitwoodi*, *M. incognita*, *M. javanica*, *Oesophagostomum* spp., *Pratylenchus penetrans*, *Steinernema feltiae*, *Trichostrongylus* spp., *Tylenchulus semipenetrans*) [9–23]. Microbivorous nematodes contribute to decomposition of organic matter and the release of nutrients for plant uptake [24], which makes them important components of the soil microfauna. A free-living, microbivorous nematode, *Panagrolaimus* sp. Fuchs, was chosen as a test organism. *Panagrolaimus* sp. is a non-target organism, easy to maintain, does not have a complex life cycle, in contrast to plant parasitic nematodes [25], and possesses some extraordinary biological properties. This nematode, known as the Antarctic nematode, is famous for its resistance to intracellular freezing and extremely cold environmental conditions [26]. *Panagrolaimus* aff. *detritophagus* is the first viable multicellular organism, isolated from 30,000–40,000-year-old permafrost deposits [27].

This study aims to: (i) investigate the in vitro impact of EOs on the psychrophilic nematode *Panagrolaimus* sp. in a direct contact bioassay, (ii) list the activity of the EOs based on median lethal concentration (LC50), (iii) determine the composition of the EOs with the best nematicidal activity, and (iv) compare the activity of EOs on the non-target panagrolaimid nematode versus plant parasitic nematodes.

2. Results

The nematicidal activity of the EOs on the *Panagrolaimus* sp. juveniles are presented in Table 1.

Table 1. LC50 (in μL/mL with 95% confidence limits and the slope values) of 50 plant essential oils investigated on *Panagrolaimus* sp. juveniles.

Species Name	Plant Part	Family	LC50 (95% CL)	Slope
Pogostemon cablin	leaves	Lamiaceae	5.641 (3.18–17.10)	2.14
Pinus pinaster	needles	Pinaceae	5.078 (3.38–9.65)	1.65
Santalum album	wood	Santalaceae	4.781 (3.45–6.34)	2.12
Azadirachta indica	seeds	Meliaceae	4.482 (2.52–10.01)	2.02
Boswellia serrata	resin	Burseraceae	4.394 (3.44–5.53)	3.03
Commiphora myrrha	resin	Burseraceae	4.301 (2.81–6.98)	2.79
Juniperus virginiana	wood	Juniperaceae	3.782 (2.43–5.48)	1.81
Cupressus sempervirens	needles	Cupressaceae	3.360 (2.47–4.82)	1.85
Abies sibirica	needles	Pinaceae	3.269 (2.14–5.41)	1.58
Cedrus atlantica	wood	Pinaceae	2.943 (2.04–4.43)	1.62
Juniperus communis	berries	Juniperaceae	2.513 (1.64–3.83)	1.68
Eucalyptus globulus	leaves	Myrtaceae	1.994 (1.47–2.56)	3.70
Myrtus communis	leaves	Myrtaceae	1.933 (1.34–2.65)	2.27
Piper nigrum	peppercorns	Piperaceae	1.775 (1.37–2.25)	2.37
Petroselinum crispum	seeds	Apiaceae	1.704 (1.14–2.38)	4.08
Zingiber officinale	roots	Zingiberaceae	1.633 (1.28–2.09)	1.98
Turnera diffusa	leaves/flowers	Passifloraceae	1.550 (1.14–2.04)	3.11
Abies alba	needles	Pinaceae	1.444 (1.03–1.95)	1.87
Taxandria fragrans	leaves	Myrtaceae	1.437 (0.99–1.93)	2.49
Melaleuca alternifolia	leaves	Myrtaceae	1.150 (0.76–1.57)	4.83
Vanilla planifolia	beans	Orchidaceae	1.135 (0.83–1.50)	2.08
Salvia rosmarinus	leaves/flowers	Lamiaceae	1.128 (0.88–1.41)	3.04
Curcuma longa	rhizomes	Zingiberaceae	1.116 (0.70–1.62)	1.50
Lavandula sp.	leaves/flowers	Lamiaceae	0.810 (0.62–1.01)	3.49
Laurus nobilis	leaves	Lauraceae	0.594 (0.31–0.92)	4.80
Melaleuca quinquenervia	leaves	Myrtaceae	0.593 (0.40–0.82)	1.76
Origanum vulgare	leaves/flowers	Lamiaceae	0.508 (0.38–0.65)	3.33
Mentha spicata	leaves/ flowers	Lamiaceae	0.505 (0.39–0.64)	2.78
Pimpinella anisum	seeds	Apiaceae	0.450 (0.26–0.63)	4.87
Salvia sclarea	leaves/ flowers	Lamiaceae	0.430 (0.31–0.57)	1.81
Anethum graveolens	seeds	Apiaceae	0.428 (0.32–0.56)	3.00
Mentha piperita	leaves/ flowers	Lamiaceae	0.405 (0.23–0.58)	3.93
Thymus vulgaris	leaves/ flowers	Lamiaceae	0.391 (0.29–0.50)	2.22

Gaultheria procumbens	leaves	Ericaceae	0.367 (0.26–0.48)	2.25
Myristica fragrans	seeds	Myristicaceae	0.345 (0.23–0.48)	4.24
Pelargonium asperum	leaves	Geraniaceae	0.279 (0.20–0.37)	3.03
Cymbopogon martini	grass blades	Poaceae	0.275 (0.20–0.35)	4.06
Syzygium aromaticum	buds	Myrtaceae	0.272 (0.20–0.34)	3.91
Ocimum basilicum	leaves/ flowers	Lamiaceae	0.263 (0.19–0.35)	2.65
Uncaria tomentosa	bark	Rubiaceae	0.222 (0.17–0.29)	2.47
Illicium verum	seeds	Schisandraceae	0.191 (0.14–0.25).	2.94
Cinnamomum verum	leaves	Lauraceae	0.172 (0.12–0.23)	5.00
Cananga odorata	flowers	Annonaceae	0.145 (0.10–0.19)	5.00
Mellisa officinalis	leaves	Lamiaceae	0.124 (0.09–0.16)	2.74
Litsea citrata	fruits	Lauraceae	0.091 (0.06–0.12)	3.14
Foeniculum vulgare	seeds	Apiaceae	0.080 (0.05–0.11)	3.68
Cymbopogon flexuosus	grass blades	Poaceae	0.071 (0.05–0.09)	2.92
Coriandrum sativum	leaves	Apiaceae	0.044 (0.02–0.04)	3.95
Cinnamomum cassia	bark	Lauraceae	0.034 (0.02–0.04)	2.00
Cinnamomum burmanii	bark	Lauraceae	0.033 (0.02–0.04)	2.73

The results based on the median lethal concentration (LC50) of 50 EOs are in range from 0.033 to 5.641 μL/mL. The list of all EOs could be divided into three groups. The first group is made up of those with LC50 values higher than 1 μL/mL, the next group with LC50 values in the range 0.1 to 1 μL/mL, and the last group with the LC50 values lower than 0.1 μL/mL. The lowest nematicidal impact is observed in the first group containing EOs from different plants with a significant content of gymnosperms, represented by the families Pinaceae and Cupressaceae. In the same group are some members of angiosperms, such as Burseraceae, Myrtaceae, etc. The second group with a moderate nematicidal effect on the panagrolaimid nematode had EOs originating mainly from the family Lamiaceae and some representatives from individual families. This study demonstrates, for the first time, the nematicidal activity of *Turnera diffusa*, *Taxandria fragrans* and *Uncaria tomentosa* EOs originating from the families *Passifloraceae*, *Myrtaceae*, and *Rubiaceae*, respectively. The best nematicidal activity among the three species was exhibited by *Uncaria tomentosa* EO, with an LC50 of 0.222 μL/mL.

The highest nematicidal impact was observed with three representatives from the family Lauraceae, namely *Litsea citrata*, *Cinnamomum cassia*, and *C. burmannii*, two representatives from the family Apiaceae—i.e., *Foeniculum vulgare* and *Coriandrum sativum*—and the single species *Cymbopogon flexuosus* from the family Poaceae, with LC50 values ranging from 0.033 to 0.091 μL/mL. The best nematicidal effect on panagrolaimid nematodes was shown by *Cinnamomum. burmannii* EO, extracted from the bark. The chemical composition of the EOs with the best nematicidal performance on *Panagrolaimus* sp., with the retention time (RT, in minutes) and the retention indices obtained experimentally and from the literature (RI^{exp} and RI^{lit}, respectively), are given in Tables 2–7.

According to the gas chromatography/mass spectrometry (GC/MS) result obtained, 24 compounds were identified, representing 99.2% of total *Litsea. citrata* EO composition. The main components belong to oxygen-containing monoterpenes (contributing 84.3%), with citral—i.e., geranial (43.4%) and neral (32.2%)—as their representatives present in the highest percentage. They are followed by monoterpene hydrocarbons with 12.2% and limonene as their representative with a contribution of 9.4%, and sesquiterpene hydrocarbons ((E)-caryophyllene, β-elemene and α-humulene) with a contribution of 1.9% to the total EO composition (Table 2).

Table 2. Chemical composition of *Litsea citrata* essential oil (EO).

RT *	Compound	Molecular Formula	RI^{exp} **	RI^{lit} ***	%
18.23	Geranial (=E–citral)	$C_{10}H_{16}O$	1274	1267	43.4
16.94	Neral (=Z-citral)	$C_{10}H_{16}O$	1243	1238	32.2
8.49	Limonene	$C_{10}H_{16}$	1028	1029	9.4
8.57	1,8-Cineole	$C_{10}H_{18}O$	1030	1031	1.9
24.27	(E)-Caryophyllene	$C_{15}H_{24}$	1419	1419	1.6
13.17	Citronellal	$C_{10}H_{18}O$	1153	1153	1.4
11.03	Linalool	$C_{10}H_{18}O$	1100	1096	1.2
5.70	α-Pinene	$C_{10}H_{16}$	933	939	1.1
14.40	(E)-Isocitral	$C_{10}H_{16}O$	1183	1180	1.1
17.42	Geraniol	$C_{10}H_{18}O$	1255	1252	0.9
6.87	β-Pinene	$C_{10}H_{16}$	977	979	0.8
7.10	6-Methyl-5-hepten-2-one	$C_8H_{14}O$	985	985	0.8
13.63	(Z)-Isocitral	$C_{10}H_{16}O$	1164	1164	0.8
6.76	Sabinene	$C_{10}H_{16}$	973	975	0.6
14.71	α-Terpineol	$C_{10}H_{18}O$	1190	1188	0.6
16.31	Nerol	$C_{10}H_{18}O$	1229	1229	0.4
6.10	Camphene	$C_{10}H_{16}$	948	954	0.3
7.25	Dehydro-1,8-cineole	$C_{10}H_{16}O$	991	991	0.2
23.10	β-Elemene	$C_{15}H_{24}$	1392	1390	0.2
12.84	exo-Isocitral	$C_{10}H_{16}O$	1144	1144	0.1
14.17	Terpinen-4-ol	$C_{10}H_{18}O$	1177	1177	0.1
25.65	α-Humulene	$C_{15}H_{24}$	1453	1454	0.1
5.51	α-Thujene	$C_{10}H_{16}$	926	930	tr
8.35	o-Cymene	$C_{10}H_{14}$	1024	1026	tr
		Total identified			99.2

** RT—retention time, ** RIexp—retention index obtained experimentally, *** RIlit—retention index from the literature, tr—traces.*

According to the results of GC/MS analysis of *Foeniculum. vulgare* EO, 18 compounds were identified, representing 99.1% of total EO composition. The main components were aromatic compounds (78.5%), with (E)-anethole as their representative present in the highest percentage (74.3%), followed by oxygen-containing monoterpenes (14.8%) and their representatives fenchone (2.1%) and carvone (2.1%), and monoterpene hydrocarbons (5.8%) with the highest amount of limonene (2.3%) (Table 3).

Table 3. Chemical composition of *Foeniculum vulgare* EO.

RT *	Compound	Molecular Formula	RI^{exp} **	RI^{lit} ***	%
18.93	(E)-Anethole	$C_{10}H_{12}O_2$	1291	1284	74.3
11.28	α-Pinene oxide	$C_{10}H_{16}O$	1106	1099	9.0
15.06	Methyl chavicol (=Estragol)	$C_{10}H_{12}O$	1199	1196	2.9
5.71	α-Pinene	$C_{10}H_{16}$	933	939	2.3
8.48	Limonene	$C_{10}H_{16}$	1028	1029	2.3
10.62	Fenchone	$C_{10}H_{16}O$	1088	1086	2.1
16.94	Carvone	$C_{10}H_{14}O$	1244	1243	2.1
17.39	p-Anis aldehyde	$C_8H_8O_2$	1254	1250	1.3
8.33	o-Cymene	$C_{10}H_{14}$	1024	1026	0.7
12.81	Camphor	$C_{10}H_{16}O$	1144	1146	0.6
8.57	1,8-Cineole	$C_{10}H_{18}O$	1030	1031	0.5
11.07	Linalool	$C_{10}H_{18}O$	1001	1096	0.4
9.53	γ-Terpinene	$C_{10}H_{16}$	1058	1059	0.3
6.10	Camphene	$C_{10}H_{16}$	948	954	0.1
6.87	β-Pinene	$C_{10}H_{16}$	977	979	0.1
11.79	α-Campholenal	$C_{10}H_{16}O$	1119	1126	0.1
6.76	Sabinene	$C_{10}H_{16}$	973	975	tr
7.24	Myrcene	$C_{10}H_{16}$	991	990	tr
		Total identified			99.1

* *RT—retention time, ** RIexp—retention index obtained experimentally, *** RIlit—retention index from the literature, tr—traces.*

The results of the GC/MS analysis of the *Cymbopogon. flexuosus* EO revealed 32 compounds, representing 97.3% of total EO composition. The main components were oxygen-containing monoterpenes (86.3%) with geranial and neral (citral), contributing 40.3% and 30.9%, respectively, geranyl acetate and geraniol (5.4% and 4.5%, respectively), followed by sesquiterpene hydrocarbons (4.0%) and their representative (E)-caryophyllene (2.1%) (Table 4).

Table 4. Chemical composition of *Cymbopogon flexuosus* EO.

RT *	Compound	Molecular Formula	RI^{exp} **	RI^{lit} ***	%
18.24	Geranial (=E-citral)	$C_{10}H_{16}O$	1274	1267	40.3
16.94	Neral (=Z-citral)	$C_{10}H_{16}O$	1244	1238	30.9
22.86	Geranyl acetate	$C_{12}H_{20}O_2$	1385	1381	5.4
17.45	Geraniol	$C_{10}H_{18}O$	1255	1252	4.5
7.10	6-Methyl-5-hepten-2-one	$C_8H_{14}O$	985	985	2.7
24.27	(E)-Caryophyllene	$C_{15}H_{24}$	1419	1419	2.1
11.03	Linalool	$C_{10}H_{18}O$	1101	1096	1.4
28.08	γ-Cadinene	$C_{15}H_{24}$	1514	1513	1.4
14.38	(E)-Isocitral	$C_{10}H_{16}O$	1182	1180	1.3
6.10	Camphene	$C_{10}H_{16}$	948	954	1.1
8.48	Limonene	$C_{10}H_{16}$	1028	1029	1.1
10.00	4-Nonanone	$C_9H_{18}O$	1071	-	0.9

RT *	Compound	Molecular Formula	RI^{exp} **	RI^{lit} ***	%
13.64	(Z)-Isocitral	$C_{10}H_{16}O$	1164	1164	0.9
5.71	α-Pinene	$C_{10}H_{16}$	933	939	0.3
7.24	1,8-Dehydro-cineole	$C_{10}H_{16}O$	991	990	0.3
8.56	1,8-Cineole	$C_{10}H_{18}O$	1030	1031	0.3
13.03	trans-α-Necrodol	$C_{10}H_{18}O$	1149	1148	0.3
13.16	Citronellal	$C_{10}H_{18}O$	1152	1153	0.3
28.45	δ-Cadinene	$C_{15}H_{24}$	1523	1523	0.3
30.72	Caryophyllene oxide	$C_{15}H_{24}O$	1582	1582	0.3
8.76	(Z)-β-Ocimene	$C_{10}H_{16}$	1036	1037	0.2
25.64	α-Humulene	$C_{15}H_{24}$	1453	1454	0.2
5.43	Tricyclene	$C_{10}H_{16}$	922	926	0.1
8.34	o-Cymene	$C_{10}H_{14}$	1024	1026	0.1
9.14	(E)-β-Ocimene	$C_{10}H_{16}$	1046	1050	0.1
10.62	Terpinolene	$C_{10}H_{16}$	1088	1088	0.1
12.84	exo-Isocitral	$C_{10}H_{16}O$	1144	1144	0.1
14.08	Rosefuran epoxide	$C_{10}H_{14}O_2$	1175	1177	0.1
14.72	α-Terpineol	$C_{10}H_{18}O$	1191	1188	0.1
16.32	Nerol	$C_{10}H_{18}O$	1229	1229	0.1
6.87	β-Pinene	$C_{10}H_{16}$	977	979	tr
7.60	n-Octanal	$C_8H_{16}O$	1003	998	tr
		Total identified			97.3

** RT—retention time, ** RIexp—retention index obtained experimentally, *** RIlit—retention index from the literature, tr—traces.*

GC/MS analysis of the *Coriandrum. sativum* EO resulted in identifying 29 compounds, representing 97.5% of total EO composition. The main components were aldehydes (contributing 51.8%), with (2E)-decenal as their representative, followed by aliphatic alcohols (among which (2E)-decen-1-ol was present in the highest percentage of 16.3%) and oxygen-containing monoterpenes (21.7%) with linalool as the most abundant compound (18.4%) (Table 5).

Table 5. Chemical composition of *Coriandrum sativum* EO.

RT *	Compound	Molecular Formula	RI^{exp} **	RI^{lit} ***	%
17.78	(2E)-Decenal	$C_{10}H_{18}O$	1263	1263	28.2
18.10	(2E)-Decen-1-ol	$C_{10}H_{20}O$	1271	1271	16.3
11.08	Linalool	$C_{10}H_{18}O$	1101	1096	18.4
26.20	(2E)-Dodecenal	$C_{12}H_{22}O$	1467	1466	8.8
15.35	n-Decanal	$C_{10}H_{20}O$	1206	1201	6.2
18.19	n-Decanol	$C_{10}H_{22}O$	1273	1269	5.4
34.07	n-Tetradecanol	$C_{14}H_{30}O$	1673	1672	4.0
21.97	(2E)-Undecenal	$C_{11}H_{20}O$	1363	1360	1.7
12.80	Camphor	$C_{10}H_{16}O$	1143	1146	1.2
23.85	Dodecanal	$C_{12}H_{24}O$	1409	1408	1.1
8.34	o-Cymene	$C_{10}H_{14}$	1024	1026	0.9
5.71	α-Pinene	$C_{10}H_{16}$	933	939	0.7
15.00	(4E)-Decenal	$C_{10}H_{18}O$	1198	1196	0.6
17.43	Geraniol	$C_{10}H_{18}O$	1255	1252	0.6
10.03	cis-Linalool oxide	$C_{10}H_{18}O_2$	1072	1072	0.4
10.62	trans-Linalool oxide	$C_{10}H_{18}O_2$	1088	1086	0.4

19.62	Undecanal	$C_{11}H_{22}O$	1307	1306	0.4
22.85	Geranyl acetate	$C_{12}H_{20}O_2$	1384	1381	0.4
7.60	n-Octanal	$C_8H_{16}O$	1003	998	0.3
8.48	Limonene	$C_{10}H_{16}$	1028	1029	0.3
31.86	Tetradecanal	$C_{14}H_{28}O$	1613	1612	0.3
8.57	1,8-Cineole	$C_{10}H_{18}O$	1030	1031	0.2
9.53	γ-Terpinene	$C_{10}H_{16}$	1058	1059	0.2
14.85	(4Z)-Decenal	$C_{10}H_{18}O$	1194	1194	0.2
6.10	Camphene	$C_{10}H_{16}$	948	954	0.1
6.88	β-Pinene	$C_{10}H_{16}$	977	979	0.1
14.71	α-Terpineol	$C_{10}H_{18}O$	1190	1188	0.1
13.69	Borneol	$C_{10}H_{18}O$	1165	1169	tr
14.17	Terpinen-4-ol	$C_{10}H_{18}O$	1177	1177	tr
		Total identified			97.5

* *RT—retention time,* ** *RIexp—retention index obtained experimentally,* *** *RIlit—retention index from the literature, tr—traces.*

The GC/MS analysis of *Cinnamomum. cassia* EO revealed 32 compounds, representing 99.3% of total EO composition. The main components belong to the group of aromatic compounds, contributing 91.8%, followed by sesquiterpene hydrocarbons (6.7%). (E)-Cinnamaldehyde and eugenol acetate were identified as the representatives of aromatic compounds contributing 76.7% and 7.4%, respectively. On the other hand, δ-cadinene with a contribution of 6.2% to the total EO composition, was identified as a representative compound from the sesquiterpene hydrocarbons group (6.7%) (Table 6).

According to the GC/MS results, 43 compounds, representing 98.1% of total *C. burmanii* EO composition, were identified. The main components belong to aromatic compounds (contributing 84.5% in the total EO composition), with (E)-cinnamaldehyde as their representative (80.5%). They are followed by sesquiterpene hydrocarbons (7.0%) with δ-cadinene and α-copaene as their members present in the amounts of 1.7% and 1.5%, respectively and oxygenated monoterpenes (5.5%) with α-terpineol (1.9%) as their main representative (Table 7).

Table 6. Chemical composition of *Cynnamomum cassia* EO.

RT *	Compound	Molecular Formula	RI^{exp} **	RI^{lit} ***	%
24.54	(E)-Cinnamaldehyde	C_9H_8O	1272	1267	76.7
33.02	Eugenol acetate	$C_{12}H_{14}O_3$	1535	1521	7.4
32.39	δ-Cadinene	$C_{15}H_{24}$	1517	1522	6.2
29.95	(E)-Cinnamyl acetate	$C_{11}H_{12}O_2$	1448	1443	4.0
17.88	Phenethyl alcohol	$C_8H_{10}O$	1126	1106	0.8
11.96	Benzaldehyde	C_7H_6O	962	952	0.7
21.91	(Z)-Cinnamaldehyde	C_9H_8O	1220	1217	0.6
22.78	Carvone	$C_{10}H_{14}O$	1243	1239	0.4
19.69	Hydrocinnamaldehyde	$C_9H_{10}O$	1163	1163	0.3
25.28	α-Methylcinnamaldehyde	$C_{10}H_{10}O$	1332	1318	0.3
30.51	Coumarin	$C_9H_6O_2$	1456	1432	0.3
15.10	γ-Terpinene	$C_{10}H_{16}$	1056	1054	0.2
23.20	2-Phenyl ethyl acetate	$C_{10}H_{12}O_2$	1259	1254	0.2

27.57	α-Copaene	$C_{15}H_{24}$	1368	1374	0.2
10.93	α-Pinene	$C_{10}H_{16}$	929	932	0.1
10.93	Camphene	$C_{10}H_{16}$	943	946	0.1
14.39	1,8-Cineole	$C_{10}H_{18}O$	1027	1026	0.1
19.88	Borneol	$C_{10}H_{18}O$	1168	1165	0.1
20.19	Terpinen-4-ol	$C_{10}H_{18}O$	1174	1174	0.1
26.98	Eugenol	$C_{10}H_{12}O_2$	1359	1356	0.1
30.99	γ-Muurolene	$C_{15}H_{24}$	1470	1478	0.1
31.11	β-Selinene	$C_{15}H_{24}$	1478	1489	0.1
32.51	trans-Cadina-1,4-diene	$C_{15}H_{24}$	1526	1533	0.1
25.90	1-epi-Cubenol	$C_{15}H_{26}O$	1622	1627	0.1
12.48	β-Pinene	$C_{10}H_{16}$	973	974	tr
14.25	β-Phellandrene	$C_{10}H_{16}$	1025	1025	tr
16.93	Terpinolene	$C_{10}H_{16}$	1087	1086	tr
27.27	Cyclosativene	$C_{15}H_{24}$	1357	1358	tr
28.10	Sativene	$C_{15}H_{24}$	1382	1374	tr
28.84	Isosativene	$C_{15}H_{24}$	1401	1417	tr
29.14	(E)-Caryophyllene	$C_{15}H_{24}$	1413	1417	tr
31.76	α-Muurolene	$C_{15}H_{24}$	1494	1500	tr
		Total identified			99.3

RT—retention time, **RIexp—retention index obtained experimentally, * RIlit—retention index from the literature, tr—traces.*

Table 7. Chemical composition of *Cynnamomum burmannii* EO.

RT *	Compound	Molecular Formula	RI^{exp} **	RI^{lit} ***	%
24.51	(E)-Cinnamaldehyde	C_9H_8O	1272	1267	80.5
20.70	α-Terpineol	$C_{10}H_{18}O$	1189	1186	1.9
32.51	δ-Cadinene	$C_{15}H_{24}$	1517	1522	1.7
27.57	α-Copaene	$C_{15}H_{24}$	1368	1374	1.5
21.82	(Z)-Cinnamaldehyde	C_9H_8O	1220	1217	1.5
19.60	Hydrocinnamaldehyde	$C_9H_{10}O$	1163	1163	1.3
19.79	Borneol	$C_{10}H_{18}O$	1168	1165	1.2
20.17	Terpinen-4-ol	$C_{10}H_{18}O$	1174	1174	1.2
31.74	α-Muurolene	$C_{15}H_{24}$	1494	1500	1.2
29.05	(E)-Caryophyllene	$C_{15}H_{24}$	1413	1417	0.8
17.10	Linalool	$C_{10}H_{18}O$	1100	1095	0.5
11.87	Benzaldehyde	C_7H_6O	962	952	0.4
24.54	Safrole	$C_{10}H_{10}O_2$	1288	1285	0.4
28.76	Isosativene	$C_{15}H_{24}$	1401	1417	0.4
31.32	α-Selinene	$C_{15}H_{24}$	1487	1498	0.4
24.75	Tridecane	$C_{13}H_{28}$	1295	1300	0.3
14.29	β-Phellandrene	$C_{10}H_{16}$	1025	1025	0.2
15.84	cis-Linalool oxide	$C_{10}H_{18}O_2$	1071	1067	0.2
20.54	Cryptone	$C_9H_{14}O$	1185	1183	0.2
28.03	Sativene	$C_{15}H_{24}$	1382	1374	0.2
30.92	γ-Muurolene	$C_{15}H_{24}$	1470	1478	0.2
31.15	β-Selinene	$C_{15}H_{24}$	1478	1489	0.2
30.21	(E)-Cinnamyl acetate	$C_{11}H_{12}O_2$	1448	1443	0.2

36.26	epi-α-Murrolol	$C_{15}H_{26}O$	1639	1640	0.2
13.37	α-Phellandrene	$C_{10}H_{16}$	1002	1002	0.1
14.42	1,8-Cineole	$C_{10}H_{18}O$	1027	1026	0.1
14.96	γ-Terpinene	$C_{10}H_{16}$	1056	1054	0.1
16.62	p-Cymenene	$C_{10}H_{14}$	1087	1089	0.1
16.99	Terpinolene	$C_{10}H_{16}$	1087	1086	0.1
18.87	trans-Limonene oxide	$C_{10}H_{16}O$	1137	1137	0.1
25.24	α-Methylcinnamaldehyde	$C_{10}H_{10}O$	1332	1318	0.1
27.18	Cyclosativene	$C_{15}H_{24}$	1357	1358	0.1
28.25	β-Elemene	$C_{15}H_{24}$	1386	1389	0.1
28.54	(Z)-Caryophyllene	$C_{15}H_{24}$	1400	1408	0.1
29.82	Humulene	$C_{15}H_{24}$	1446	1452	0.1
30.51	Coumarin	$C_9H_6O_2$	1456	1432	0.1
36.35	α-Muurolol (Torreyol)	$C_{15}H_{26}O$	1643	1644	0.1
12.83	Myrcene	$C_{10}H_{16}$	994	998	tr
17.70	Phenethyl alcohol	$C_8H_{10}O$	1126	1106	tr
19.45	Isoborneol	$C_{10}H_{18}O$	1154	1155	tr
32.80	trans-Cadina-1,4-diene	$C_{15}H_{24}$	1526	1533	tr
33.15	α-Calacorene	$C_{15}H_{20}$	1538	1544	tr
32.92	Benzyl benzoate	$C_{14}H_{12}O_2$	1772	1759	tr
		Total identified			98.1

RT—retention time, ** RIexp—retention index obtained experimentally, * RIlit—retention index from the literature, tr—traces.*

3. Discussion

Essential oils with the highest nematicidal activity, demonstrated in this study, have been reported to be efficient against wide spectra of pathogens, diseases, and parasites.

The *Litsea citrata* EO showed antibacterial, antifungal, acaricidal, and nematicidal activities. The fruit essential oil of *Litsea cubeba* (*syn. Litsea citrata*) exhibited antibacterial activity against *Bacillus cereus*, *Staphylococcus aureus*, *Vibrio parahaemolyticus*, and *Klebsiella pneumoniae* [28]. As an antifungal agent it was effective against *Candida krusei* and *C. guilliermondii* but did not act against *C. albicans*, *C. tropicalis* and *C. parapsilosis* [29]. The *Litsea cubeba* EO had acaricidal activity against house dust mites, *Dermatophagoides farinae* and *D. pteronyssinus*, and stored food mites, *Tyrophagus putrescentiae* [30]. The LC50 values of ajowan, allspice and litsea were 0.431, 0.609 and 0.504 mg/mL, respectively, and exhibited good nematicidal activity against *B. xylophilus* [31]. Citral, i.e., geranial and neral, were the main compounds in the *Litsea citrata* EO in this study. Citral (3,7-dimethyl-2,6-octadienal) is the monoterpene aldehyde representing natural mixture of the two geometric isomers: geranial (trans-isomer) with a strong lemon odor and neral (cis-isomer) with a lemon odor that is less intense and sweeter than geranial [32].

The *Foeniculum vulgare* EO exhibited antifungal, antibacterial, antiviral, and nematicidal activities. In the inverted petriplate method, the volatile oil showed complete zone inhibition against *Aspergillus niger*, *A. flavus*, *Fusarium graminearum*, and *F. moniliforme* at a 6-µL dose [33]. Hot water extracts of fennel seeds was effective against *Enterococcus faecalis*, *S. aureus*, *Escherichia coli*, *Pseudomonas aeruginosa*, *Salmonella typhimurium*, *S. typhi*, and *Shigella flexneri* [34]. The DNA virus *Herpes simplex* type-1 (HSV-1) and the RNA virus parainfluenza type-3 (PI-3) were inhibited by the *Foeniculum. vulgare* EO

[35]. Essential oils of *Carum carvi, F. vulgare, Mentha rotundifolia*, and *M. spicata* showed the highest nematicidal activity against *M. javanica* juveniles [9].

Trans-anethole was the most abundant compound in the *Foeniculum vulgare* EO, reaching 74% of the total identified constituents. Propenylbenzenes, such as anethole, were reported to be mutagenic for *Salmonella* tester strains and also carcinogenic in the induction of hepatomas in B6C3F1 mice and skin papillomas in CD-1 mice [36].

The *Cymbopogon flexuosus* lemongrass EO was reported to have antibacterial, antifungal and anti-inflammatory activity. The EO from *C. flexuosus* exhibited an antimicrobial effect against *B. subtilis, Staphylococcus aureus, A. flavus* and *A. fumigatus* [37]. The lemongrass (*C. flexuosus*) inflorescence EO was inhibitory to *Pyricularia oryzae, Dreshlera oryzae, A. niger* and *Penicillium italicum* [38]. Lemongrass EO, which has citral as its main component, has exhibited an anti-inflammatory effect in both animal and human cells [39]. In this study, the content of the lemongrass EO's major compounds, geranial and neral, was similar to their content in the *L. citrata* EO with slightly lower amounts—40.3% and 30.9%, respectively.

The *Coriandrum sativum* EO exhibited antifungal, antibacterial, insecticidal, and nematicidal activities. The *Coriandrum sativum* EO showed excellent antifungal activity against seedborne pathogens *P. oryzae, Bipolaris oryzae, Alternaria alternata, Tricoconis padwickii, Drechslera tetramera, D. halodes, Curvularia lunata, F. moniliforme*, and *F. oxysporum* [40]. The methanolic extract of *C. sativum* showed antibacterial activity against *E. coli, P. aeruginosa, S. aureus*, and *K. pneumoniae* [41]. The leaf oil had significant toxic effects against the larvae of *Aedes aegypti* with an LC50 value of 26.93 ppm and an LC90 value of 37.69 ppm, and the stem oil has toxic effects against the larvae of *A. aegypti* with an LC50 value of 29.39 ppm and an LC90 value of 39.95 ppm [42]. Among the 28 plant EOs tested for their nematicidal activities against the pine wood nematode, *B. xylophilus*, the best nematicidal activity was achieved with the EO of coriander [43]. In this study, the major compound in the *C. sativum* EO was trans-2-decenal with 28.2%, affiliated to the group of medium-chain aldehydes. Aliphatic aldehydes (mainly mainly C10–C16 aldehydes), with their unpleasant odor, are the main components of the volatile oil from the fresh herb [44]. Aldehydes present in the coriander EO are important biologically active substances due to their possible toxic activity against tropical mosquitoes transmitting dangerous illnesses [45].

The *Cinnamomum cassia* EO exhibited antimicrobial, antiviral, insecticidal, and nematicidal activities. The cassia EO acted as fungal growth inhibitor against *A. flavus* and *A. oryzae* [46] and as a bacterial inhibitor of *S. aureus* and *E. coli* [47]. The silver nanoparticles derived from cinnamon extract enhanced the antiviral activity and were found to be effective against highly pathogenic avian influenza virus subtype H7N3, when incubated with the virus prior to infection and introduced to cells after infection [48]. The chloroform extract from *C. cassia* was the most effective against *Dermestes maculatus* larvae, the pest of Egyptian mummies [8]. Cassia oil was efficient against *Sitophilus zeamais* [49], and the booklice *Liposcelis bostrychophila* [50]. As judged by the 24-h LC50 values, two cassia oils (0.084–0.085 mg/mL) and four cinnamon oils (0.064–0.113 mg/mL) were toxic toward adult *B. xylophilus* [51].

As opposed to cassia, *Cinnamomum burmannii* EO has been less studied. The *Cinnamomum burmannii* EO exhibited significant antibacterial properties against five common foodborne pathogenic bacteria, namely, *B. cereus*, *L. monocytogenes*, *S. aureus*, *E. coli*, and *Salmonella anatum* [52].

The major component of cinnamon bark EO is (E)-cinnamaldehyde. In the contact with bacterial membrane, cinnamaldehyde causes the loss of membrane functionality or the loss of channel proteins in the membrane, resulting in death of bacterial cells [53]. Besides this, (*E*)-cinnamaldehyde was significantly more effective than its corresponding acid (cinnamic acid) and alcohol (cinnamyl alcohol) and could be used as a fumigant with contact action in the control of house dust mites, *D. farinae* and *D. pteronyssinus* [54].

It has been emphasized that the major components play important roles in the toxicity of EOs [31,42,55] and the majority of them belong to the class of terpenes. Terpenes are the largest class of secondary metabolites and basically consist of five carbon isoprene units, which are assembled to each other (many isoprene units) by thousands of ways. Terpenes are simple hydrocarbons, while terpenoids (monoterpenes, sesquiterpenes, diterpenes, sesterpenes, and triterpenes) are a modified class of terpenes with different functional groups and an oxidized methyl group moved or removed at various positions [56].

Organic compounds that contain the group -CHO (the aldehyde group; i.e., a carbonyl group (C=O) with a hydrogen atom bound to the carbon atom) are known as aldehydes. In systematic chemical nomenclature, aldehyde names end with the suffix -al [57].

In this study, the major components and presumably the most active components (geranial, neral, trans-2-decenal, and trans-cinnamaldehyde) of *Litsea citrata*, *Cymbopogon flexuosus*, *Coriandrum sativum*, *Cinnamomum cassia*, and *C. burmannii* EOs are aldehydes. Aldehydes are highly reactive molecules that may have a variety of effects on biological systems. Although some aldehyde-mediated effects are beneficial, many effects are deleterious, including cytotoxicity, mutagenicity, and carcinogenicity [58], and generally, they are toxic to the human body [59] and evidently, to nematodes. Despite the potential risks of aldehyde exposure, the toxic mechanisms are only understood in general terms. Human exposure to aldehydes represents a significant toxicological concern and, therefore, understanding the corresponding molecular mechanism of toxicity is important for accurate risk assessment and remediation. In this perspective, it has been shown that environmental and endogenous aldehydes can be described by their relative softness and electrophilicity, which are important electronic determinants of the respective second order reaction rates with nucleophilic targets on macromolecules. These soft-soft and hard-hard adduct reactions appear to mediate toxicity by impairing the function of macromolecules (e.g., proteins, DNA, and RNA) that play critical roles in cytophysiological processes. However, more research is needed to broaden our understanding of how these specific covalent reactions disable macromolecular targets [60].

Comparing the results for the toxicity of EOs for nematodes with a different oral apparatus, they are mostly in agreement. However, some results for *Panagrolaimus* sp. deviate from those obtained for plant parasitic nematodes.

The *Rosmarinus officinalis* (*syn. Salvia rosmarinus*) EO at2 μg/mL induced 100% mortality of *Xiphinema index* adults [61], while in this study, the same EO was characterized as having low toxicity and classified into the first group of EOs.

The LC50 of the *M. spicata* EO was 0.2 mg/mL, for *M. javanica* juveniles [62], while in this study the LC50 of the same oil was 0.505 μL/mL against *Panagrolaimus* sp. juveniles. Good nematicidal activity against male, female and juvenile nematodes of *B. xylophilus* was achieved, among other EOs, with the essential oils of *Boswellia carterii* [31]. In this study, *Boswellia serata* was classified into the group of EOs with low toxicity. The *Pinus pinea* EO was found to be toxic against *M. incognita* juveniles with an estimated LC50 of 44 ppm [63], while in this study EOs from gymnosperms, e.g., *Pinus pinaster* (LC50: 5.078 μL/mL), generally showed low toxicity to *Panagrolaimus* species.

Variations in acute toxicity among EOs of the same plant species are greatly influenced by production, storage conditions, climatic or edaphic factors [64]. The chemical content varies even within the same crop. Significant variations were found in many EO components, both across years and throughout harvest dates within locations [65]. However, the different impact of the same EO on free-living versus plant parasitic nematodes may be due to different feeding behaviors, different dimensions, and different metabolic activities and demonstrate a possible direction in the search for active compounds that will be at the same time toxic to plant parasitic nematodes and not have unacceptable effects on the environment and non-target species.

4. Materials and Methods

4.1 Nematode Culture and Direct Contact Bioassay

A culture of *Panagrolaimus* sp. was grown monoxenically on previously frozen agricultural compost and extracted from it with a Baerman funnel [66] over 24 h. Using a compound microscope and a micropipette, juveniles were separated from adult nematodes, counted in aliquots of 50 in 20 μL of water suspension and the live specimens were used in the experiments. The 50 commercial plant EOs from 22 families were purchased from the market and used to investigate their in vitro nematicidal activity against the panagrolaimid nematode *Panagrolaimus* sp. (Table 1). Serial dilutions starting from 0.2 μL/mL, in a double decreasing range up to 0.00975 μL/mL of EOs, were made and stabilized with 0.1-μL/mL Break-Thru® 446 oil enhancer. The direct contact bioassay was performed in small glass petri dishes containing 2 mL of solution and 50 nematodes incubated at 18 °C in the dark. The experiments were performed in five replicates. The lethal effect was monitored after 24 h. An aqueous solution of the emulsifier without EO served as the control. Prior to the assessment of the EOs, the mortality of panagrolaimid nematodes in the aqueous solution was compared with the mortality of nematodes in 0.1-μL/mL emulsifier and no significant differences between the treatments were observed. The nematodes were considered dead if they did not react on touching with a small needle.

4.2 Chemical Analyses

The gas chromatography/mass spectrometry (GC/MS) analysis was performed on an Agilent 6890N network gas chromatograph attached to a mass spectrometer (Agilent 5975B) equipped with a fused silica capillary column (HP-5ms) with dimensions as follows: 30-m length, 0.25-mm internal diameter, 0.25-μm film thickness, coated with 5% diphenyl- and 95% dimethyl-polysiloxane. The samples were diluted in diethyl ether (1:10) and a volume of 1.0 μL was injected. The injector was set at 220 °C and performed in the split mode at a ratio of 1:20. Helium was used as the carrier gas at a flow rate of 0.9 mL/min. The oven temperature increased from 60 to 246 °C at a rate of 3 °C/min. Temperatures of the mass selective detector (MSD) transfer line, ion source and quadruple mass analyzer were set at 280, 230 and 150 °C, respectively. The ionization voltage was 70 eV and the scan range was 35–400 *m/z*. Compound identifications were based on comparisons of their mass spectra with the mass spectra obtained from the National Institute of Standards and Technology database and by comparisons of the retention indices with values reported in the literature (RIlit) [67]. A homologous series of *n*-alkanes (C_8–C_{34}) was run under the same operating conditions as the EO to determine the experimental retention indices (RIexp). The relative amounts of individual components (expressed in percentages) were calculated via peak area normalization, without the use of correction factors. Compounds present in traces (tr) with their amounts less than 0.05% are indicated (Tables 2–7).

4.3 Statistical Data Analysis

In order to evaluate the nematicidal activity of the EOs, median lethal concentration (LC50) was calculated using the Probit Analysis program [68]. The *Panagrolaimus* mortality was corrected using Abbott's formula [69]. The nematicidal activity, i.e., acute toxicity of the examined EOs based on the median lethal concentration, was designated as high (LC50: <0.1 μL/mL), moderate (LC50: 0.1–1 μL/mL) and low (LC50: >1 μL/mL) (Table 1).

References

1. Koul, O.; Walia, S.; Dhaliwal, G.S. Essential oils as green pesticides: Potential and constraints. *Biopestic. Int.* **2008**, *4*, 63–84.

2. Santos, A.; Flores, M. Effects of glyphosate on nitrogen fixation of free-living heterotrophic bacteria. *Lett. Appl. Microbiol.* **1995**, *20*, 349–352. [CrossRef]

3. Saygideger, S.D.; Okkay, O. Effect of 2,4-dichlorophenoxyacetic acid on growth, protein and chlorophyll— A content of *Chlorella vulgaris* and *Spirulina platensis* cells. *J. Environ. Biol.* **2008**, *29*, 175–178. [PubMed]

4. Zaller, J.G.; Heigl, F.; Ruess, L.; Grabmaier, A. Glyphosate herbicide affects belowground interactions between earthworms and symbiotic mycorrhizal fungi in a model ecosystem. *Sci. Rep.* **2014**, *4*, 5634. [CrossRef] [PubMed]

5. Ibrahim, L.; Preuss, T.G.; Ratte, H.T.; Hommen, U. A list of fish species that are potentially exposed to pesticides in edge-of-field water bodies in the European Union—A first step towards identifying vulnerable representatives for risk assessment. *Environ. Sci. Pollut. Res.* **2013**, *20*, 2679–2687. [CrossRef]

6. Leonards, P.E.; Zierikzee, Y.; Brinkman, U.A.T.; Cofino, W.P.; van Straalen, N.M.; van Hattum, B. The selective dietary accumulation of planar polychlorinated biphenyls in the otter (*Lutra lutra*). *Environ. Toxicol. Chem.* **1997**, *16*, 1807–1815. [CrossRef]

7. Brain, R.A.; Anderson, J.C. The agro-enabled urban revolution, pesticides, politics, and popular culture: A case study of land use, birds, and insecticides in the USA. *Environ. Sci. Pollut. Res.* **2019**, *26*, 21717–21735. [CrossRef]

8. Abdel-Maksoud, G.; El-Amin, A.-R.; Afifi, F. Insecticidal activity of *Cinnamomum cassia* extractions against the common Egyptian mummies' insect pest (*Dermestes maculatus*). *Int. J. Conserv. Sci.* **2014**, *5*, 355–368.
9. Oka, Y.; Nacar, S.; Putievsky, E.; Ravid, U.; Yaniv, Z.; Spiegel, Y. Nematicidal activity of essential oils and their components against the root-knot nematode. *Phytopathology* **2000**, *90*, 710–715. [CrossRef]
10. Pérez, M.P.; Navas-Cortés, J.A.; Pascual-Villalobos, M.J.; Castillo, P. Nematicidal activity of essential oils and organic amendments from Asteraceae against root-knot nematodes. *Plant Path* **2003**, *52*, 395–401. [CrossRef]
11. Park, I.; Park, J.; Kim, K.; Choi, K.; Choi, I.; Kim, C. Nematicidal activity of plant EOs and components from garlic (*Allium sativum*) and cinnamon (*Cinnamomum verum*) oils against the pine wood nematode (*Bursaphelenchus xylophilus*). *Nematology* **2005**, *7*, 767–774. [CrossRef]
12. Elbadri, G.A.; Lee, D.W.; Park, J.C.; Yu, H.B.; Choo, H.Y.; Lee, S.M.; Lim, T.H. Nematocidal screening of essential oils and herbal extracts against *Bursaphelenchus xylophilus*. *Plant Pathol. J.* **2008**, *24*, 178–182. [CrossRef]
13. Barbosa, P.; Lima, A.S.; Vieira, P.; Dias, L.S.; Tinoco, M.T.; Barroso, J.G.; Pedro, J.G.; Figueriedo, A.C.; Mota, M. Nematicidal activity of essential oils and volatiles derived from Portuguese aromatic flora against the pinewood nematode, *Bursaphelenchus xylophilus*. *J. Nematol.* **2010**, *42*, 8–16. [PubMed]
14. Ntalli, N.G.; Ferrari, F.; Giannakou, I.; Menkissoglu-Spiroudi, U. Phytochemistry and nematicidal activity of the essential oils from 8 Greek Lamiaceae aromatic plants and 13 terpene components. *J. Agric. Food Chem.* **2010**, *58*, 7856–7863. [CrossRef]
15. Andrés, M.F.; González-Coloma, A.; Sanz, J.; Burillo, J.; Sainz, P. Nematicidal activity of essential oils: A review. *Phytochem. Rev.* **2012**, *11*, 371–390. [CrossRef]
16. Caboni, P.; Saba, M.; Tocco, G.; Casu, L.; Murgia, A.; Maxia, A.; Spiroudi, U.M.; Ntalli, N. Nematicidal activity of mint aqueous extracts against the root-knot nematode *Meloidogyne incognita*. *J. Agric. Food Chem.* **2013**, *61*, 9784–9788. [CrossRef]
17. Kim, S.I.; Lee, J.K.; Na, Y.E.; Yoon, S.T.; Oh, Y.J. Nematicidal and ovicidal activities of *Dryobalanops aromatica* and *Mentha haplocalyx* var. piperascens-derived materials and their formulations against *Meloidogyne incognita* second-stage juveniles and eggs. *Nematology* **2014**, *16*, 193–200. [CrossRef]
18. Faria, J.M.S.; Sena, I.; Ribeiro, B.; Rodrigues, A.M.; Maleita, C.M.N.; Abrantes, I.; Bennett, R.; Mota, M.; Da Silva Figueiredo, A.C. First report on *Meloidogyne chitwoodi* hatching inhibition activity of essential oils and essential oils fractions. *J. Pest Sci.* **2016**, *89*, 207–217. [CrossRef]
19. André, W.P.P.; Ribeiro, W.L.C.; Oliveira, L.M.B.D.; Macedo, I.T.F.; Rondon, F.C.M.; Bevilaqua, C.M.L. Óleos essenciais e seus compostos bioativos no controle de nematoides gastrintestinais de pequenos ruminantes. *Acta Sci. Vet.* **2018**, *46*, 1514–1522. [CrossRef]

20. Ferreira, L.E.; Benincasa, B.I.; Fachin, A.L.; Contini, S.H.T.; França, S.C.; Chagas, A.C.S.; Beleboni, R.O. Essential oils of *Citrus aurantifolia, Anthemis nobile* and *Lavandula officinalis*: In Vitro anthelmintic activities against *Haemonchus contortus. Parasites Vectors* **2018**, *11*, 269. [CrossRef]

21. Saha, S.; Lachance, S. Effect of essential oils on cattle gastrointestinal nematodes assessed by egg hatch, larval migration and mortality testing. *J. Helminthol.* **2020**, *94*, e111. [CrossRef] [PubMed]

22. Pardavella, I.; Nasiou, E.; Daferera, D.; Trigas, P.; Giannakou, I. The use of essential oil and hydrosol extracted from *Satureja hellenica* for the Control of *Meloidogyne incognita* and *M. javanica. Plants* **2020**, *7*, 856. [CrossRef]

23. Oka, Y.; Ben-Daniel, B.; Cohen, Y. Nematicidal activity of the leaf powder and extracts of *Myrtus communis* against the root-knot nematode *Meloidogyne javanica. Plant Path* **2012**, 1–9. [CrossRef]

24. Ferris, H.; Venette, R.C.; Van der Meulen, H.R.; Lau, S.S. Nitrogen mineralization by bacterial–feeding nematodes: Verification and measurement. *Plant Soil* **1998**, *203*, 159–171. [CrossRef]

25. Oro, V.; Knezevic, M.; Dinic, Z.; Delic, D. Bacterial microbiota isolated from cysts of *Globodera rostochiensis* (Nematoda: Heteroderidae). *Plants* **2020**, *9*, 1146. [CrossRef]

26. Seybold, A.C.; Wharton, D.A.; Thorne, M.A.; Marshall, C.J. Establishing RNAi in a non–model organism: The Antarctic nematode *Panagrolaimus* sp. DAW1. *PLoS ONE* **2016**, *11*, e0166228. [CrossRef]

27. Shatilovich, A.V.; Tchesunov, A.V.; Neretina, T.V.; Grabarnik, I.P.; Gubin, S.V.; Vishnivetskaya, T.A.; Onstott, T.C.; Rivkina, E.M. Viable nematodes from late Pleistocene permafrost of the Kolyma river lowland. *Dokl. Biol. Sci.* **2018**, *480*, 100–102. [CrossRef]

28. Su, Y.-C.; Ho, C.-L. Essential oil compositions and antimicrobial activities of various parts of *Litsea cubeba* from Taiwan. *Nat. Prod. Comm.* **2016**, *11*, 4515–4518. [CrossRef]

29. Ebani, V.V.; Nardoni, S.; Bertelloni, F.; Giovanelli, S.; Rocchigiani, G.; Pistelli, L.; Mancianti, F. Antibacterial and antifungal activity of essential oils against some pathogenic bacteria and yeasts shed from poultry. *Flavour Fragr. J.* **2016**, *31*, 302–309. [CrossRef]

30. Jeon, Y.-J.; Lee, H.-S. Chemical composition and acaricidal activities of essential oils of *Litsea cubeba* fruits and *Mentha arvensis* leaves against house dust and stored food mites. *J. Essent. Oil Bear.* **2016**, *19*, 1721–1728. [CrossRef]

31. Park, I.-K.; Kim, J.; Lee, S.-G.; Shin, S.-C. Nematicidal activity of plant essential oils and components from ajowan (*Trachyspermum ammi*), allspice (*Pimenta dioica*) and litsea (*Litsea cubeba*) essential oils against pine wood nematode (*Bursaphelenchus xylophilus*). *J. Nematol.* **2007**, *39*, 275–279. [PubMed]

32. Di Mola, A.; Massa, A.; De Feo, V.; Basile, A.; Pascale, M.; Aquino, R.P.; De Caprariis, P. Effect of citral and citral related compounds on viability of pancreatic and human B-lymphoma cell lines. *Med. Chem. Res.* **2017**, *26*, 631–639. [CrossRef]

33. Singh, G.; Maurya, S.; De Lampasona, M.P.; Catalan, C. Chemical constituents, antifungal and antioxidative potential of *Foeniculum vulgare* volatile oil and its acetone extract. *Food Control* **2006**, *17*, 745–752. [CrossRef]

34. Kaur, G.J.; Arora, D.S. Antibacterial and phytochemical screening of *Anethum graveolens, Foeniculum vulgare* and *Trachyspermum ammi*. *BMC Complement. Altern. Med.* **2009**, *9*, 30. [CrossRef]

35. Orhan, I.E.; Ozcelik, B.; Kartal, M.; Kan, Y. Antimicrobial and antiviral effects of essential oils from selected Umbelliferae and Labiatae plants and individual essential oil components. *Turk. J. Biol.* **2012**, *36*, 239–246.

36. Kim, S.G.; Liem, A.; Stewart, B.C.; Miller, J.A. New studies on *trans*-anethole oxide and *trans*-asarone oxide. *Carcinogenesis* **1999**, *20*, 1303–1307. [CrossRef]

37. Kakarla, S.; Ganjewala, D. Antimicrobial activity of essential oils of four lemongrass (*Cymbopogon flexuosus* Steud) varieties. *Med. Aromat. Plant Sci. Biotechnol.* **2009**, *3*, 107–109.

38. Sarma, A.; Saikia, R.; Sarma, T.C.; Boruah, P. Lemongrass [*Cymbopogon flexuosus* (Steud) Wats] inflorescence oil: A potent anti fungal agent. *J. Essent. Oil Bear.* **2004**, *7*, 87–90. [CrossRef]

39. Han, X.; Parker, T.L. Lemongrass (*Cymbopogon flexuosus*) essential oil demonstrated anti–inflammatory effect in pre-inflamed human dermal fibroblasts. *Biochim. Open* **2017**, *4*, 107–111. [CrossRef]

40. Mandal, S.; Mandal, M. Coriander (*Coriandrum sativum* L.) essential oil: Chemistry and biological activity. *Asian Pac. J. Trop. Biomed.* **2015**, *5*, 421–428. [CrossRef]

41. Ratha bai, V.; Kanimozhi, D. Evaluation of antimicrobial activity of *Coriandrum sativum*. *IJSRR* **2012**, *1*, 1–10.

42. Chung, I.M.; Ahmad, A.; Kim, S.J.; Naik, P.M.; Nagella, P. Composition of the essential oil constituents from leaves and stems of Korean *Coriandrum sativum* and their immunotoxicity activity on the *Aedes aegypti* L. *Immunopharmacol. Immunotoxicol.* **2012**, *34*, 152–156. [CrossRef] [PubMed]

43. Kim, J.; Seo, S.M.; Lee, S.G.; Shin, S.C.; Park, I.K. Nematicidal activity of plant essential oils and components from coriander (*Coriandrum sativum*), oriental sweetgum (*Liquidambar orientalis*), and valerian (*Valeriana wallichii*) essential oils against pinewood nematode (*Bursaphelenchus xylophilus*). *J. Agric. Food Chem.* **2008**, *56*, 7316–7320. [CrossRef] [PubMed]

44. Potter, T.L.; Fagerson, I.S. Composition of coriander leaf volatiles. *J. Agric. Food Chem.* **1990**, *38*, 2054–2056. [CrossRef]

45. Bhuiyan, N.I.; Begum, J.; Sultana, M. Chemical composition of leaf and seed essential oil of *Coriandrum sativum* L. from Bangladesh. *Bangladesh J. Pharmacol.* **2009**, *4*, 150–153. [CrossRef]

46. Kocevski, D.; Du, M.; Kan, J.; Jing, C.; Lacanin, I.; Pavlovic, H. Antifungal effect of *Allium tuberosum, Cinnamomum cassia,* and *Pogostemon cablin* essential oils and their components against population of *Aspergillus* species. *J. Food Sci.* **2013**, *78*, M731–M737. [CrossRef]

47. Huang, D.F.; Xu, J.-G.; Liu, J.-X.; Zhang, H.; Hu, Q.P. Chemical constituents, antibacterial activity and mechanism of action of the essential oil from *Cinnamomum cassia* bark against four food related bacteria. *Microbiology* **2014**, *83*, 357–365. [CrossRef]

48. Munazza, F.; Najam-us-Sahar, S.Z.; Deeba, A.; Farhan, A. In vitro antiviral activity of *Cinnamomum cassia* and its nanoparticles against H7N3 influenza A virus. *J. Microbiol. Biotechnol.* **2016**, *26*, 151–159. [CrossRef]

49. Yang, Y.; Isman, M.B.; Tak, J.-H. Insecticidal activity of 28 essential oils and a commercial product containing *Cinnamomum cassia* bark essential oil against *Sitophilus zeamais* Motschulsky. *Insects* **2020**, *11*, 474. [CrossRef]

50. Liu, X.C.; Cheng, J.; Zhao, N.N.; Liu, Z.L. Insecticidal activity of essential oil of *Cinnamomum cassia* and its main constituent, trans-cinnamaldehyde, against the booklice, *Liposcelis bostrychophila*. *Trop. J. Pharm. Res.* **2014**, *13*, 1697–1702. [CrossRef]

51. Kong, J.-O.; Lee, S.-M.; Moon, Y.-S.; Lee, S.-G.; Ahn, Y.-J. Nematicidal activity of cassia and cinnamon oil compounds and related compounds toward *Bursaphelenchus xylophilus* (Nematoda: Parasitaphelenchidae). *J. Nematol.* **2007**, *39*, 31–36. [PubMed]

52. Shan, B.; Cai, Y.-Z.; Brooks, J.D.; Corke, H. Antibacterial properties and major bioactive components of cinnamon stick (*Cinnamomum burmannii*): Activity against foodborne pathogenic bacteria. *J. Agric. Food Chem.* **2007**, *55*, 5484–5490. [CrossRef] [PubMed]

53. Song, Y.R.; Choi, M.S.; Choi, G.W.; Park, I.K.; Oh, C.S. Antibacterial activity of cinnamaldehyde and estragole extracted from plant essential oils against *Pseudomonas syringae* pv. actinidiae causing bacterial canker disease in kiwifruit. *Plant Pathol. J.* **2016**, *32*, 363–370. [CrossRef] [PubMed]

54. Wang, Z.; Kim, H.K.; Tao, W.; Wang, M.; Ahn, Y.-J. Contact and fumigant toxicity of cinnamaldehyde and cinnamic acid and related compounds to *Dermatophagoides farinae* and *Dermatophagoides pteronyssinus* (Acari: Pyroglyphidae). *J. Med. Entomol.* **2011**, *48*, 366–371. [CrossRef]

55. Echeverrigaray, S.; Zacaria, J.; Beltrão, R. Nematicidal activity of monoterpenoids against the root-knot nematode *Meloidogyne incognita*. *Phytopathology* **2010**, *100*, 199–203. [CrossRef]

56. Perveen, S. *Terpenes and Terpenoids*; IntechOpen: Rijeka, Croatia, 2018; pp. 1–13. [CrossRef]

57. Daintith, J. *A Dictionary of Chemistry*, 6th ed.; Oxford University Press: Oxford, UK, 2008; pp. 16–17.

58. Lindahl, R. Aldehyde dehydrogenases and their role in carcinogenesis. *Crit. Rev. Biochem. Mol. Biol.* **1992**, *27*, 283–335. [CrossRef]

59. Laskar, A.A.; Younus, H. Aldehyde toxicity and metabolism: The role of aldehyde dehydrogenases in detoxification, drug resistance and carcinogenesis. *Drug Metab. Rev.* **2019**, 1–23. [CrossRef]

60. Lopachin, R.M.; Gavin, T. Molecular mechanisms of aldehyde toxicity: A chemical perspective. *Chem. Res. Toxicol.* **2014**, *27*, 1081–1091. [CrossRef]

61. Avato, P.; Laquale, S.; Argentieri, M.P.; Lamiri, A.; Radicci, V.; D'Addabbo, T. Nematicidal activity of essential oils from aromatic plants of Morocco. *J. Pest Sci.* **2017**, *90*, 711–722. [CrossRef]

62. Santana, O.; Andrés, M.F.; Sanz, J.; Errahmani, N.; Abdeslam, L.; González-Coloma, A. Valorization of essential oils from Moroccan aromatic plants. *NPC* **2014**, *9*, 1109–1114. [CrossRef]

63. Ibrahim, S.K.; Traboulsi, A.F.; El-Haj, S. Effect of essential oils and plant extracts on hatching, migration and mortality of *Meloidogyne incognita*. *Phytopathol. Mediterr.* **2006**, *45*, 238–246. [CrossRef]

64. Kim, J.; Seo, S.-M.; Park, I.-K. Nematicidal activity of plant essential oils and components from *Gaultheria fragrantissima* and *Zanthoxylum alatum* against the pine wood nematode, *Bursaphelenchus xylophilus*. *Nematology* **2011**, *13*, 87–93. [CrossRef]

65. Napoli, E.; Giovino, A.; Carrubba, A.; Siong, V.H.Y.; Rinoldo, C.; Nina, O.; Ruberto, G. Variations of essential oil constituents in oregano (*Origanum vulgare* subsp. *viridulum* (= *O. heracleoticum*) over cultivation cycles. *Plants* **2020**, *9*, 1174. [CrossRef] [PubMed]

66. Hooper, D.J. Extraction of free-living stages from soil. In *Laboratory Methods for Work with Plant and Soil Nematodes*, 6th ed.; Southey, J.F., Ed.; Ministry of Agriculture, Fisheries and Food: London, UK, 1986; pp. 5–30.

67. Adams, R.P. *Identification of Essential Oil Components by Gas Chromatography/Mass Spectrometry*, 4th ed.; Allured Publishing Corporation: Carol Stream, IL, USA, 2007; pp. 1–804.

68. Daum, R.J. Revision of two computer programs for Probit analysis. *Bull. Entomol. Soc. Am.* **1970**, *16*, 10–15. [CrossRef]

69. Abbott, W.S. A method of computing the effectiveness of an insecticide. *J. Am. Mosq. Control Assoc.* **1987**, *3*, 302–303. [CrossRef]

Chapter-9

Popular Biofortified Cassava Cultivars Are Heavily Impacted by Plant Parasitic Nematodes, Especially *Meloidogyne* Spp.

1. Introduction

Cassava is a major staple food crop in tropical and subtropical Africa, Asia, and Latin America, where approximately 500 million people depend on it as a major carbohydrate (energy) source [1]. It is an important crop for food security in these regions, partly because it yields more energy per hectare than many other major crop. In Africa, cassava is the most important of all root and tuber crops as a source of calories for human and livestock needs, ranking 4th, after rice, sugarcane and maize [2]. To meet rising demands for cassava but also to improve its nutritional value as a food, there has been considerable investment to breed improved cultivars, including for higher mineral and vitamin contents, a process referred to as biofortification [3]. The specific enhancement of nutritional elements through genetic improvement is referred to as biofortification [4]. The Bill and Melinda Gates Foundation have supported a global effort to develop cassava germplasm enriched with bioavailable nutrients since 2005.

The BioCassava Plus initiative has six major objectives, including reducing cyanogen content, delaying postharvest deterioration and developing disease-resistant cultivars [5]. Using hybridization and selective breeding, researchers in Nigeria have developed new yellow cultivars of cassava that naturally produce a higher level of beta-carotene, which will help in reducing malnutrition caused by vitamin A deficiency in the region [6]. A total of seven biofortified cassava cultivars with total carotenoid content in the range of 8–12 μg/g fresh weight and 30–33% dry matter have been released [5], which compare with white cultivar total carotenoid content and dry matter in the range of 0.05–0.09 μg/g fresh weight and 35–37%, respectively. The dry matter content for provitamin A cultivars, therefore, is relatively lower compared to locally used cultivars and is a priority for improvement [5]. Cassava biofortification has largely been aimed at addressing vitamin A deficiency [7], an important public health problem in sub-Saharan Africa.

However, while raising the nutritional value of storage roots is a worthy objective, with great prospects for impacting the lives of millions, it must be additionally coordinated and associated with other valuable traits, such as pest and disease resistance. A number of biotic constraints affect the production of cassava, especially diseases and in particular virus diseases [8]. Major efforts have focused on breeding resistance against these threats into new, improved cultivars [8]. Less recognized threats, however, such as PPN have received much less attention. Although not well recognized, there is considerable, and growing, evidence of the damage that PPN inflict on cassava production [9–12]. In many cases, however, nematode damage often goes unnoticed. Traditionally, cassava is considered a hardy crop and generally viewed as immune to PPN. The naturally 'knobbly' and rough texture of the roots, which can disguise nematode damage to casual observation, partly aids this perception, while there may also be few roots present at harvest, especially if affected by nematode infection and have become necrotic and died off. Among the PPN that infect cassava, root-knot nematodes (*Meloidogyne* spp.) are the most prominent [13,14]. A number of studies have demonstrated the damaging impact of root-knot nematode infection, including their association with an increased incidence of rots [14]. Other PPN are also associated with cassava losses, such as *Pratylenchus brachyurus* and *Scutellonema bradys*, but the root-knot nematodes *M. incognita* and *M. javanica* are the most commonly reported and the most important nematode pests [12]. Some reports have documented almost total losses due to *Meloidogyne* spp. [15], although the association is sometimes not always so clear [16]. The association with rots can also disguise nematode damage and losses attributable to PPN but is an important aspect. This indirect consequence can lead to greater losses due to secondary fungal and bacterial rots and indeed has been shown to be strongly associated [17]. Given that improving the storability of cassava storage roots is a key breeding objective, the appropriate management of PPN will undoubtedly contribute to improving this objective [15,17]. Earlier studies have illustrated the variable susceptibility of cassava cultivars to *Meloidogyne* spp. [14,16], including biofortified cultivars [14,18,19]. The current study was undertaken to evaluate the effect of PPN on officially released biofortified cultivars in the field. These cultivars, released in Kenya and Nigeria, had been reported in a previous study to be susceptible to *M. incognita* in pots [19]. Our study builds on the pot evaluation study; using a field naturally infested with PPN, we assessed the impact of PPN infection on these cultivars and the implication of this for cassava farmers and consumers in the region.

2. Results

2.1 Plant-Parasitic Nematode Identification and Densities

Eleven genera of PPN were recorded: Meloidogyne, Pratylenchus, Helicotylenchus, Scutellonema, Hoplolaimus, Tylenchus, Longidorus, Aphelenchus, Xiphinema, Rotylenchulus and Radopholus (Figure 1). The initial population densities (Pi) were relatively low for seven nematode genera, while the more prominent genera were Meloidogyne, Pratylenchus, Helicotylenchus and Scutellonema (Table 1). Carbofuran application significantly ($p \leq 0.05$) suppressed PPN densities in all treated plots, compared with the untreated plots (Figure 1). The genera Meloidogyne, Pratylenchus, Helicotylenchus and Scutellonema remained prominent over the duration of the trials with Pf's significantly ($p \leq 0.05$) higher than other genera; Meloidogyne spp. had higher soil densities than all other genera (Figure 1).

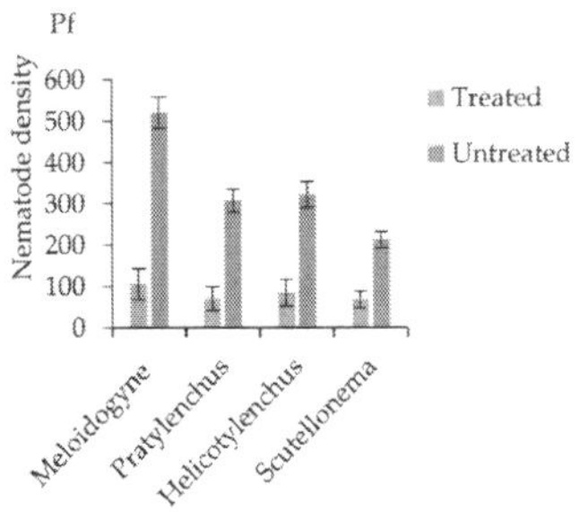

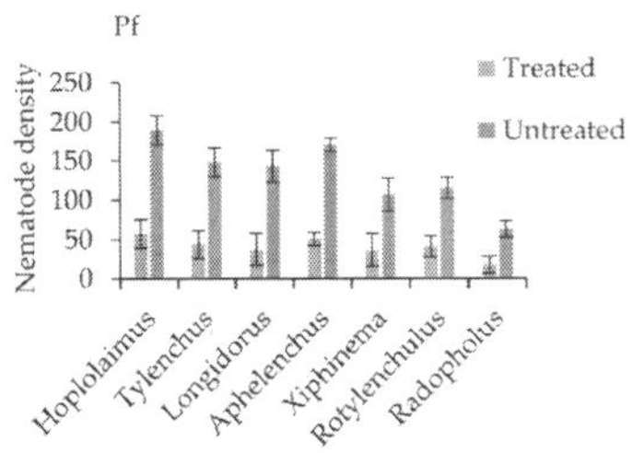

Figure 1. Effect of carbofuran on the final population of four major plant-parasitic nematodes encountered in field trials in Nigeria. P*f* = final density at harvest—12 months after planting; treatments: treated = 3G Carbofuran at 60 g/plot twice; untreated = untreated control. Error bars = Least Significant Difference ($p \leq 0.05$).

Table 1. Population of four major plant-parasitic nematodes encountered at planting in field trials in Nigeria [1].

Treatment[2]	Meloidogyne[3]	Pratylenchus [3]	Helicotylenchus [3]	Scutellonema [3]
Total Initial Population (Pi)				
Carbofuran	59.86[a]	23.71 [a]	36.79 [a]	19.43 [a]
Untreated	59.07[a]	25.07 [a]	36.71 [a]	18.43 [a]

[1]n = 8: means of four replications x two experiments; 2 3G Carbofuran was applied at a rate of 60 g/plot twice; [3]soil densities from 250 g/soil; for each treatment group values within a column followed by a different letter are significantly ($p \leq 0.05$) different.

2.2. Root Galling Damage and Host Suitability of Biofortified Cassava Cultivars Due to Meloidogyne Spp., 12 Months after Planting

All cultivars in the untreated plots reacted to *Meloidogyne* species infection with varying intensity, ranging from a gall index of 3.6–5.0 (Table 2). No galling of feeder roots was recorded in carbofuran-treated plots, but a low *Meloidogyne* population was recorded in the soil. A gall index of 5.0 was recorded on the check cultivar only, IITA-TMS-IBA30572, which also recorded the highest number of galls. Of the biofortified cultivars, IITA-TMS-IBA011368, IITA-TMS-IBA011371 and IITA-TMS-IBA070593 recorded the highest number of galls. In the untreated plots, all biofortified cassava cultivars reacted to *Meloidogyne* spp. infection and were all rated as good hosts, based upon a reproduction factor (RF) greater than 5.0 (Table 2).

Table 2. Root galling damage and host suitability of six biofortified cassava cultivars due to *Meloidogyne* spp, 12 months after planting in field trials in Nigeria[1].

Cultivar	Treatment [2]	Total *Meloidogyne* Density (root + soil) [3]	Gall Index [4]	RF (Pf/Pi) [5]	Host Status [6]
IITA-TMS-IBA011368	Carbofuran	411 ^{a}	0.0 ^{a}		
	Untreated	2087 ^{b}	4.3 ^{b}	36.6	G
IITA-TMS-IBA011412	Carbofuran	455 ^{a}	0.0 ^{a}		
	Untreated	1688 ^{b}	3.8 ^{b}	27	G
IITA-TMS-IBA011371	Carbofuran	485 ^{a}	0.0 ^{a}		
	Untreated	2194	4.3 ^{b}	43.4	G
IITA-TMS-IBA070593	Carbofuran	525 ^{a}	0.0 ^{a}		
	Untreated	2064 ^{b}	4.3 ^{b}	35.7	G
IITA-TMS-IBA070539	Carbofuran	599 ^{a}	0.0 ^{a}		
	Untreated	1781 ^{b}	3.8 ^{b}	29.7	G
NR 07/0220	Carbofuran	483 ^{a}	0.0 ^{a}		
	Untreated	1800 ^{b}	3.6 ^{b}	28.2	G
IITA-TMS-IBA30572 (check)	Carbofuran	251 ^{a}	0.0 ^{a}		
	Untreated	2361 ^{b}	5.0 ^{b}	42.6	G

[1]n = 8: means of four replications x two experiments; 2 3G Carbofuran was applied at a rate of 60 g/plot twice; [3] root and soil densities combined from 10 g/roots and 250 g/soil; 4 treated plots recorded no galling;

Table 3. (a). Mean squares for growth and development of six biofortified cassava and one white cultivar in field trials in Nigeria. (b) Growth and development of six biofortified cassava cultivars in field trials in Nigeria[1].

(a)

Source	Df	Plant Height (cm)				Stem Girth (cm)			
		3 MAP	6 MAP	9 MAP	12 MAP	3 MAP	6 MAP	9 MAP	12 MAP
Year	1	124.95	410.81	84.88	393.38	0.0015	0.00009	1.10 *	2.26 *
Replicate	3	121.96	513.91	1095.76	502.9	0.1600 *	0.44104 *	0.09	0.05
Cultivar	6	3564.50 ***	5790.16 ***	4530.87 ***	9944.81 ***	0.7353 ***	0.94247 ***	0.32	0.47
Treatment	1	12,539.84 ***	38,498.68 ***	82,889.28 ***	127,419.80 ***	0.7873 **	3.97509 ***	5.99 ***	6.56 ***
Cultivar * treatment	6	263.1	762.68	367.09	2058.23	0.0451	0.24967 *	0.17	0.04
Cultivar * year	6	105.58	892.17	383.49	1343.34	0.1689 *	0.23217	0.19	0.62 *
Treatment * year	1	13.23	805.18	324.02	2047.73	0.5144 *	0.07509	0.21	0.02
Error	87	151.04	467.32	819.3	1200.16	0.0576	0.11483	0.17	0.22

(b)

Cultivar	Treatment [2]	Plant Height (cm)				Stem Girth (cm)			
		3 MAP	6 MAP	9 MAP	12 MAP	3 MAP	6 MAP	9 MAP	12 MAP
IITA-TMS-IBA011368	Carbofuran	89.58 ^{a}	180.21 ^{a}	198.30 ^{a}	279.45 ^{a}	1.28 ^{a}	2.12 ^{a}	2.46 ^{a}	2.86 ^{a}
	Untreated	61.30 ^{b}	125.21 ^{b}	154.16 ^{b}	188.10 ^{b}	1.14 ^{a}	1.86 ^{a}	1.93 ^{b}	2.26 ^{b}
IITA-TMS-IBA011412	Carbofuran	97.29 ^{a}	154.33 ^{a}	196.06 ^{a}	261.43 ^{a}	1.53 ^{a}	2.34 ^{a}	2.68 ^{a}	2.94 ^{a}
	Untreated	65.13 ^{b}	99.46 ^{b}	132.21 ^{b}	197.45 ^{b}	1.29 ^{a}	1.51 ^{b}	1.83 ^{b}	2.40 ^{b}
IITA-TMS-IBA011371	Carbofuran	83.69 ^{a}	169.83 ^{a}	221.08 ^{a}	266.39 ^{a}	1.31 ^{a}	2.25 ^{a}	2.55 ^{a}	2.91 ^{a}
	Untreated	61.95 ^{b}	138.66 ^{b}	154.70 ^{b}	203.90 ^{b}	1.19 ^{a}	1.94 ^{a}	2.28 ^{b}	2.59 ^{b}
IITA-TMS-IBA070593	Carbofuran	62.95 ^{a}	146.20 ^{a}	185.55 ^{a}	214.50 ^{a}	1.04 ^{a}	1.94 ^{a}	2.51 ^{a}	2.96 ^{a}
	Untreated	52.96 ^{a}	125.44 ^{a}	144.78 ^{b}	190.73 ^{a}	1.03 ^{a}	1.60 ^{a}	2.24 ^{a}	2.56 ^{b}
IITA-TMS-IBA070539	Carbofuran	66.33 ^{a}	144.23 ^{a}	181.95 ^{a}	235.83 ^{a}	1.46 ^{a}	1.71 ^{a}	2.53 ^{a}	3.03 ^{a}
	Untreated	50.60 ^{b}	109.41 ^{b}	129.30 ^{b}	167.90 ^{b}	1.18 ^{a}	1.48 ^{a}	2.04 ^{b}	2.51 ^{b}
NR 07/0220	Carbofuran	44.19 ^{a}	109.68 ^{a}	162.99 ^{a}	201.60 ^{a}	0.83 ^{a}	1.43 ^{a}	2.14 ^{a}	2.50 ^{a}
	Untreated	29.81 ^{b}	86.49 ^{a}	110.10 ^{b}	129.88 ^{b}	0.70a	1.35 ^{a}	1.84 ^{b}	2.08 ^{b}
IITA-TMS-IBA30572 (check)	Carbofuran	83.24 ^{a}	146.34 ^{a}	191.30 ^{a}	266.95 ^{a}	1.45 ^{a}	2.11 ^{a}	2.44 ^{a}	2.94 ^{a}
	Untreated	57.36 ^{b}	106.56 ^{b}	126.11 ^{b}	175.98 ^{b}	1.33 ^{a}	1.53 ^{b}	1.93 ^{a}	2.35 ^{a}
Overall mean	Carbofuran	75.32 ^{a}	150.12 ^{a}	191.03 ^{a}	246.59 ^{a}	1.27 ^{a}	1.99 ^{a}	2.47 ^{a}	2.88 ^{a}
	Untreated	54.16 ^{b}	113.03 ^{b}	135.91 ^{b}	179.13 ^{b}	1.12 ^{a}	1.61 ^{b}	2.01 ^{b}	2.39 ^{b}
SE	Carbofuran	3.69	8.94	7.35	9.51	0.1	0.1	0.09	0.12
	Untreated	3.91	5.61	10.08	11.93	0.08	0.14	0.16	0.20

, **, * = mean squares significant at $p \leq 0.05$, 0.01 and 0.0001 probability levels, respectively; MAP = months after planting. [1] n = 8: means of four replications x two experiments; [2] 3G Carbofuran was applied at a rate of 60 g/plot twice; MAP = months after planting; SE = standard error; for each treatment group values within a column followed by a different letter are significantly ($p \leq 0.05$) different.*

gall index: 1 = 1–2 galls,2 = 3–10 galls, 3 = 11–30 galls, 4 = 31–100 galls, 5 => 100 galls [20]; 5 RF = nematode reproduction factor [21]; 6 host status was categorized as good (G) when Pf/Pi > 5.0, fair (F) if 5.0 Pf/Pi > 1, poor (P) if 1 Pf/Pi > 0, and nonhost (N) when Pf/Pi = 0 [22]; for each treatment group values within a column followed by a different letter are significantly (p ≤ 0.05) different.

2.3 Growth and Development of Biofortified Cassava Cultivars

The analysis of the data showed that there was no cultivar or treatment interaction per year at 6, 9 and 12 months after planting (MAP) (Table 3a). Therefore, the data from the experiments were pooled together for analysis (Table 3b).

Table 3. (a). Mean squares for growth and development of six biofortified cassava and one white cultivar in field trials in Nigeria. (b) Growth and development of six biofortified cassava cultivars in field trials in Nigeria 1.

Source	Df	Plant Height (cm)				Stem Girth (cm)			
		3MAP	6MAP	9MAP	12MAP	3MAP	6MAP	9MAP	12MAP
Year	1	124.95	410.81	84.88	393.38	0.0015	0.00009	1.10*	2.26*
Replicate	3	121.96	513.91	1095.76	502.9	0.1600*	0.44104*	0.09	0.05
Cultivar	6	3564.50***	5790.16***	4530.87***	9944.81***	0.7353***	0.94247***	0.32	0.47
Treatment	1	12,539.84 ***	38,498.68 ***	82,889.28 ***	127,419.80 ***	0.7873**	3.97509***	5.99***	6.56***
Cultivar*treatment	6	263.1	762.68	367.09	2058.23	0.0451	0.24967*	0.17	0.04
Cultivar*year	6	105.58	892.17	383.49	1343.34	0.1689*	0.23217	0.19	0.62*
Treatment*year	1	13.23	805.18	324.02	2047.73	0.5144*	0.07509	0.21	0.02
Error	87	151.04	467.32	819.3	1200.16	0.0576	0.11483	0.17	0.22
				(b)					
Cultivar	**Treatment**	**Plant Height (cm)**					**Stem Girth (cm)**		
		3MAP	6MAP	9MAP	12MAP	3MAP	6MAP	9MAP	12MAP
IITA-TMS-IBA011368	Carbofuran	89.58^{a}	180.21^{a}	198.30^{a}	279.45^{a}	1.28^{a}	2.12^{a}	2.46^{a}	2.86^{a}
	Untreated	61.30^{b}	125.21^{b}	154.16^{b}	188.10^{b}	1.14^{a}	1.86^{a}	1.93^{b}	2.26^{b}
IITA-TMS-IBA011412	Carbofuran	97.29^{a}	154.33^{a}	196.06^{a}	261.43^{a}	1.53^{a}	2.34^{a}	2.68^{a}	2.94^{a}
	Untreated	65.13^{b}	99.46^{b}	132.21^{b}	197.45^{b}	1.29^{a}	1.51^{b}	1.83^{b}	2.40^{b}
IITA-TMS-IBA011371	Carbofuran	83.69^{a}	169.83^{a}	221.08^{a}	266.39^{a}	1.31^{a}	2.25^{a}	2.55^{a}	2.91^{a}
	Untreated	61.95^{b}	138.66^{b}	154.70^{b}	203.90^{b}	1.19^{a}	1.94^{a}	2.28^{b}	2.59^{b}
IITA-TMS-IBA070593	Carbofuran	62.95^{a}	146.20^{a}	185.55^{a}	214.50^{a}	1.04^{a}	1.94^{a}	2.51^{a}	2.96^{a}
	Untreated	52.96^{a}	125.44^{a}	144.78^{b}	190.73^{a}	1.03^{a}	1.60^{a}	2.24^{a}	2.56^{b}
IITA-TMS-IBA070539	Carbofuran	66.33^{a}	144.23^{a}	181.95^{a}	235.83^{a}	1.46^{a}	1.71^{a}	2.53^{a}	3.03^{a}
	Untreated	50.60^{b}	109.41^{b}	129.30^{b}	167.90^{b}	1.18^{a}	1.48^{a}	2.04^{b}	2.51^{b}
NR07/0220	Carbofuran	44.19^{a}	109.68^{a}	162.99^{a}	201.60^{a}	0.83^{a}	1.43^{a}	2.14^{a}	2.50^{a}
	Untreated	29.81^{b}	86.49^{a}	110.10^{b}	129.88^{b}	0.70a	1.35^{a}	1.84^{b}	2.08^{b}
IITA-TMS-IBA30572 (check)	Carbofuran	83.24^{a}	146.34^{a}	191.30^{a}	266.95^{a}	1.45^{a}	2.11^{a}	2.44^{a}	2.94^{a}
	Untreated	57.36^{b}	106.56^{b}	126.11^{b}	175.98^{b}	1.33^{a}	1.53^{b}	1.93^{a}	2.35^{a}
Overallmean	Carbofuran	75.32^{a}	150.12^{a}	191.03^{a}	246.59^{a}	1.27^{a}	1.99^{a}	2.47^{a}	2.88^{a}
	Untreated	54.16^{b}	113.03^{b}	135.91^{b}	179.13^{b}	1.12^{a}	1.61^{b}	2.01^{b}	2.39^{b}
SE	Carbofuran	3.69	8.94	7.35	9.51	0.1	0.1	0.09	0.12
	Untreated	3.91	5.61	10.08	11.93	0.08	0.14	0.16	0.20

, **, * = mean squares significant at p 0.05, 0.01 and 0.0001 probability levels, respectively; MAP = months after planting. 1 n = 8: means of four replications x two experiments; 2 3G Carbofuran was applied at a rate of 60 g/plot twice; MAP = months after planting; SE = standard error; for each treatment group values within a column followed by a different letter are significantly ($p \leq 0.05$) different.*

The application of carbofuran improved the growth of all cassava cultivars at some point during the growing cycle of the experiments; treated plants were generally significantly ($p \leq 0.05$) taller and sturdier (Table 3b). Generally, stunting of aerial growth was observed on untreated plants at 3 MAP, which became more pronounced at 6, 9 and 12 MAP, when compared with treated plots. Cultivars were significantly ($p \leq 0.05$) shorter in untreated plots compared with treated plots, except for IITA-TMS-IBA070593. Significant ($p \leq 0.05$) reduction was also recorded in the stem girth of untreated plots in some of the cultivars, when compared with the treated plots. The overall mean showed that the growth and development of all cultivars in the untreated plots were significantly ($p \leq 0.05$) suppressed at 3, 6, 9 and 12 MAP, when compared with treated plots, and the standard error (SE) was mostly higher in the untreated plots when compared with treated plots and increased at 3, 6, 9 and 12 MAP (Table 3b).

2.4 Yield Evaluation of Biofortified Cassava Cultivars

The results showed that there was cultivar and treatment interaction per year (Table 4a). The non-marketable storage yields showed no interaction per year (Table 4b), and these data were pooled together for analysis (Table 4e).

(a)

Source	Df	Plant Weight [1]	Marketable Storage Roots [1]		Non-Marketable Storage Roots [1]		Total Yield/Plot
		Fresh Weight (kg)	Number	Weight (kg)	Number	Weight (kg)	Storage Roots (kg)
Year	1	10.26 *	24.89 *	23.87 **	8.58 **	3.16 **	10.38 *
Replicate	3	0.26	4.64	2.32	2.53	0.24	4.4
Cultivar	6	13.41 ***	27.71 ***	14.93 ***	4.07 **	0.71 *	16.50 ***
Treatment	1	142.88 ***	183.09 ***	223.74 ***	40.08 ***	9.37 ***	162.48 ***
Cultivar * treatment	6	2.39	3.07	3.53 *	3.73	0.82 *	2.44
Cultivar * year	6	0.54	4.99	0.7	3.64**	0.24	1.16
Treatment * year	1	0.84	10.81	0.01	0.44	1.16 *	1.35
Error	87	2.01	3.96	1.55	1.12	0.25	1.98

(b)

Source	Df	Rotted Storage Roots [1]		Deformed Storage Roots [1]	
		Number	Weight (kg)	Number	Weight (kg)
Year	1	1.75	0.97 *	2.89 *	0.79 **
Replicate	3	0.67	0.29	1.5	0.09
Cultivar	6	1.36	0.36	0.82	0.04
Treatment	1	22.32 ***	6.51 ***	2.29	0.17
Cultivar * treatment	6	1.99 *	0.48 *	0.58	0.1
Cultivar * year	6	0.33	0.12	2.18 *	0.11
Treatment * year	1	0.14	0.79 *	0.32	0.12
Error	87	0.74	0.18	0.61	0.07

Table 4. (a) Mean squares for yield evaluation of six biofortified cassava and one white cultivar in field trials in Nigeria. (b) Mean squares for non-marketable yield evaluation of six biofortified cassava and one white cultivar in field trials in Nigeria. (c) Yield evaluation of six biofortified cassava cultivars in first field trial in Nigeria 1. (d) Yield evaluation of six biofortified cassava cultivars in second field trial in Nigeria 1. (e) Non-marketable yield evaluation of six biofortified cassava cultivars in field trials in Nigeria 1.

Table 4. *Cont.*

(c)

Cultivar	Treatment [2]	Plant Weight [3]	Marketable Storage Roots [3]		Non-Marketable Storage Roots [3]		Total Yield/Plot
		Fresh Weight (kg)	Number	Weight (kg)	Number	Weight (kg)	Storage Roots (kg)
First trial							
IITA-TMS-IBA011368	Carbofuran	6.68^{a}	9.00^{a}	7.05^{a}	1.25^{a}	0.25^{a}	7.30^{a}
	Untreated	3.50^{b}	4.50^{b}	2.23^{b}	4.50^{b}	1.93^{b}	3.70^{b}
IITA-TMS-IBA011412	Carbofuran	4.58^{a}	7.23^{a}	5.65^{a}	1.25^{a}	0.23^{a}	5.88^{a}
	Untreated	2.88^{a}	3.25^{b}	2.43^{b}	5.00^{b}	1.85^{b}	3.60^{b}
IITA-TMS-IBA011371	Carbofuran	5.40^{a}	5.00^{a}	6.08^{a}	1.00^{a}	0.18^{a}	6.25^{a}
	Untreated	3.30^{b}	3.25^{a}	2.28^{b}	1.50^{a}	0.73^{a}	3.00^{b}
IITA-TMS-IBA070593	Carbofuran	3.95^{a}	4.75^{a}	3.50^{a}	0.75^{a}	0.30^{a}	3.65^{a}
	Untreated	2.00^{a}	2.50^{a}	2.80^{a}	0.50^{a}	0.15^{a}	3.10^{a}
IITA-TMS-IBA070539	Carbofuran	4.48^{a}	8.25^{a}	4.88^{a}	0.75^{a}	0.23^{a}	5.10^{a}
	Untreated	2.30^{b}	5.25^{a}	2.10^{b}	1.75^{a}	0.75^{a}	2.85^{a}
NR 07/0220	Carbofuran	2.73^{a}	6.00^{a}	3.05^{a}	0.75^{a}	0.30^{a}	3.35^{a}
	Untreated	1.78^{a}	1.75^{b}	0.70^{a}	1.75^{a}	1.00^{a}	1.68^{a}
IITA-TMS-IBA30572	Carbofuran	5.32^{a}	8.75^{a}	4.93^{a}	1.25^{a}	0.60^{a}	5.53^{a}
(check)	Untreated	3.28^{b}	6.25^{a}	2.95^{a}	1.25^{a}	0.85^{a}	3.80^{b}
Overall mean	Carbofuran	4.73^{a}	7.00^{a}	5.02^{a}	1.00^{a}	0.30^{a}	5.29^{a}
	Untreated	2.72^{b}	3.82^{b}	2.21^{b}	2.32^{b}	1.04^{a}	3.10^{b}
SE	Carbofuran	0.4	0.58	0.28	0.28	0.05	0.36
	Untreated	0.43	0.78	0.7	0.48	0.22	0.48

(d)

Cultivar	Treatment [2]	Plant Weight [3]	Marketable Storage Roots [3]		Non-marketable Storage Roots [3]		Total Yield/Plot
		Fresh weight (kg)	Number	Weight (kg)	Number	Weight (kg)	Storage Roots (kg)
Second trial							
IITA-TMS-IBA011368	Carbofuran	7.55^{a}	9.00^{a}	8.33^{a}	0.00^{a}	0.00^{a}	8.33^{a}
	Untreated	3.75^{b}	6.75^{b}	4.20^{b}	2.00^{b}	1.05^{a}	5.25^{b}
IITA-TMS-IBA011412	Carbofuran	5.60^{a}	6.00^{a}	6.43^{a}	0.75^{a}	0.20^{a}	6.63^{a}
	Untreated	3.40^{b}	6.25^{a}	3.65^{b}	1.75^{a}	0.68^{b}	3.33^{b}
IITA-TMS-IBA011371	Carbofuran	6.23^{b}	5.50^{a}	6.05^{a}	1.00^{a}	0.10^{a}	6.28^{a}
	Untreated	3.50^{a}	3.70^{b}	3.70^{b}	1.25^{a}	0.20^{a}	3.93^{b}
IITA-TMS-IBA070593	Carbofuran	5.03^{a}	6.75^{a}	5.75^{a}	0.50^{a}	0.10^{a}	5.85^{a}
	Untreated	3.10^{a}	4.50^{a}	2.95^{b}	1.50^{a}	0.38^{a}	3.33^{a}

Table 4. *Cont.*

IITA-TMS-IBA070539	Carbofuran	5.43^{a}	8.25^{a}	5.55^{a}	1.00^{a}	0.33^{a}	5.88^{a}
	Untreated	3.25^{a}	8.00^{a}	3.03^{b}	1.75^{a}	0.55^{a}	3.58^{b}
NR 07/0220	Carbofuran	3.13^{a}	7.50^{a}	3.28^{a}	0.50^{a}	0.15^{a}	3.43^{a}
	Untreated	2.03^{a}	4.25^{b}	2.18^{a}	2.25^{b}	0.53^{a}	2.70^{a}
IITA-TMS-IBA30572 (check)	Carbofuran	5.63^{a}	8.25^{a}	6.35^{a}	0.25^{a}	0.13^{a}	6.48^{a}
	Untreated	2.53^{b}	4.25^{a}	2.10^{b}	0.10^{a}	0.25^{a}	2.35^{b}
Overall mean	Carbofuran	5.51^{a}	7.32^{a}	5.96^{a}	0.51^{a}	0.14^{a}	6.13^{a}
	Untreated	3.08^{b}	5.39^{b}	3.12^{b}	1.51^{a}	0.52^{a}	3.50^{b}
SE	Carbofuran	0.45	0.6	0.23	0.22	0.09	0.42
	Untreated	0.52	0.81	0.92	0.54	0.26	0.64

(e)

Cultivar	Treatment [2]	Rotted Storage Roots [3]		Deformed Storage Roots [3]	
		Number	Weight (kg)	Number	Weight (kg)
IITA-TMS-IBA011368	Carbofuran	0.00^{a}	0.00^{a}	0.63^{a}	0.13^{a}
	Untreated	2.13^{b}	1.09^{b}	0.88^{a}	0.28^{a}
IITA-TMS-IBA011412	Carbofuran	0.38^{a}	0.06^{a}	0.63^{a}	0.15^{a}
	Untreated	1.75^{a}	0.79^{a}	1.50^{b}	0.41^{b}
IITA-TMS-IBA011371	Carbofuran	0.50^{a}	0.06^{a}	0.25^{a}	0.11^{a}
	Untreated	1.00^{a}	0.49^{a}	0.50^{a}	0.15^{a}
IITA-TMS-IBA070593	Carbofuran	0.38^{a}	0.05^{a}	0.15^{a}	0.05^{a}
	Untreated	0.38^{a}	0.03^{a}	0.51^{a}	0.24^{a}
IITA-TMS-IBA070539	Carbofuran	0.25^{a}	0.11^{a}	0.63^{a}	0.16^{a}
	Untreated	0.88^{a}	0.33^{a}	0.75^{a}	0.29^{a}
NR 07/0220	Carbofuran	0.00^{a}	0.00^{a}	0.63^{a}	0.20^{a}
	Untreated	1.13^{b}	0.51^{b}	1.00^{a}	0.21^{a}
IITA-TMS-IBA30572 (check)	Carbofuran	0.13^{a}	0.03^{a}	0.25^{a}	0.11^{a}
	Untreated	0.63^{a}	0.44^{b}	0.50^{a}	0.34^{a}
Overall mean	Carbofuran	0.23^{a}	0.04^{a}	0.45^{a}	0.13^{a}
	Untreated	1.13^{b}	0.53^{b}	0.81^{b}	0.27^{b}
SE	Carbofuran	0.12	0.03	0.23	0.07
	Untreated	0.37	0.2	0.35	0.12

(a) *, **, *** = mean squares significant at $p \leq 0.05$, 0.01 and 0.0001 probability levels, respectively; [1] fresh storage roots number and weight in 5 plants/plot. (b) *, **, *** = mean squares significant at $p \leq 0.05$, 0.01 and 0.0001 probability levels, respectively; [1] fresh storage roots number and weight in 5 plants/plot. (c) [1] n = 4: means of four replications; [2] 3G Carbofuran was applied at a rate of 60 g/plot twice; [3] fresh storage roots number and weight in 5 plants/plot; SE = standard error; for each treatment group values within a column followed by a different letter are significantly ($p \leq 0.05$) different. (d) [1] n = 4: means of four replications; [2] 3G Carbofuran was applied at a rate of 60 g/plot twice; [3] fresh storage roots number and weight in 5 plants/plot; SE = standard error; for each treatment group values within a column followed by a different letter are significantly ($p \leq 0.05$) different. (e) [1] n = 8: means of four replications x two experiments; [2] 3G Carbofuran was applied at a rate of 60 g/plot twice; [3] fresh storage roots number and weight in 5 plants/plot; SE = standard error; for each treatment group values within a column followed by a different letter are significantly ($P \leq 0.05$) different.

Plant weight and storage root yields were largely improved ($p \leq 0.001$) across cassava cultivars in the two trials (Table 4a,c,d). Nematicide treatment led to $_{\text{higher}}$ ($p \leq 0.05$) numbers and fresh weights of marketable storage roots in most cultivars, compared with untreated plots (Figure 2). The number and weight of non-marketable storage roots were significantly ($p \leq 0.05$) lowered in cultivars IITA-TMS-IBA011368 and IITA-TMS-IBA011412 when compared with untreated plots in the first trial (Table 4c), while in the second trial (Table 4d), these cultivars in addition to NR 07/0220 were significantly ($p \leq 0.05$) lowered. A significant ($p \leq 0.05$) reduction in the number of rotted storage roots was also observed in treated plots, compared with untreated, with lower numbers for cultivars IITA-TMS-IBA011368 and NR 07/0220 (Table 4e). Fresh weights of rotted storage roots were similarly lower in treated plots, with cultivars IITA-TMS-IBA011368, NR 07/0220 and IITA-TMS-IBS30572 having significantly ($p \leq 0.05$) less. The number and weight of deformed storage roots of cultivar IITA-TMS-IBA011412 were also less ($p \leq 0.05$) in the treated plots, compared with the untreated (Table 4e). The overall mean showed that the aerial fresh weight of plants and the number and weight of marketable storage roots in the untreated plots were significantly ($p \leq 0.05$) lower, when compared with treated plots in the two trials, while SE rates were higher in the untreated plots when compared with treated plots (Table 4c,d), while the number and weight of non-marketable (rotted and deformed) storage roots were significantly ($p \leq 0.05$) lower in treated plots, with higher SE in the untreated plots when compared with treated plots (Table 4e).

(a) (b)

Figure 2. Storage roots of treated and untreated biofortified cassava cultivar IITA-TMS- IBA011368 at 12 months after planting; (a) treated with 3G Carbofuran; (b) untreated storage roots.

Results showed that two biofortified cassava cultivars, IITA-TMS-IBA070593 and IITA-TMS- IBA070539, had significantly ($p \leq 0.05$) lower total carotenoid contents of roots from untreated plants when compared with treated plants (Table 5). Similarly, dry matter content from untreated plants was lower in cultivars IITA-TMS-IBA011368 and IITA-TMS-IBA070593. When assessing total carotenoid and dry matter contents at the plot scale, however, taking into consideration the contents and yields, all the biofortified cultivars had significantly ($p \leq 0.05$) lower values, compared with treated plots. The overall mean showed that total carotenoids per plot and dry matter per plant and per plot were significantly ($p \leq 0.05$) lower in untreated plots, while SE rates were higher in the untreated plots when compared with treated plots (Table 5).

Table 5. Nutritional quality of six biofortified cassava cultivars in field trial in Nigeria 1.

Cultivar	Treatment 2	Storage Roots Yield (kg/plot 3)	Total Carotenoid (µg/g fr.wt./plant 3)	Total Carotenoid (kg/plot 3)	Dry Matter (%/plant 3)	Dry Matter (kg/plot 3)
IITA-TMS-IBA011368	Carbofuran	7.30 ^{a}	6.38 ^{a}	0.23 ^{a}	35.75 ^{a}	2.61 ^{a}
	Untreated	3.70 ^{b}	7.38 ^{a}	0.14 ^{b}	30.25 ^{b}	1.12 ^{b}
IITA-TMS-IBA011412	Carbofuran	5.88 ^{a}	6.58 ^{a}	0.19 ^{a}	33.25 ^{a}	1.96 ^{a}
	Untreated	3.60 ^{b}	6.45 ^{a}	0.12 ^{a}	30.50 ^{a}	1.10 ^{b}
IITA-TMS-IBA011371	Carbofuran	6.25 ^{a}	7.80 ^{a}	0.24 ^{a}	29.50 ^{a}	1.88 ^{a}
	Untreated	3.00 ^{b}	8.13 ^{a}	0.12 ^{b}	30.00 ^{a}	0.89 ^{b}
IITA-TMS-IBA070593	Carbofuran	3.65 ^{a}	9.81 ^{a}	0.18 ^{a}	34.50 ^{a}	1.26 ^{a}
	Untreated	3.10 ^{a}	6.37 ^{b}	0.10 ^{a}	19.75 ^{b}	0.61 ^{b}
IITA-TMS-IBA070539	Carbofuran	5.10 ^{a}	11.71 ^{a}	0.30 ^{a}	35.75 ^{a}	1.82 ^{a}
	Untreated	2.85 ^{a}	7.93 ^{b}	0.11 ^{b}	37.75 ^{a}	1.08 ^{b}
NR 07/0220	Carbofuran	3.35 ^{a}	5.06 ^{a}	0.08 ^{a}	24.25 ^{a}	0.81 ^{a}
	Untreated	1.68 ^{a}	7.03 ^{a}	0.06 ^{a}	24.50 ^{a}	0.41 ^{b}
IITA-TMS-IBA30572	Carbofuran	5.53 ^{a}	0.04 ^{a}	0.00 ^{a}	38.75 ^{a}	2.14 ^{a}
(check)	Untreated	3.80 ^{b}	0.00 ^{a}	0.00 ^{a}	36.25 ^{a}	1.38 ^{b}
Overall mean	Carbofuran	5.29 ^{a}	6.77 ^{a}	0.18 ^{a}	33.11 ^{a}	1.78 ^{a}
	Untreated	3.10 ^{b}	6.18 ^{a}	0.09 ^{b}	29.85 ^{b}	0.94 ^{b}
SE	Carbofuran	0.36	0.77	0.05	1.49	0.12
	Untreated	0.48	0.76	0.08	2.17	0.19

1 n = 4: means of four replications; 2 3G Carbofuran was applied at a rate of 60 g/plot twice; 3 fresh storage weight in 5 plants/plot; SE = standard error; for each treatment group values within a column followed by a different letter are significantly ($p \leq 0.05$) different.

3. Discussion

A pot study using the same six biofortified cultivars as in the current study found them all to be good hosts to the root-knot nematode *M. incognita*, which reduced growth and development after six months in the screenhouse [19]. Although a number of other PPN genera were encountered, the majority were in relatively low densities and likely posed little threat to the cassava. Four genera were more prominent, of which *Meloidogyne*, the most important nematode genera attacking cassava, dominated the PPN community. The focus of the current study therefore centered on *Meloidogyne* spp., although it is understood that *Pratylenchus, Helicotylenchus* and *Scutellonema* spp. could have had some influence on cassava growth, which can become important when they are present in large densities [23]. The effect of *M. incognita* on the nutritional content of these biofortified cassava was less conclusive, but the study provided an indication that *M. incognita* infection can negatively impact cassava quality. The current study clearly supports the pot study findings but now also demonstrates that *Meloidogyne* spp. infection will reduce the nutritional value of improved, biofortified cassava under field conditions. Although the effect varied across cultivar and quality was not consistently reduced proportionally (per unit weight), by taking the yield impact into account, the overall damaging effect of *Meloidogyne* spp. can be better appreciated. All the tested cassava cultivars were susceptible to *Meloidogyne* spp. infection, resulting in significant ($p \leq 0.05$) root galling damage and a reduction in plant growth and storage root yield of all but one of the six biofortified cultivars. Along with PPN densities, rotted storage roots were also much reduced in nematicide-treated plots. Rotting of storage roots directly affects their in-field storability, as well as their post-harvest longevity. Therefore, placing more emphasis on the management of PPN may be well justified, especially given the emphasis placed on nutritional biofortification and that increasing the storability and longevity of storage roots is a key breeding trait [16,24]. Carbofuran, however, is a toxic carbamate pesticide, which affects a wider range of pests and diseases than PPN alone. The reduction of rot-causing pathogens therefore is likely an additional effect, which would additionally reduce potential rots of cassava roots and should be considered. The pesticide did, however, enable a suitable comparison of PPN field densities, creating a differential against which to assess their impact on cassava. Other studies that have sought to assess the effect of PPN on cassava yield have used similar techniques, in addition to other methods,

such as solarization [12,18,25]. From these studies some sizeable yield reductions have been associated with PPN, in particular *Meloidogyne* spp., demonstrating the need for their management if cassava production systems are to be sustainably intensified. Low yields have consistently characterized cassava production in Nigeria and other sub-Saharan countries. Nematode management may provide a major way forward in improving yields in farmers' fields. The association between root rot incidence and *Meloidogyne* spp. infection has also been well demonstrated on cassava [18,25], as well as other root and tuber crops [16,26]. There is no doubting therefore the value of investing in PPN management and root-knot nematodes in particular, towards improving cassava productivity [12,14,15].

The current study showed that high PPN densities were associated with reduced crop performance following treatment with carbofuran, resulting in significant ($p \leq 0.001$) yield loss of biofortified cassava. *Meloidogyne* spp. in the untreated plots caused galling on feeder roots of all biofortified cultivars. In Nigeria, *Meloidogyne* spp. infection caused significant ($p \leq 0.05$) suppression in the growth and yield of elite cassava cultivars after 12 months in the field, despite relatively low observed levels of the nematode [14]. The loss in cassava yield was, however, mainly attributed to direct damage of the root system by the feeding activities of *Meloidogyne* spp. Although the current study was conducted at the International Institute of Tropical Agriculture (IITA) station, no inoculation was undertaken, natural PPN infestation levels were used and the trials were managed relative to farmer conditions. It is assumed, therefore, that these results provide a relatively fair reflection of the likely losses that farmers would experience. Elsewhere in Nigeria, significant cassava yield losses have also been recorded in farmer field trials naturally infested with *Meloidogyne* spp. Up to 200% yield increases were observed following the reduction of *Meloidogyne* spp. using solarization [25]. In Uganda, severe galling due to *Meloidogyne* spp. was reported in farmers' fields [27]. Separately, 94% of fields examined in Uganda presented galling damage, with 17% severely affected, indicating substantial yield losses [28]. The impact of *Meloidogyne* spp. on cassava production is a threat to production that is likely to become increasingly acute and more intense under more intensified cropping conditions [11,14,15]. Besides reducing crop growth, vigor and productivity, PPN can reduce the quality and nutritional value of crop products. This is not surprising as PPN infect the root system, disrupting nutrient uptake and reducing their distribution within the plant [29–31]. PPN parasitize plants, changing the nutrient apportioning and cause disturbance in water and nutrient relations necessary for optimal plant growth [32]. Although a number of studies on various crops have indicated or demonstrated this, the empirical evidence is relatively limited. In our study, the total carotenoid and dry matter contents per plot of all biofortified cultivars were significantly ($p \leq 0.05$) lower in untreated plots with higher PPN densities than treated plots. The current study and the preliminary pot study [15] now clearly show the impact on nutrition that *Meloidogyne* spp. can have, both on an individual plant, but especially when multiplied at scale. For example, the carotenoid content of cultivar IITA-TMS-IBA070539 was less by 0.19 kg per plot in untreated plots. This equates to a loss of 63% of total carotenoid content in the yield and quality of biofortified cassava, seriously undermining the efforts and investment to develop these high content biofortified cultivars.

Our study further confirms earlier reports that *Meloidogyne* spp. are the most prevalent and abundant PPN affecting cassava in Southwestern Nigeria. In the current study, the *Meloidogyne* spp. were not identified to species level, although it is likely that *M. javanica* and/or *M. incognita* were present, both of which are common to the region [33] and are the two most commonly occurring *Meloidogyne* spp. found infecting cassava [23]. As resistance against *Meloidogyne* spp. can be bred for in cassava, it would appear a useful mechanism for improving cassava for more intensive cultivation. The presence and infection of cassava by *Meloidogyne* spp. will reduce the yield and quality of cassava, including the nutritional content of biofortified cassava. Furthermore, *Meloidogyne* spp. infection is additionally associated with higher levels and incidence of rots, reducing the storability of cassava. PPN infection and damage to cassava has largely been overshadowed by other pests and diseases but is, however, a considerable threat to both yield, quality and storability. Breeding or actively selecting for nematode resistance during the evaluation process may therefore be more warranted than generally acknowledged or appreciated. In addition to creating more durable cultivars, more suitable to intensified cropping conditions, indirectly, this is likely to improve in-ground storability.

4. Materials and Methods

4.1 Experimental Details and Layout

Two field trials were planted in June 2017 and May 2018 in a well-drained sandy loam soil after ploughing and harrowing once each, at the IITA Ibadan, Nigeria (120 km north of Lagos). Cassava stems~15 cm long were planted at an angle into the ground, spaced 1 × 1 m in a line 7 m long for each cultivar, representing a plot of 8 plants. Trials were maintained for 12 months after planting (MAP) before harvesting. The study consisted of two factors, cassava genotype (seven cultivars) and nematicide treatment (two levels). Six biofortified cassava cultivars (IITA-TMS-IBA011368, IITA-TMS-IBA011412, IITA-TMS-IBA011371, IITA-TMS-IBA070593, IITA-TMS-IBA070539 and NR 07/0220) and a check cultivar of white cassava (IITA-TMS- IBA30572) were obtained from the IITA. The nematicide 3G Carbofuran was applied at the rate of 3 kg a.i./ha (60 g/plot) at planting and repeated at 3 MAP and compared with a control receiving no nematicide. The experiment was laid out in a randomized complete block design with four replicates each per cultivar per treatment.

4.2 Assessment of Nematode Population Density and Damage

Soil samples were collected from 8 points per plot using a soil auger to a depth of 30 cm at planting to obtain initial nematode population density (P*i*) and at harvest to obtain final nematode population density (P*f*). Nematodes were extracted from 250 g soil sub-samples using the tray method [34], after removing all stones and debris and thoroughly mixing the bulked soil from each plot. At harvest, roots from 5 plants per plot were combined, gently tapped free of soil, chopped finely, thoroughly mixed and a 10 g sub-sample removed for nematode extraction using the same method as for soil. Nematode extracts were removed after 24 h, allowed to settle for 5 h and the volume adjusted to 30 mL by siphoning off the excess [35]. The mean nematode density was assessed under a compound microscope from 5 x 1 mL aliquots pipetted into a Doncaster counting dish [36]. Nematodes were identified to genus level using Bell's Key [37] and a multiple tally counter used to count the different nematode genera. Total number of nematodes per plot from soil and root data was used to calculate the nematode reproduction factor (RF) [21]:

$$\frac{Pf \times 250\,\mathrm{g/soil} + Pf \times 10\,\mathrm{g/root}}{Pi} \qquad (1)$$

At harvest, the number of galls on 5 cm feeder roots per plant, removed randomly from 5 plants per plot, was counted and galling index (GI) per plant root assessed using the 1–5 gall index scale [20] (1 = 1–2 galls, 2 = 3–10 galls, 3 = 11–30 galls, 4 = 31–100 galls, 5 => 100 galls).

Host Status

Host suitability was categorized as good when $Pf/Pi > 5.0$, fair if $5.0 \geq Pf/Pi > 1$, poor if $1 \geq Pf/Pi > 0$ and nonhost when $Pf/Pi = 0$ based on a study method [22].

4.3 Measurement of Crop Growth Parameters

Crop growth parameters were collected at 3, 6, 9 and 12 MAP for plant height and girth from five randomly selected cassava plants per plot. At harvest, the five selected plants per plot were additionally assessed as a bulk (plot) for aerial plant weight, number and weight of marketable and non-marketable storage roots. Plant height was measured to the tallest point of pre-harvested plants using a wooden ruler; girth was measured at 10 cm above the soil surface using a Vernier caliper. Stem and leaf material per plant was weighed together per plot and recorded as plant fresh weight. Harvested storage roots were sorted into non-marketable (small) and marketable storage roots. Deformed storage roots (physically twisted) and those affected by root rot were counted and weighed separately. Total yield was computed from all harvested marketable and non-marketable storage roots per plot.

4.4 Carotenoid and Dry Matter Analysis

The nutritional content of storage roots was assessed using total carotenoid nutrient and dry matter content following the procedure outlined in [15]. Cassava storage roots from each plot were randomly divided, one for fresh and the other for dried analysis for total carotenoid and dry matter content, respectively. The roots were chopped into ~0.5 cm^3 cubes and 100 g sub-samples for each plot were randomly removed to determine the total carotenoid content using the iCheck™ method (BioAnalyt GmbH, Teltow, Germany). Total carotenoid content and dry matter were conducted for the first trial only due to the high cost of this procedure. For dry matter analysis, the 100 g fresh storage root cubes were oven-dried at 70 °C for 72 h, then milled to obtain a homogeneous powder, stored in moisture-free plastic containers and dry matter calculated for each cultivar [38]:

$$\text{Dry matter (\%)} = \frac{\text{Final weight}}{\text{Fresh weight}} \times 100 \quad (2)$$

4.5. Statistical Analysis of Data

Data were subjected to a factorial analysis of variance (ANOVA) using SAS 9.4 [39] statistical package and means separated using least significant difference (LSD) at 5% level of probability. The data from the two experiments were pooled for analysis for those that recorded no cultivar or treatment interaction per year.

5. Conclusions

It is abundantly clear from the results that nematodes are a major constraint to cassava production. Root-knot nematode *Meloidogyne* spp. and the lesion nematode *Pratylenchus* spp. were the most common and important nematodes encountered from the study while *Helicotylenchus* spp., *Scutellonema* spp. and *Hoplolaimus* spp. could also become important when present in large numbers. All the biofortified cassava cultivars were susceptible and reacted to *Meloidogyne* spp. with varying intensity of root galling, ranging between 3.50 and 5.00 index. This was associated with a significant ($p \leq 0.05$) reduction in above-ground fresh weight, plant height, stem girth, marketable storage root weight and number in most cultivars. The nutrient analysis clearly demonstrates the negative impact of PPN on the nutrient quality of biofortified cassava. Therefore, breeding and/or selecting for resistance against PPN, especially *Meloidogyne* spp., is here highlighted as highly necessary to achieve good yields and maintain nutrient quality in biofortified cassava. This has particular relevance under more intensified cropping conditions, which exaggerate soil and root borne constraints. Furthermore, the effect of root-knot nematode infection on the reduction of total carotenoid and dry matter contents should be investigated. Carbofuran was used to effectively manage PPN densities in the field in the current study, but it is an environmentally hazardous product that has been removed from the market in many places, even if it is systemic and not toxic to plants [40,41]. Synthetic pesticides are also often out of reach for resource-poor African farmers due to their high cost. Consequently, there is the need to work out effective and sustainable nematode control strategies in order to improve growth, yield and quality of biofortified cassava. Root-knot nematodes are highly pervasive pests, which are becoming increasingly problematic across tropical cropping systems and as such require particular attention from breeders.

References

1. Montagnac, J.A.; Davis, C.R.; Tanumihardjo, S.A. Nutritional value of cassava for use as a staple food and recent advances for improvement. *Compr. Rev. Food Sci. Food Saf.* **2009**, *8*, 181–194. [CrossRef]

2. Dapaah, S.K. Contributions of root and tuber crops to socio-economic changes in the developing world: The case of Africa, with special emphasis on Ghana. In *Tropical Root Crops in a Developing Economy, Proceedings of the Ninth Symposium of the International Society for Tropical Root Crops, Accra, Ghana, 20–26 October 1991*; Ofori, F., Hahn, S.K., Eds.; IDRC: Ottawa, ON, Canada, 1991; pp. 18–24.

3. Centro International de Agricultura Tropical. *Cassava Program: Annual Report for 1987–1991*; Centro International de Agricultura Tropical: Cali, Colombia, 1996.
4. Tanumihardjo, S.A.; Bouis, H.; Hotz, C.; Meenakshi, J.V.; McClafferty, B. Biofortification of staple crops: An emerging strategy to combat hidden hunger. *Compr. Rev. Food Sci. Food Saf.* **2008**, *7*, 329–334.
5. HarvestPlus2014. Available online: http://www.harvestplus.org/content/vitamin-cassava (accessed on 12 June 2014).
6. Busani, B. Nigeria: Not Everyone Pleased with New Vitamin A Fortified Cassava. Available online: www.globalissues.org/news/2011/12/30/12339 (accessed on 13 January 2012).
7. Saltzman, A.; Birol, E.; Bouis, H.E.; Boy, E.; De Moura, F.F.; Islam, Y.; Pfeiffer, W.H. Biofortification: Progress toward a more nourishing future. *Glob. Food Secur.* **2013**, *2*, 9–17. [CrossRef]
8. Okigbo, B.N. Nutritional implications of projects giving high priority to the production of staples of low nutritive quality. In the case for cassava (*Manihot esculenta* Crantz) in the humid tropics of West Africa. *Food Nutr. Bull.* **1980**, *2*, 1–10. [CrossRef]
9. IITA. Cassava in Tropical Africa. In *A Reference Manual*; IITA: Ibadan, Nigeria, 1990; p. 35.
10. Legg, J.; Okonya, J.; Wyckhuys, K.; Coyne, D.L. Integrated pest management of root and tuber crops in the tropics. In *Integrated Pest Management in Tropical Regions*; Rapisarda, C., Cocuzza, G.E.M., Eds.; CABI Publishing: Wallingford, UK, 2019; pp. 90–112.
11. Coyne, D.L.; Cortada, L.; Danzell, J.J.; Claudius-Cole, A.O.; Haukeland, S.; Luambano, N.; Talwana, H. Plant-parasitic nematodes and food security in Sub-Saharan Africa. *Annu. Rev. Phytopathol.* **2018**, *56*, 381–403. [CrossRef] [PubMed]
12. Coyne, D.L.; Affokpon, A. Nematode parasites of tropical root and tuber crops. In *Plant Parasitic Nematodes in Subtropical and Tropical Agriculture*, 3rd ed.; Sikora, R.A., Coyne, D.L., Hallman, J., Timper, P., Eds.; CAB International: Wallingford, UK, 2018; pp. 252–289.
13. Coyne, D.L.; Talwana, L.A.H. Reaction of cassava cultivars to root-knot nematode (*Meloidogyne* spp.) in pot experiments and farmer-managed field trials in Uganda. *Int. J. Nematol.* **2000**, *10*, 153–158.
14. Akinsanya, A.K.; Afolami, S.O. Effect of seven elite cassava (*Manihot esculenta* Crantz) varieties to infection by *Meloidogyne* spp. and other nematodes in the field. *Nematropica* **2018**, *48*, 50–58.
15. Akinsanya, A.K.; Afolami, S.O.; Kulakow, P.; Coyne, D.L. The root knot nematode, *Meloidogyne incognita*, profoundly affects the production of popular biofortified cassava cultivars. *Nematology* **2020**, *22*. [CrossRef]
16. Theberge, R.L. *Common Africa Pest and Disease of Cassava, Yam, Sweet Potato and Cocoyam*; IITA: Norwich, UK; Balding and Mansel Ltd.: Norwich, UK, 1985; pp. 26–28.

17. Coyne, D.; Kagoda, A.; Wambugu, E.; Ragama, P. Response of cassava to nematicide application and plant parasitic nematode infection in East Africa with emphasis on root knot nematodes. *Int. J. Pest Manag.* **2006**, *52*, 215–223. [CrossRef]

18. Akinlesi, R.A. Interaction of *Meloidogyne incognita* with *Botryodiplodia theobromae* on *Manihot esculenta* (cassava) and Its Biocontrol. Ph.D. Thesis, University of Ibadan, Ibadan, Nigeria, 2014; pp. 144–153.

19. Akinsanya, A.K.; Afolami, S.O. Screenhouse response of seven elite Cassava (*Manihot esculenta* Crantz) varieties to *Meloidogyne incognita* infection. *Nematropica* **2019**, *49*, 91–98.

20. Taylor, A.L.; Sasser, J.N. *Biology, Identification and Control of Root-Knot Nematodes (Meloidogyne Species)*; Department of Plant Pathology, North Carolina State University Graphics and the USAID: Raleigh, NC, USA, 1978; p. 111.

21. Oostenbrink, M. Major characteristics of the relationship between nematodes and plants. *Meded. Landbouwhogesch. Wagening.* **1966**, *66*, 44–46.

22. Fengru, Z.; Schmitt, D.P. Host status of 32 species of *Meloidogyne konaensis*. *Nematology* **1994**, *26*, 744–748.

23. Coyne, D.L.; Talwana, H.A.L.; Maslen, N.R. Plant parasitic nematodes associated with root and tuber crops in Uganda. *Afr. Crops Prot.* **2003**, *9*, 87–98.

24. Sánchez, H.; Ceballos, F.; Calle, J.C.; Pérez, C.; Egesi, C.E.; Cuambe, A.F.; Escobar, D.O.; Chávez, A.L.; Fregene, M. Tolerance to Postharvest Physiological Deterioration in Cassava Roots. *Crops Sci.* **2010**, *50*, 1333–1338.

25. Abidemi, M.O. Plant-Parasitic Nematodes Associated with Cassava in Southwestern Nigeria and Control of Meloidogyne Incognita with Tithonia Diversifolia and Plastic Mulch. Ph.D. Thesis, University of Ibadan, Ibadan, Nigeria, 2014; pp. 185–200.

26. Ray, S.; Sahoo, N.K.; Mohanty, K. Plant parasitic nematodes associated with tuber crops in Orissa. *J. Root Crops* **1992**, *18*, 70–72.

27. Bridge, J.; Otim, N.; Namaganda, J. The root knot nematode, *Meloidogyne incognita*, causing damage to cassava in Uganda. *Afro-Asian J. Nematol.* **1991**, *1*, 116–117.

28. Coyne, D.L.; Namaganada, J.M. Root-knot nematodes, *Meloidogyne* spp. Incidence on cassava in two area of Uganda. *Roots Tuber Newslett.* **1994**, *1*, 2–3.

29. Sijmons, P.C.; Grundler, F.M.W.; von Mende, N.; Burrows, P.R.; Wyss, U. *Arabidopsis thaliana* as a new model host for plant-parasitic nematodes. *Plant J.* **1991**, *1*, 245–254. [CrossRef]

30. Fatemy, F.; Evans, K. Effects of *Globodera rostochiensis* and water stress on shoot and root growth and nutrient uptake of potatoes. *Reuue Nematol.* **1986**, *9*, 181–184.

31. Melakeberhan, H.; Ferris, H.; Dias, J.M. Physiological response to resistance and susceptible *Vitis vinifera* to *M. incognita*. *J. Nematol.* **1990**, *22*, 224–230.

32. Hussey, R.S.; Williamson, V.M. Physiological and molecular aspects of nematode parasitism. In *Plant and Nematode Interactions*; Barker, K.R., Pederson, G.A., Windham, G.L., Eds.; American Society of Agronomy: Madison, WI, USA, 1996; pp. 87–108.

33. Dos Santos, M.F.A.; da Silva Mattos, V.; Monteiro, J.M.S.; Almeidaa, M.R.A.; Jorge, A.S., Jr.; Cares, J.E.; Castagnone-Sereno, P.; Coyne, D.; Carneiro, R.M.D.G. Diversity of *Meloidogyne* spp. from peri-urban areas of sub-Saharan Africa and their genetic similarity with populations from the Latin America. *Physiol. Mol. Plant Pathol.* **2019**, *105*, 110–118. [CrossRef]

34. Whitehead, A.G.; Hemming, J.R. A comparison of some quantitative methods of extracting small vermiform nematodes from soil. *Ann. App. Biol.* **1965**, *55*, 25–38. [CrossRef]

35. Caveness, F.E. A simple siphon method of separating nematodes from excess water. *Nematropica* **1975**, *5*, 30–32.

36. Doncaster, C.C. A counting dish for nematodes. *Nematologica* **1962**, *7*, 334–336. [CrossRef]

37. Bell, M. Plant-Parasitic Nematodes: Lucid Key to 30 Genera of Plant-Parasitic Nematodes. 2004. Available online: http://www.lucidcentral.com/keys/nematodes (accessed on 21 May 2005).

38. Rodriguez-Amaya, D.B.; Kimura, M. *Harvestplus Handbook for Carotenoid Analysis*; HarvestPlus Technical Monograph 2; IFPRI: Washington, DC, USA, 2004.

39. SAS Institute. *SAS User's Guide: Statistics, Version 9.4*; SAS Institute: Cary, NC, USA, 2012.

40. Sasser, J.C.; Carter, C.C.; Hartman, K.M. *Standardization of Host Suitability Studies and Reporting of Resistance to Root-Knot Nematode*; United States Agency for International Development (USAID): Raleigh, NC, USA, 1985; p. 7.

41. Daramola, F.Y. Nematodes Associated with Pineapple (*Ananas comosus* Meer) in Southern Nigeria and Their Management with Poultry Manure and Carbofuran for Improved Productivity. Ph.D. Thesis, Federal University of Agriculture, Abeokuta, Nigeria, 2012; p. 13.

Chapter-10

Changes in the Plant β-Sitosterol/ Stigmasterol Ratio Caused by the Plant Parasitic Nematode Meloidogyne incognita

1. Introduction

Plants are consistently exposed to numerous pests and pathogens, which leads to variations in plant metabolism, including sterol profiles. Sterols are biomolecules which play important roles in various biological processes. Besides their essential function in cell membrane support and fluidity, they are also important as hormone precursors and are involved in biotic and abiotic stress responses [1–5]. Sterols belong to the large group of isoprenoid synthesized via the lanosterol (animals and fungi) or cycloartenol (plants) pathway (Figure 1), sharing a basic structure with a four-cyclic hydrocarbon ring, called gonane, and a hydroxyl group at position C-3. Depending on the organism, sterols are differently modified in the ring structure or in the side chain at position C-17, by methylations or double bonds [4,6]. Cholesterol, arguably the most studied sterol, is mainly synthesized in animals. In contrast, plants largely contain a mixture of C-24 sterols, such as β-sitosterol, campesterol and stigmasterol (collectively known as phytosterols). Nevertheless, they also synthesize minor amounts of cholesterol (Figure 1).

Remarkably not all multicellular organisms that require sterols for growth and re-production are able to synthesize these molecules de novo [7]. Plant parasitic nematodes (PPN), for instance, are among the sterol auxotrophic parasites that rely on host plant sterols for growth and reproduction [7–9]. Several PPN are sedentary endoparasites that burrow inside plant roots and induce the formation of feeding sites, such as the root-knot nematodes, *Meloidogyne* spp. These nematodes induce the formation of giant cells in the differentiating vascular tissue that act as nutrient sinks, for example, sterols, which the nematode feeds on [10,11].

Figure 1. Plant sterol synthesis pathway starting with the conversion of 2,3 oxidosqualene to cycloartenol by oxidosqualene cyclase (OSC). OSC enzymes are classed as cycloartenol synthase (CAS) or lanosterol synthase (LAS) depending on their first cyclic product. The main sterol synthesis pathway in plants is indicated by multiple arrows representing several enzymatic steps with detailed information on β-sitosterol conversion to stigmasterol by a C22-desaturase. The most common end sterols in plants are highlighted in gray. The lanosterol synthesis pathway, well known for animal and fungi is indicated by dotted lines as lanosterol synthesis has been reported in plants, although lanosterol was not detected in this study.

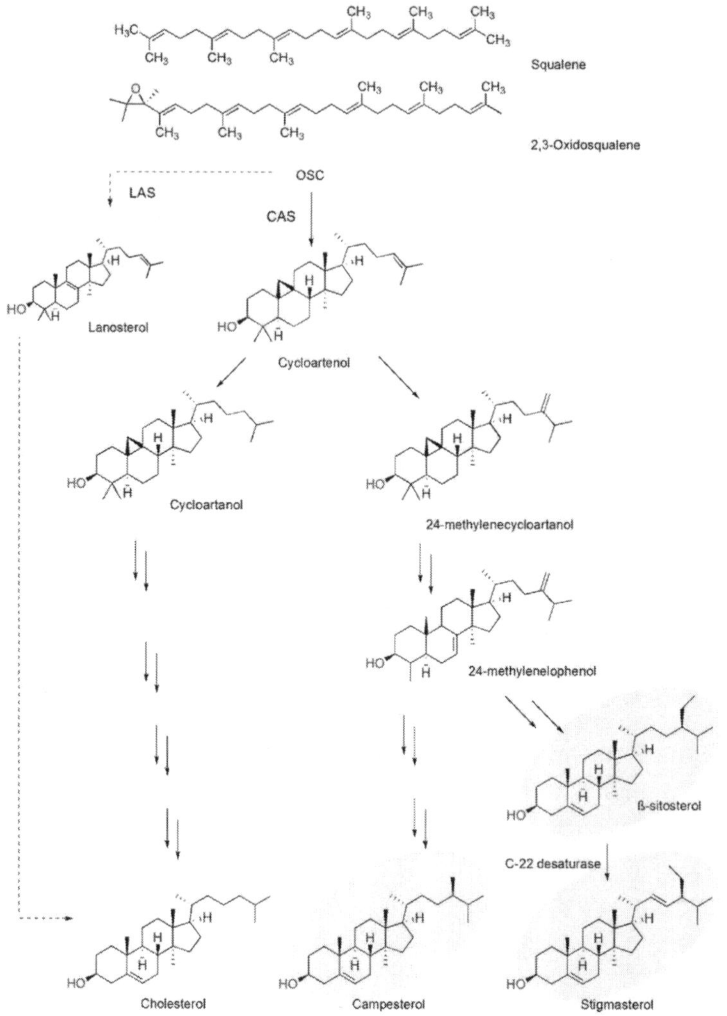

Biotic and abiotic factors have been reported to cause changes in plant sterol lev- els. Metabolic changes in β-sitosterol and stigmasterol levels have also been associated with fungal or bacterial infection and were related to the induction of signaling pathways leading to the synthesis of antimicrobial molecules and changes in membrane permeability [2,5,12–15]. Differences in the β-sitosterol/stigmasterol ratio have also been associated with resistance and susceptibility of tomato plants to *Meloidogyne incognita* [16]. Furthermore, studies of Hedin et al. [17] show changes in β-sitosterol/stigmasterol levels after *M. incognita* infection of cotton plant roots. Besides these biotic factors, abiotic stresses, such as drought and temperature, have also been reported to affect plant β-sitosterol and stigmasterol levels [5,18].

Stigmasterol is synthesized from β-sitosterol by a single desaturase reaction that occurs at position C22 of the sterol side chain, catalyzed by the enzyme sterol C22-desaturase that belongs to the cytochrome P450 710 family (EC 1.14.19.41; Figure 1) [19,20]. Little is known about the regulation of β-sitosterol and stigmasterol levels in roots during plant defense against PPN. Thus, understanding how plant sterols change after PPN infection and how these changes influence plant defense might help designing nematode-resistant or tolerant crops, possibly with an altered sterol profile. In this way, to better understand the role of plant sterol composition during nematode infection, we investigated the sterol composition of *Brassica juncea* (brown mustard), *Cucumis sativus* (cucumber), *Glycine max* (soybean), *Solanum lycopersicum* (tomato cv. Moneymaker and cv. Oskar) and *Zea mays* (corn), after infection with *M. incognita*. Furthermore, changes in sterol composition were tracked over time and expression levels of sterol C22-desaturase gene followed in tomato cv. Moneymaker.

2. Results and Discussion

2.1 Plant Sterol Composition

First, we investigated the profiles of free sterols in the roots of five different agricultural crops, brown mustard, corn, cucumber, soybean and two tomato cultivars (cv. Moneymaker and cv. Oskar) (Figure 2, Table 1). Notably, the cholesterol levels were significantly higher in both tomato cultivars than in the other four crop species. Brown mustard (*B. juncea*) had higher levels of β-sitosterol and lower levels of stigmasterol than all the other species. Significant sterol variations among vegetables, fruits, berries and medicinal plants have been reported [21–23]. However, data available for comparisons of plant root sterol composition are limited. With 80.7% stigmasterol in corn root systems, our data are in agreement with previous reports of Bladocha and Benveniste [24], which showed that sterol composition of corn roots and leaves differed strongly in the ratio of β-sitosterol to stigmasterol. Stigmasterol was the most abundant root sterol and β-sitosterol the most abundant sterol in leaves. In the medicinal plant Cannabis, significant differences in campesterol, β-sitosterol and stigmasterol have been observed between organs, with β-sitosterol as the most abundant sterol in stem bark and roots and stigmasterol being most abundant in leaves. Campesterol had the lowest concentration in roots and stem bark compared to β-sitosterol and stigmasterol [23].

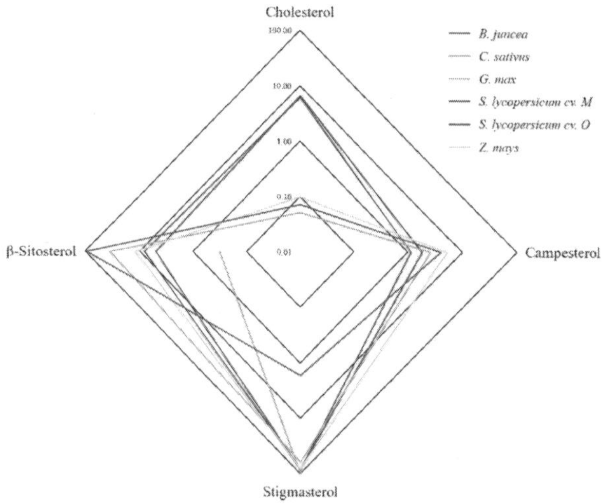

Figure 2. Free sterol composition in percentage of *Brassica juncea, Cucumis sativus, Glycine max, Solanum lycopersicum* cv. Moneymaker (M) and cv. Oskar (O) and *Zea mays.*

Table 1. Average percentage of free and total (in brackets) sterols of *Meloidogyne incognita* infected (Inf.) and non-infected brown mustard (*Brassica juncea*), cucumber (*Cucumis sativus*), soybean (*Glycine max*), tomato (*Solanum lycopersicum* cv. Moneymaker and cv. Oskar) and corn (*Zea mays*) roots.

Plant Species	Sample	Cholesterol	Campesterol	Stigmasterol	β-Sitosterol
	Root	0.1 (0.1)	4.1 (5.6)	1.7 (1.6)	94.1 (92.7)
B. juncea	Inf. root	0.1 (0.2)	5.6 (7.3)	1.9 (1.9)	92.5 (90.7)
	p value	0.79	0.07	0.72	0.12
	Root	ND (0.1)	ND (ND)	99.7 (99.5)	0.3 (0.5)
C. sativus	Inf. root	ND (0.2)	ND (ND)	99.0 (99.1)	1.0 (0.7)
	p value	NA	NA	0.03 *	0.03 *
	Root	0.1 (0.3)	2.6 (2.3)	62.4 (56.5)	34.9 (40.9)
G. max	Inf. root	0.1 (0.2)	2.6 (2.5)	61.7 (59.3)	35.7 (37.8)
	p value	0.95	0.07	0.45	0.41
	Root	6.5 (9.1)	1.9 (2.3)	86.7 (75.5)	5.0 (13.1)
S. lycopersicum cv. Moneymaker	Inf. root	7.5 (11.4)	1.9 (3.0)	75.0 (65.6)	15.6 (20.0)
	p value	0.15	0.28	9.4 × 10−4 ***	5.1 × 10−5 ***
	Root	6.1 (7.3)	1.1 (1.5)	84.7 (80.3)	8.0 (10.9)
S. lycopersicum cv. Oskar	Inf. root	8.2 (9.8)	1.5 (1.7)	78.7 (75.4)	11.6 (13.2)
	p value	0.1	0.02 *	0.07	0.09
	Root	0.1 (0.2)	5.3 (5.3)	82.7 (81.2)	11.9 (13.3)
Z. mays	Inf. root	0.2 (0.3)	5.8 (5.5)	80.7 (80.7)	13.3 (13.5)
	p value	0.05 *	0.09	0.1	0.003 **

*Student's t-test was used for comparisons of uninfected vs. infected root systems. ***, $p < 0.001$; **, $p < 0.01$; *, $p < 0.05$. ND = not detected. n = minimum of 3 samples.*

Similar to our study where *B. juncea* sterols were composed of 94.1 % β-sitosterol (Figure 2; Table 1), the sterol composition from roots and leaves of the close relative *Brassica napus* is dominated by β-sitosterol [25]. On the other hand, Surjus and Durand [26] reported that β-sitosterol is the prominent plant sterol in roots of soybean cv. Hodgson, which does not match our findings where stigmasterol is the most abundant sterol with 62.4% in soybean cv. Aveline Bio.

C. sativus was the only species in this study where no campesterol was detected in the root sterol fraction, which was mainly composed of stigmasterol (Figure 2; Table 1). A study on the sterol composition of selected grains, legumes and seeds has shown that campesterol was also not detected in pumpkin seeds [27], whose sterols were mainly made up of β-sitosterol. In another study, neither campesterol, stigmasterol nor β-sitosterol were detected in *C. sativus* fruits, however other sterols were present [21]. Altogether, sterol compositions differ between organs of a plant, and even the same organs of different cultivars of the same species can differ significantly in their sterol composition and abundance [28].

Within plants, conjugated sterols are ubiquitous. However, their profile and relative content can differ among organs, plant developmental stage and environmental signals [29]. The analysis of total sterols (sterol ester and free sterols) and free sterol fraction is included in Table 1. When comparing the total sterol fraction to the free sterol fraction, the abundance of cholesterol and β-sitosterol increased, campesterol maintained a similar relative abundance, while the abundance of stigmasterol decreased. These results indicate that more cholesterol and β-sitosterol are present as steryl esters compared to stigmasterol. Overall, sterol profile changes have been reported for different tissues and conjugated forms [29] and even if a plant sterol, such as cholesterol represents a minor amount of the total sterol fraction of the plant, it can be the most abundant phytosterol in some tissue. For example, the sterol fraction of the phloem exudate of bean and tobacco plants contains over 88% of cholesterol [30].

2.2 Plant Sterol Composition after Meloidogyne Incognita Infection

The sterol compositions of *M. incognita*-infected *B. juncea, C. sativus, G. max, S. lycopersicum* cvs. Oskar and Moneymaker, and *Z. mays* roots were determined 21 dpi (Table 1), to allow nematodes to establish and expand feeding sites [10]. Compared to uninfected tomato roots, sterols of cv. Moneymaker and cv. Oskar were composed of 6.5% and 6.1% free cholesterol, 86.7% and 84.7% stigmasterol, 5.0% and 8.0% β-sitosterol and 1.9% and 1.1% campesterol, respectively (Table 1). That means, infection with *M. incognita* led to an overall increase in cholesterol and β-sitosterol and a decrease in stigmasterol. Cholesterol levels increased up to 7.5% in cv. Moneymaker roots and up to 8.2% in cv. Oskar. The highest contribution of cholesterol to the sterol pool was determined in the galls, i.e., the nematode feeding sites, with 12.3% (cv. Oskar) and 10.3% (cv. Moneymaker; Table S2). Yet, the most pronounced sterol change observed 21 days post *M. incognita* inoculation was in the relative abundance of β-sitosterol and stigmasterol. In both tomato cultivars, levels of free β-sitosterol increased from 5.0% to 15.6% and 8.0% to 11.6% in cv. Moneymaker and cv. Oskar, respectively. At the same time, stigmasterol levels decreased from 86.7% to 75% and from 84.7% to 78.7% in infected roots of cv. Moneymaker and cv. Oskar, respectively. These changes in the β-sitosterol/stigmasterol ratio were even more pronounced when the sterol composition of the galls was evaluated (Figure 3; Table S2).

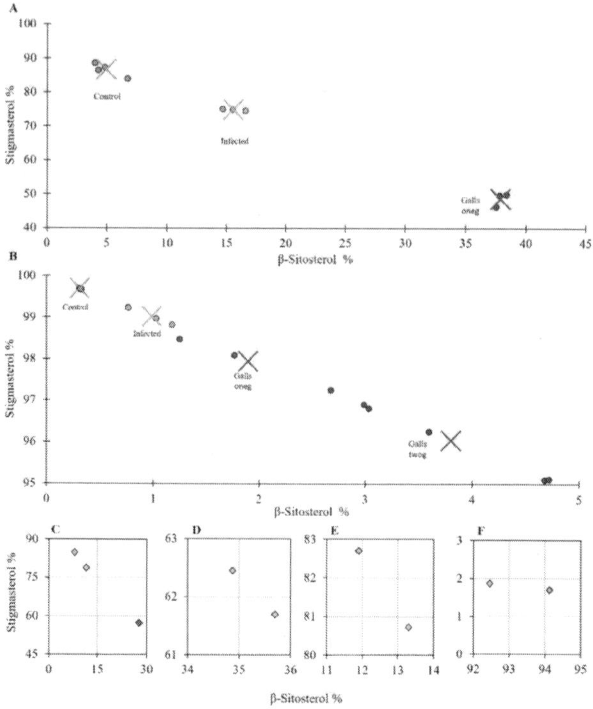

Figure 3. Relative stigmasterol to β-sitosterol abundance of uninfected (green), M. incognita infected (yellow) and galls (red) for one generation (oneg) and (brown) for second generation (twog) of M. incognita, samples of the plants: Solanum lycopersicum cv. Moneymaker (A) and Cucumis sativus (B). For S. lycopersicum cv. Oskar (C), Glycine max (D), Zea mays (E) and Brassica juncea (F) the results are presented as the mean of the 3 replicates. For A and B the average is marked by X. n = ≥3 replicates.

The analysis of the free and the total sterol fraction of *C. sativus*, *G. max* and *Z. mays* roots infected by *M. incognita* showed similar β-sitosterol, stigmasterol and cholesterol changes compared to the control plants (Table 1). Infection with *M. incognita* resulted in a relative increase in cholesterol and β-sitosterol combined with a relative decrease in stigmasterol levels compared to uninfected plants. However, such changes were not observed in *B. juncea*, where infection resulted in an increase in campesterol.

It seems plausible that the observed changes in the sterol pool are linked to a metabolic reaction against the infection by *M. incognita*. For example, in solanaceous plants, cholesterol can make up a significant portion of the overall sterol pool and has been suggested as a precursor of toxic steroidal alkaloids and glycoalkaloids [31]. Campesterol is used in numerous plants as precursor for the synthesis of brassinosteroid phytohormones, essential for the regulation of numerous plant processes, such as cell expansion and elongation, senescence and protection against drought and chilling [32]. The conversion of β-sitosterol to stigmasterol has been linked to biotic and abiotic stress [2,19,20] and has previously been linked to resistance against *M. incognita* in tomato cultivars [16].

As *M. incognita* induces a formation of giant cells, the galls sterol composition might be influenced by the lipid bilayer reorganization of these cells. Studies on the lipid bilayer revealed that β-sitosterol is slightly more efficient in ordering a fluid membrane of 2-dipalmitoyl-sn-glycero-3-phosphocholine than stigmasterol, resulting in a more packed membrane liquid ordered phase [33]. Furthermore, simulations have shown that cholesterol was slightly more efficient in packing the lipid bilayer than β-sitosterol [34].

Since the β-sitosterol to stigmasterol ratio is regulated by a single C22 desaturation step and strong changes in this ratio were observed, scatter plots were prepared to compare β-sitosterol/stigmasterol changes after nematode infection in the different plant species (Figure 3). All plant species analyzed displayed an increase of β-sitosterol and a decrease in stigmasterol after nematode infection, with the exception of *B. juncea*, which showed a decrease of β-sitosterol levels. β-Sitosterol accounted for 94.1% and stigmasterol for only 1.7% of free sterols in non-infected *B. juncea* plants (Table 1). This might be the reason why *B. juncea* displayed a completely different alteration on the sterol profile in response to nema- tode infection than the other plant species investigated (Table 1, Figures 2 and 3). Anyhow, similar β-sitosterol/stigmasterol observations can be seen for other sterol analyses, e.g., of two cotton cultivars, cv. ST-213 and cv. 81-249 where the β-sitosterol/ stigmasterol ratio changed from 32.6/53.1% (cv. St213) and 30.0/43.8% (cv. 81-249) to 36.8/43.8% (cv. ST-213) and 33.8/47.3% (cv. 81-249) after *M. incognita* infection [17].

A reason for the different sterol response in *B. juncea* compared to the other plant species might be that *Brassica* species contain a particular sterol, brassicasterol. Brassicast- erol synthesis belongs to the same sterol branch as campesterol (Figure 1). The campesterol precursor 24-methyldesmosterol is converted to 24-epi-campesterol and then to brassi-casterol. This final enzymatic step described in *Arabidopsis thaliana* is catalyzed by a C22 desaturase [19]. In this context, it is also important to note that *Brassica* species can produce isothiocyanates (ITCs) the glycosides of which are hydrolyzed by myrosinases in response to herbivory [35]. ITCs are highly toxic, leading to a suppressive effect of *Brassica* species on soil-borne pathogens and herbivores [36]. Therefore, *Brassica* species including *B. juncea* are used as cover crops in PPN management via so-called bio-fumigation [37,38]. Nevertheless, *B. juncea* is a host of *M. incognita* [39].

2.3 β-Sitosterol/Stigmasterol Conversion in Tomato after Meloidogyne Incognita Infection

The β-sitosterol to stigmasterol conversion requires the creation of a double bond at position C22, which is catalyzed by a monooxygenase of the Cytochrome P450 enzyme family 710A (CYP710A), the only family in the CYP710 clan (Figure 1) [19,40]. The observed increase of β-sitosterol and decrease of stigmasterol led us to investigate the expression of the tomato gene *SlCYP710A11* during *M. incognita* infection. This gene encodes the enzyme previously characterized as a C22 desaturase in tomato sterol biosynthesis [19]. Temporal gene expression analysis of the *SlCYP710A11* gene in uninfected tomato cv. Mon- eymaker showed only small variations in gene expression levels during a time course of 21 days (Figure 4A). However, in tomato plants of the same developmental stage infected by *M. incognita*, the expression of *SlCYP710A11* was downregulated significantly in the samples taken at 14 and 21 dpi (Figure 4B). At the same time, the tomato sterol profile of β-sitosterol and stigmasterol reflected the gene expression levels (Figure 4C) in that the β-sitosterol/stigmasterol ratio gradually increased over the course of 21 days due to a rela- tive increase of β-sitosterol and a corresponding decrease of stigmasterol (Figure 4C).

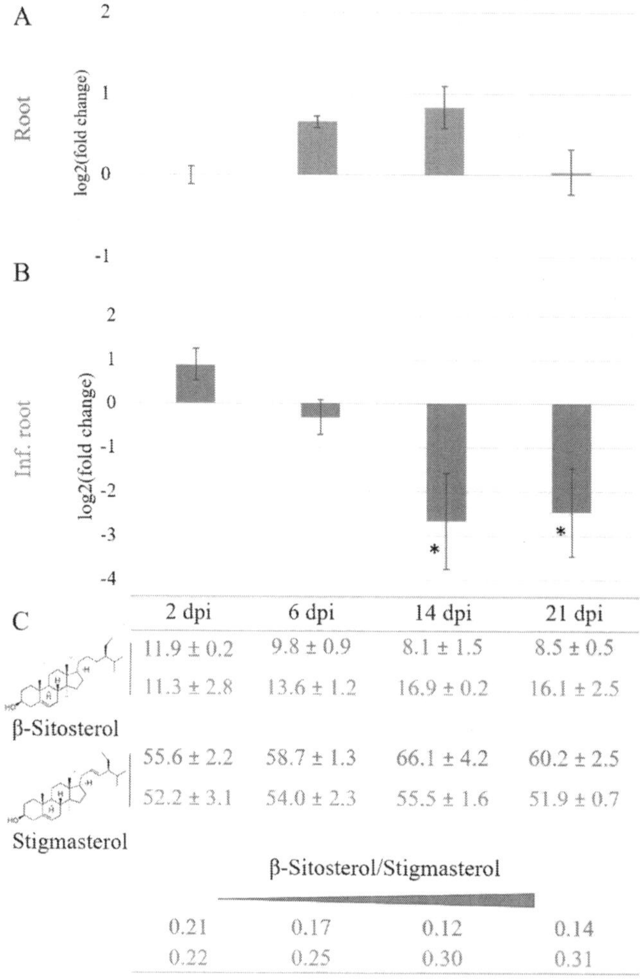

Figure 4. Temporal gene expression analysis of the C-22 desaturase gene CYP710A11 and changes in the β-sitosterol/stigmasterol ratio in Solanum lycopersicum cv. Moneymaker 2, 6, 14 and 21 days post inoculation (dpi). SlCYP710A11 gene expression is presented as fold change. (A) Data on uninfected roots are marked in blue. (B) Data on M. incognita-infected roots (Inf. roots) in orange. N = 4 biological replicates of 2 pooled plants per analysis. (C) Changes in β-sitosterol/stigmasterol ratios are displayed as percentage of total sterols extracted. ANOVA was used for comparisons of gene expression levels in uninfected vs. infected root systems. *, $p < 0.05$.

The β-sitosterol/stigmasterol change was most pronounced at 21 dpi, confirming the previous results on plants infected with *M. incognita* that displayed reduced relative levels of stig- masterol and increased levels of β-sitosterol compared to the uninfected plants, most easily to explain by a decrease in C22 desaturase activity (Figure 4B). Interestingly, the change in the β-sitosterol/stigmasterol ratio was already visible at 6 dpi when transcriptional repression was not apparent yet, suggesting additional regulatory mechanisms (Figure 4B). Altogether, both gene expression and sterol profile data support the finding that the synthe- sis of stigmasterol from β-sitosterol is downregulated as an effect of *M. incognita* infection in *S. lycopersicum*.

Since the reaction to *M. incognita* infection is a modulation of C22 desaturase activity on behalf of the plants, it is important to note that the enzyme responsible for the conversion of 24-epi-campesterol to brassicasterol also represents a C22 desaturase; indeed, it was found for *Arabidopsis* that the enzyme encoded by *CYP710A2* was responsible for both brassicasterol and stigmasterol production [19]. However, *M. incognita* infection did not lead to a significant change in the sterol pattern of *B. juncea* (Table 1). Hence, in spite of the fact that brassicasterol was not analyzed, we can conclude that it is unlikely that the expression of the *CYP710A2* orthologue was affected.

Changes in the β-sitosterol/stigmasterol equilibrium might represent a general plant response to environmental cues as reviewed by Zhang et al. [28], and not a specific response to *M. incognita*. For example, an increase in stigmasterol levels has generally been observed as response to cold stress [5,41]. An increase in C22 desaturase expression levels has been reported as response of *Arabidopsis thaliana* plants to biotic and abiotic factors: to inducers of PAMP-triggered immunity like flagellin and lipopolysaccharides, to reactive oxygen species (ROS) and osmotic stress as well as to infections with bacterial and fungal pathogens [3,5,14,15,42]. Other than in *Arabidopsis*, a relative increase in stigmasterol has also been observed in leaves of *Triticum aestivum* infected by a biotrophic fungus, and in *Z. mays* leaves infected by a necrotrophic fungus ([43,44].

While our results seemed to show *CYP710A* gene induction at the first two time points of *M. incognita* infection, these changes were not significant. However, the repression of *SlCYP710A11* expression at 14 and 21 dpi, and the corresponding changes in the β-sitosterol/stigmasterol ratio, contrast with the abovementioned studies, where *CYP710A* expression was induced, β-sitosterol levels decreased and stigmasterol levels increased. It has to be kept in mind that most previous studies on plant sterol abundance during plant defense focused on shorter time intervals after exposure to pathogens, above-ground plant organs and were conducted mainly on *Arabidopsis* plants, where β-sitosterol is the most abundant sterol and brassicasterols make up part of the end sterols [14,15,43,44]. Furthermore, *Arabidopsis*, like *B. juncea*, is a member of the Brassicaceae and can produce nematocidal ITCs, which might affect its additional responses to PPN [45]. Altogether, the finding of an increase in the β-sitosterol/stigmasterol ratio in response to PPN infection in a diverse group of plants that do not produce nematocidal toxins might indeed represent a specific response. However, given that this response takes some time to establish, it is possible that it is not part of the defense against PPN, but of the supply of PPN with suitable sterols by the plant.

Given that plant-pathogen interactions are processes with different stages, in which gene expression levels often vary, it is not surprising to see changes in profiles of metabolites, such as sterols that could play a critical role in a plant-nematode interaction. It would also not be surprising that different pathogens/herbivores trigger similar or different plant responses. At this point, additional investigations have to be conducted to (a) compare the effects of PPN vs. other root pathogens/herbivores, and (b) evaluate the impact of the initial plant sterol composition on sterol changes after pathogen attack. After all, in the current study, *B. juncea* had the highest β-sitosterol abundance and was the only outlier in the sterol response to *M. incognita*, presumably due to the fact that Brassicaceae have particular sterol profiles including brassicasterol.

2.4 CYP710A

CYP710A represents the plant cytochrome P450 monooxygenase family encoding the sterol C22 desaturase, which is converting β-sitosterol to stigmasterol [40]. Like plants, fungi possess C22 desaturase enzymes known as CYP61 family of P450 enzymes, which are experimentally characterized and phylogenetically represent orthologues of the plant CYP710 protein family. Phylogenetic analysis of P450 diversity suggests that the CYP710 family is conserved from green algae to higher plants throughout evolution [46] and that the biochemical function can be traced back to plant-fungal divergence but was lost in animals [40]. During evolution, sterol 14-demethylase (CYP51) gene is assumed to have given rise to the *CYP710/CYP61* genes as their function in sterol biosynthesis is downstream of that of CYP51 [40]. CYP51 enzymes are present in plants, fungi and animals synthesizing sterols.

While the phylogeny of P450 monooxygenases is well researched, only limited phylogenetic information is available for CYP710 [28,40]. Overall, CYP710 enzyme activity and/or gene expression has only been studied in few plants, such as *A. thaliana* [2,14,19], *S. lycopersicum* [19], *Physcomitrella patens* [47] and *Calotropis procera* [48]. Therefore, we conducted a phylogenetic analysis of our studied tomato SlCYP710A11 protein and other plant CYP710 enzymes (Figure 5; Table S3). The well-studied AtCYP710A1 (*A. thaliana*) and SlCYP710A11 (*S. lycopersicum*) amino acid sequences were used as queries to mine for plant homologues. Four hits were scored in *A. thaliana*: Cytochrome P450 proteins 710A1, 710A2, 710A3 and 710A4 (NCBI accessions NP_180997.1, NP_180996.1, NP_180451.1 and

NP_180452.1). It is worth mentioning that in *A. thaliana* both 710A1 and 710A2 can convert β-sitosterol to stigmasterol [19]. For *Z. mays*, two protein sequences were found in the NCBI database, from two different studies, one annotated as 'uncharacterized protein' and one as CYP710A11 (NP_001307723.1 and PWZ33314.1, respectively). For *G. max*, two proteins were identified, one annotated as CYP710A1 (XP_003542931.1) and one as CYP710A11 (XP_003546088.1). Only one homologous protein was found in *C. sativus* (XP_004134602.1), also annotated as CYP710A11 (Table S3). Since *B. juncea* sequences were not present in the NCBI or UniProt databases, *Brassica rapa* was used as a close relative.

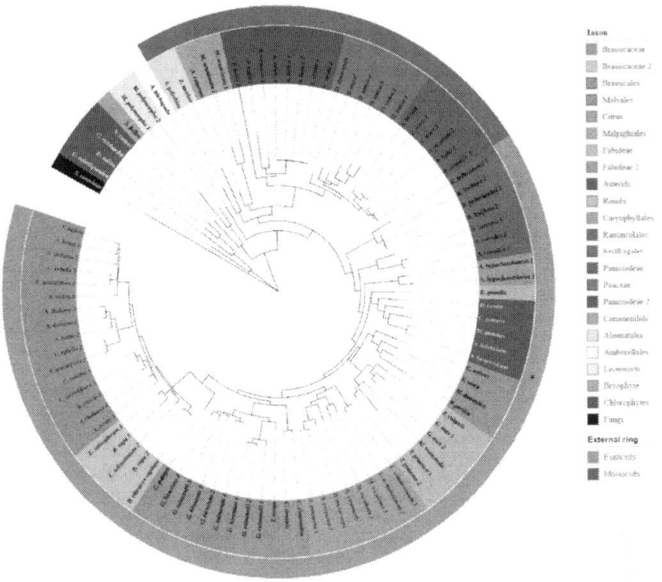

Figure 5. Phylogenetic maximum likelihood tree of the CYP710 enzyme family. The tree is rooted at ERG5, which is the Saccharomyces cerevisiae protein from which all CYP710 proteins originated [20]. Multiple sequences of the same plant species are numbered, and accession numbers of all selected proteins are reported in Table S2. Tree branches are colored and grouped by taxon. External ring shows the eudicots apart from the monocots. * Tomato CYP710A11 enzyme.

During the blast search, multiple gene duplication events were observed, mostly at the species level (data not shown). The only duplication observed at the family level was found in the Brassicaceae family (whole genome duplication [49]). The phylogenetic analysis showed the divergence of eudicot and monocot CYP710 enzymes and basically followed plant phylogeny (Figure 5).

Based on the sterol analysis of the selected plants, the phylogenetic analysis, and recent studies (e.g., where *C. procera CYP710A* gene expression did not respond to abiotic factors [48]), we cannot conclude that in all plants C22 desaturase gene expression responds the same way to PPN infection. Moreover, not all CYP710A enzymes function the same way in sterol biosynthesis, and there might be undiscovered members of the CYP710A family catalyzing the same, or a different reaction (like the desaturation of 24-epi-campesterol to brassicasterol as reviewed by Zhang et al. [28]). Generally, among plant sterol synthesis enzymes, sterol methyl transferase (SMT), delta (24)-sterol reductase (DWF1) and CYP710A are assumed to adjust end sterol composition [28]. Altogether, further studies are required to address the questions if the observed β-sitosterol/stigmasterol changes are species- specific and how additional sterol related genes are involved in the activation of *CYP710A* and changes of the β-sitosterol/stigmasterol equilibrium, and to evaluate their impact on nematode performance. These data might help to develop new nematode-resistant cultivars able to maintain a sterol equilibrium that is not suitable for nematode development.

3. Materials and Methods

3.1 Nematode Inoculation and Plant Material

The root-knot nematodes, *Meloidogyne incognita* (isolate Reichenau 2, R2) were maintained at Agroscope (Wädenswil, Switzerland) on *S. lycopersicum* cv. Oskar. Greenhouse conditions were set at 22 2 °C, 60% relative humidity (RH) and 16 h/8 h light/dark rhythm. Second-stage juveniles (J2) were extracted from heavily galled root systems using a mist chamber (PM 7/119). J2 were stored at 6 °C prior to use [50]. For sterol profiling a minimum of three biological replicates were used per treatment (negative and positive controls) and species:, *Brassica juncea* cv. Sareptasenf (P. H. Petersen), *Cucumis sativus* cv. Landgurken (Bigler Samen) *Glycine max* cv. Aveline Bio (UFA), *Solanum lycopersicum* culti- vars (cvs.) Moneymaker (HILDA) and Oskar (Syngenta) and *Zea mays* cv. Grünschnittmais (UFA) were used. Seeds were pre-germinated (*B. juncea* 3–5 days, *C. sativus* 2–3 days, *G. max* 4–6 days, *S. lycopersicum* 4–6 days and *Z. mays* 5–6 days) in Petri dishes with 5 mm of tap water and then planted into 14 cm diameter plastic pots, using a 3:1 (vol/vol) silver sand:steamed soil mixture (sieved field soil from Cadenazzo, Switzerland). Greenhouse conditions were set to 22 4 °C, 60% RH and 16 h:8 h light:dark rhythm. Three four-week-old plants of each species/cultivar were inoculated with 10,000 *M. incognita* (R2) J2 per pot.

3.2 Sterol Extraction and GC-MS Analysis

Infected and uninfected (control) plant roots were washed free of soil 21 days post inoculation (dpi). For "galls" sterol analysis, galled uproot systems were manually separated with a scalpel. Roots and galls were washed and the separated materials shock-frozen in liquid nitrogen, and ground to powder using mortar and pestle. Sterols were extracted according to Bligh and Dyer [51]. Each root-powder sample (1 g) was separated into two equal parts and total lipids were extracted in chloroform:methanol (2:1 *v*/*v*) for 1 h at 60 °C. One of the two lipid fractions was further saponified for extraction of free and esterified sterols. Saponification was performed as described by Dahlin et al. [52] (alkaline saponification with 2M KOH in 95% ethanol). Both lipid fractions (saponified and total lipid extract) of each root sample were dried under nitrogen and processed for sterol separation by suspending the dried samples in hexane and using a silica solid phase extraction (SPE) column (6 mL SiOH columns, Chromabond, Macherey Nagel, Düren, Germany) as described by Azadmard-Damirchi and Dutta [53]. Eluted sterols were dried under nitrogen and suspended in chloroform for sterol analysis on the Varian 450-GC coupled to a Varian 240-MS Ion Trap (GC-MS) (Darmstadt, Germany). The software VARIAN MS Workstation v. 6.9.3 was used for instrument control and data acquisition. A VARIANT FactorFour Capillary column VF-5 ms of 30 m length, 0.25 mm inner diameter, and 0.25 µm film thickness was used as stationary phase. Helium was used as carrier gas at a flow rate of 1.0 mL/min. Inlet temperature was set at 320 °C. 10 µL of the chloroform sample were injected. Initial GC temperature was set at 225 °C and ramped up to 300 °C at 1.5 °C/min. Temperature was maintained at 300 °C for 10 min before ramping to 320 °C with 5 °C/min, and finally remaining stable at 320 °C for 6 min. Transfer line was set to 270 °C and ion trap temperature was 150 °C. Ion trap was operated with electron ionization (EI) set at an ionization energy of 70 eV and scan mode selection (*m*/*z* 50–900) started after 5 min solvent delay. Sterol standards (cholesterol, campesterol, β-sitosterol and stigmasterol) were obtained from Sigma-Aldrich (St. Louis, MO, USA) and used to compare retention times, sterol fragmentation and for relative sterol quantification. The software R (v. 3.6.2; R core team, 2018) was used to perform Student's *t*-tests (*t*-tests) and ANOVA (analysis of variance) tests on the data obtained to investigate the statistical differences between samples. *T*-tests were used when only infected and uninfected samples were compared, ANOVA was performed when gall samples were included in the comparison.

3.3 CYP710A11 Temporal Gene Expression Analysis

Tomato cv. Moneymaker plants were grown as described above. 4000 *M. incognita* J2/ plant were inoculated by pipetting equal amounts of nematodes into four 5 cm deep holes next to three-week-old tomato plants. 8 Plants were used per time point and pooled in 4 groups of 2 plants each. Plant roots were harvested from infected and uninfected plants at 2, 6, 14 and 21 dpi, frozen in liquid nitrogen and stored at 20 °C before RNA extraction in liquid nitrogen using the Thermo Scientific GeneJET Plant RNA Purification Mini Kit (Waltham, MA, USA). Genomic DNA was removed from the isolated RNA using iScript DNase, followed by RNA quality testing by agarose gel electrophoresis and NanoDrop One One/OneC Microvolume UV-Vis Spectrophotometer measurements (Thermo Fisher Scientific, Reinach, Switzerland). cDNA synthesis was performed using the iScript cDNA synthesis kit (Bio-Rad, Hercules, CA, USA). The tomato gene coding sequence of *SlCYP710A11* was used to design qPCR primers with the online tool Primer3 (v. 4.1.0, Whitehead Institute for Biomedical Research), with the setting of 20 nt primer sequence length, 110 to 130 bases of amplified fragments, 50% GC content and 60 °C melting temperature. Primer sequences (Table S1) were BLASTed against WormBase and NCBI databases to check target specificity. The same parameters were used to design qPCR primers for the reference genes. NormFinder statistical algorithms were used to evaluate the housekeeping gene stability of actin, α-tubulin, SlCBL1, GADPH and eEF1-α. Primer efficiency was determined using the program Real-time PCR Miner [54]. qPCR analyses were carried out according to the 480 SYBR Green 1 Master mix (Roche, Basel, Switzerland) protocol and optimized to the primer melting temperature of 60 °C on the Roch LightCycler 480. For each qPCR run, the Roche LightCycler 480 program was used for melting peak and temperature evaluation. Each experiment was normalized according to the reference gene expression of actin and α-tubulin. Relative fold-changes in expression levels were analysed in Excel using $2^{(-\Delta\Delta Ct)}$ [55].

3.4 Phylogenetic Analysis of Cytochrome P450 Proteins

The protein sequences of *A. thaliana* AtCYP710A1 and *S. lycopersicum* SlCYP710A11, retrieved from the UniProtKB (UniProt) database, were used as queries in a sequence simi- larity search, performed on the UniProt and National Center for Biotechnology Information (NCBI) databases. The number of CYP710A1 proteins and their accession numbers were recorded for the plant species used in the sterol analysis. Protein sequences were searched for conserved protein domains using the Pfam (v. 32, European Bioinformatics Institute) and PANTHER protein databases. AtCYP710A1 was also used as query in a BLAST on Phytozome database (v12.1.5) [56]. Retrieved cytochrome P450 710 protein sequences were aligned using MUSCLE with the software MegaX (Molecular Evolutionary Genetics Analysis X). Aligned sequences were used in MegaX for phylogenetic analysis using the Maximum Likelihood approach, with 1000 bootstraps.

The online tool iTOL (interactive Tree Of Life, v. 5.6) was used to finalize the phylogenetic tree.

4. Conclusions

In this study, we report changes in plant sterol profiles, in response to infection by the plant parasitic nematode *M. incognita*. The β-sitosterol/stigmasterol ratio in *C. sativus*, *G. max*, *S. lycopersicum* cv. Moneymaker and cv. Oskar and *Z. mays* were strongly affected by *M. incognita*. Interestingly, *B. juncea* revealed a sterol response different from that in the other plants examined. Since the conversion of β-sitosterol to stigmasterol is mediated by a single desaturation reaction at position C22 of the sterol side chain catalyzed by CYP710A, we investigated the transcriptional response of tomato *SlCYP710A11*. Infection of *S. lycopersicum* cv. Moneymaker with *M. incognita* led to repression of *SlCYP710A11* transcription that paralleled the change in the β-sitosterol/stigmasterol ratio. However, a detailed comparison indicates that the change in expression levels was not the only factor changing the sterol profile. Further studies are required to investigate whether the changes in plant sterol composition were specific to the response to *M. incognita* infection, if other nematode species generate the same changes in plant sterol composition, and whether they can represent a resistance mechanism.

References

1. London, E. Insights into lipid raft structure and formation from experiments in model membranes. *Curr. Opin. Struct. Biol.* **2002**, *12*, 480–486. [CrossRef]

2. Arnqvist, L.; Persson, M.; Jonsson, L.; Dutta, P.C.; Sitbon, F. Overexpression of *CYP710A1* and *CYP710A4* in transgenic Arabidopsis plants increases the level of stigmasterol at the expense of sitosterol. *Planta* **2007**, *227*, 309–317. [CrossRef] [PubMed]

3. Sewelam, N.; Jaspert, N.; Van Der Kelen, K.; Tognetti, V.B.; Scmitz, J.; Frerigmann, H.; Stahl, E.; Zeier, J.; Van Breusege, F.; Maurino, V.G. Spatial $|_2|_2$ signaling specificity: $|_2|_2$ from chloroplasts and peroxisomes modulates the plant transcriptome differentially. *Mol. Plant* **2014**, *7*, 1191–1210. [CrossRef] [PubMed]

4. Valitova, J.N.; Sulkarnayeva, A.G.; Minibayeva, F.V. Plant sterols: Diversity, biosynthesis, and physiological functions. *Biochemistry* **2016**, *81*, 1050–1068. [CrossRef] [PubMed]

5. Aboobucker, S.I.; Suza, W.P. Why do plants convert sitosterol to stigmasterol? *Front. Plant Sci.* **2019**, *10*, 354. [CrossRef] [PubMed]

6. Desmond, E.; Gribaldo, S. Phylogenomics of sterol synthesis: Insights into the origin, evolution and diversity of a key eukaryotic feature. *Genome Biol. Evol.* **2009**, *1*, 364–381. [CrossRef] [PubMed]

7. Lebedev, R.; Trabelcy, B.; Goncalves, I.L.; Gerchman, Y.; Sapir, A. Metabolic reconfiguration in *C. elegans* suggests a pathway for widespread sterol auxotrophy in the animal kingdom. *Curr. Biol.* **2020**, *30*, 3031–3038. [CrossRef]

8. Chitwood, D.J.; Lusby, W.R. Metabolism of plant sterols by nematodes. *Lipids* **1991**, *26*, 619–627. [CrossRef]

9. Chitwood, D.J. Biochemistry and functions of nematode steroids. *Crit. Rev. Biochem. Mol. Biol.* **1999**, *34*, 273–284. [CrossRef]

10. Shukla, N.; Yadav, R.; Kaur, P.; Rasmussen, S.; Goel, S.; Agarwal, M.; Jagannath, A.; Gupta, R.; Kumar, A. Transcriptome analysis of root-knot nematode (*Meloidogyne incognita*)-infected tomato (*Solanum lycopersicum*) roots reveal complex gene expression profiles and metabolic networks of both host and nematode during susceptible and resistance responses. *Mol. Plant Pathol.* **2018**, *19*, 615–633. [CrossRef]

11. Sato, K.; Kadota, Y.; Shirasu, K. Plant immune responses to parasitic nematodes. *Front. Plant Sci.* **2019**, *10*, 1165. [CrossRef] [PubMed]

12. Hartmann, M.A. Plant sterols and the membrane environment. *Trends Plant Sci.* **1998**, *3*, 170–175. [CrossRef]

13. Hodzic, A.; Rappolt, M.; Amenitsch, H.; Laggner, P.; Pabst, G. Differential modulation of membrane structure and fluctuations by plant sterols and cholesterol. *Biophys. J.* **2008**, *94*, 3935–3944. [CrossRef] [PubMed]

14. Griebel, T.; Zeier, J. A role for β-sitosterol to stigmasterol conversion in plant-pathogen interaction. *Plant J.* **2010**, *63*, 254–568. [CrossRef] [PubMed]

15. Wang, K.; Senthil-Kumar, M.; Ryu, C.; Kang, L.; Mysore, K.S. Phytosterols play a key role in plant innate immunity against bacterial pathogens by regulating nutrient efflux into the apoplast. *Plant Physiol.* **2012**, *158*, 1789–1802. [CrossRef] [PubMed]

16. Zinovieva, S.V.; Vasyukova, N.I.; Ozeretskovskaya, I.L. Involvement of plant sterols in the system tomatoes-nematode *Meloidogyne incognita*. *Helminthologia* **1990**, *27*, 211–216.

17. Hedin, P.A.; Callahan, F.E.; Dollar, D.A.; Greech, R.G. Total sterols in root-knot nematode *Meloidogyne incognita* infected cotton *Gossypium hirsutum* (L.) plant roots. *Comp. Biochem. Physiol.* **1995**, *111*, 447–452. [CrossRef]

18. Sawai, S.; Ohyama, K.; Yasumoto, S.; Seki, H.; Sakuma, T.; Yamamoto, T.; Takebayashi, Y.; Kojima, M.; Sakakibara, H.; Aoki, T.; et al. Sterol side chain reductase 2 is a key enzyme in the biosynthesis of cholesterol, the common precursor of toxic steroidal glycoalkaloids in potato. *Plant Cell* **2014**, *26*, 3763–3774. [CrossRef]

19. Morikawa, T.; Mizutani, M.; Aoki, N.; Watanabe, B.; Saga, H.; Saito, S.; Oikawa, A.; Suzuki, H.; Sakurai, N.; Shibata, D.; et al. Cytochrome P450 *CYP710A* encodes the sterol C-22 desaturase in Arabidopsis and tomato. *Plant Cell* **2006**, *18*, 1008–1022. [CrossRef]

20. Nelson, D.R. Plant cytochrome P450s from moss to poplar. *Phytochem. Rev.* **2006**, *5*, 193–204. [CrossRef]

21. Piironen, V.; Toivo, J.; Puupponen-Pimiä, R.; Lampi, A.M. Plant sterols in vegetables, fruits and berries. *J. Sci. Food Agric.* **2003**, *83*, 330–337. [CrossRef]

22. Jun-Hua, H.A.N.; Yue-Xin, Y.A.N.G.; Mei-Yuan, F.E.N.G. Contents of phytosterols in vegetables and fruits commonly consumed in China. *Biomed. Environ. Sci.* **2008**, *21*, 449–453. [CrossRef]

23. Jin, D.; Dai, K.; Xie, Z.; Chen, J. Secondary metabolites profiled in cannabis inflorescences, leaves, stem barks, and roots for medicinal purposes. *Sci. Rep.* **2020**, *10*, 1–14. [CrossRef]

24. Bladocha, M.; Benveniste, P. Manipulation by tridemorph, a systemic fungicide, of the sterol composition of maize leaves and roots. *Plant Physiol.* **1983**, *71*, 756–762. [CrossRef] [PubMed]

25. Chalbi, N.; Martínez-Ballesta, M.C.; Youssef, N.B.; Carvajal, M. Intrinsic stability of Brassicaceae plasma membrane in relation to changes in proteins and lipids as a response to salinity. *J. Plant Physiol.* **2015**, *175*, 148–156. [CrossRef]

26. Surjus, A.; Durand, M. Lipid changes in soybean root membranes in response to salt treatment. *J. Exp. Bot.* **1996**, *47*, 17–23. [CrossRef]

27. Ryan, E.; Galvin, K.; O'Connor, T.P.; Maguire, A.R.; O'Brien, N.M. Phytosterol, squalene, tocopherol content and fatty acid profile of selected seeds, grains, and legumes. *Plant Foods Hum. Nutr.* **2007**, *62*, 85–91. [CrossRef]

28. Zhang, X.; Lin, K.; Li, Y. Highlights to phytosterols accumulation and equilibrium in plants: Biosynthetic pathway and feedback regulation. *Plant Physiol. Biochem.* **2020**, *155*, 637–649. [CrossRef]

29. Ferrer, A.; Altabella, T.; Arró, M.; Boronat, A. Emerging roles for conjugated sterols in plants. *Prog. Lipid Res.* **2017**, *67*, 27–37. [CrossRef]

30. Behmer, S.; Olszewski, N.; Sebastiani, J.; Palka, S.; Sparacino, G.; Grebenok, R.J. Plant phloem sterol content: Forms, putative functions, and implications for phloem-feeding insects. *Front. Plant Sci.* **2013**, *4*, 370. [CrossRef]

31. Petersson, E.V.; Nahar, N.; Dahlin, P.; Broberg, A.; Tröger, R.; Dutta, P.C.; Jonsson, L.; Sitbon, F. Conversion of exogenous cholesterol into glycoalkaloids in potato shoots, using two methods for sterol solubilisation. *PLoS ONE* **2013**, *8*, e82955. [CrossRef] [PubMed]

32. Bajguz, A. Metabolism of brassinosteroids in plants. *Plant Physiol. Biochem.* **2007**, *45*, 95–107. [CrossRef] [PubMed]

33. Silva, C.; Aranda, F.J.; Ortiz, A.; Martínez, V.; Carvajal, M.; Teruel, J.A. Molecular aspects of the interaction between plants sterols and DPPC bilayers: An experimental and theoretical approach. *J. Colloid Interface Sci.* **2011**, *358*, 192–201. [CrossRef] [PubMed]

34. Emami, S.; Azadmard-Damirchi, S.; Peighambardoust, S.H.; Hesari, J.; Valizadeh, H.; Faller, R. Molecular dynamics simulations of ternary lipid bilayers containing plant sterol and glucosylceramide. *Chem. Phys. Lipids* **2017**, *203*, 24–32. [CrossRef] [PubMed]

35. Larsen, P.O. *Secondary Plant Products*; Conn, E.E., Ed.; Academic Press: New York, NY, USA, 1981; Volume 7, pp. 501–525, ISBN 978-0-12-675407-0.

36. Matthiessen, J.N.; Kirkegaard, J.A. Biofumigation and biodegradation: Opportunity and challenge in soil-borne pest and disease management. *Crit. Plant Sci.* **2006**, *25*, 235–265. [CrossRef]

37. Morris, E.K.; Fletcher, R.; Veresoglou, S.D. Effective methods of biofumigation: A meta-analysis. *Plant Soil* **2019**, *446*, 379–392. [CrossRef]

38. Dahlin, P.; Hallmann, J. New insights on the role of allyl isothiocyanate in controlling the root knot nematode *Meloidogyne hapla*. *Plants* **2020**, *9*, 603. [CrossRef] [PubMed]

39. Edwards, S.; Ploeg, A. Evaluation of 31 potential biofumigant brassicaceous plants as hosts for three *Meloiodogyne* species. *J. Nematol.* **2014**, *46*, 287–295.

40. Nelson, D.R. Cytochrome P450 diversity in the tree of life. *Biochim. Biophys. Acta Proteins Proteom.* **2018**, *1866*, 141–154. [CrossRef]

41. Rogowska, A.; Szakiel, A. The role of sterols in plant response to abiotic stress. *Phytochem. Rev.* **2020**, *19*, 1525–1538. [CrossRef]

42. Jones, J.D.G.; Dangl, J.L. The plant immune system. *Nat. Rev.* **2006**, *444*, 323–328. [CrossRef] [PubMed]

43. Jennings, P.H.; Zscheile, F.P., Jr.; Brannaman, B.L. Sterol changes in maize leaves infected with *Helminthosporium carbonum*. *Plant Physiol.* **1970**, *45*, 634–635. [CrossRef] [PubMed]

44. Nowak, R.; Kim, W.K.; Rohringer, R. Sterols of healthy and rust-infected primary leaves of wheat and of non-germinated and germinated uredospores of wheat stem rust. *Can. J. Bot.* **1972**, *50*, 185–190. [CrossRef]

45. Kissen, R.; Pope, T.W.; Grant, M.; Pickett, J.A.; Rossiter, J.T.; Powell, G. Modifying the alkylglucosinolate profile in *Arabidopsis thaliana* alters the tritrophic interaction with the herbivore Brevicoryne brassicae and parasitoid Diaeretiella rapae. *J. Chem. Ecol.* **2009**, *35*, 958–969. [CrossRef] [PubMed]

46. Ghosh, S. Triterpene structural diversification by plant cytochrome P450 enzymes. *Front. Plant Sci.* **2017**, *8*, 1886. [CrossRef] [PubMed]

47. Morikawa, T.; Saga, H.; Hashizume, H.; Ohta, D. *CYP710A* genes encoding sterol C22-desaturase in *Physcomitrella patens* as molecular evidence for the evolutionary conservation of a sterol biosynthetic pathway in plants. *Planta* **2009**, *229*, 1311–1322. [CrossRef]

48. Ramadan, A.M.; Azeiz, A.A.; Baabad, S.; Hassanein, S.; Gadalla, N.O.; Hassan, S.; Algandaby, M.; Bakr, S.; Khan, T.; Abouseadaa, H.H.; et al. Control of β-sitosterol biosynthesis under light and watering in desert plant *Calotropis procera*. *Steroids* **2019**, *141*, 1–8. [CrossRef]

49. Yu, J.; Tehrim, S.; Wang, L.; Dossa, K.; Zhang, X.; Ke, T.; Liao, B. Evolutionary history and functional divergence of the cytochrome P450 gene superfamily between *Arabidopsis thaliana* and *Brassica* species uncover effects of whole genome and tandem duplications. *BMC Genom.* **2017**, *18*, 733. [CrossRef]

50. PM 7/119 (1). Nematode extraction. *EPPO Bull.* **2013**, *43*, 471–495. [CrossRef]

51. Blight, E.G.; Dyer, W.J. A rapid method of total lipid extraction and purification. *Can. J. Biochem. Physiol.* **1959**, *37*, 911–917. [CrossRef]

52. Dahlin, P.; Srivastava, V.; Ekengren, S.; McKee, L.S.; Bulone, V. Comparative analysis of sterol acquisition in the oomycetes *Saprolegnia parasitica* and *Phytophthora infestans*. *PLoS ONE* **2017**, *12*, e0170873. [CrossRef] [PubMed]

53. Azadmard-Damirchi, S.; Dutta, P.C. Novel solid-phase extraction method to separate 4-desmethyl-, 4-monomethyl- and 4,4′- dimethylsterols in vegetable oils. *J. Chromatogr. A* **2006**, *1108*, 183–187. [CrossRef] [PubMed]

54. Zhao, S.; Fernald, R.D. Comprehensive algorithm for quantitative real-time polymerase chain reaction. *J. Comput. Biol.* **2005**, *12*, 1047–1064. [CrossRef] [PubMed]

55. Livak, K.J.; Schmittgen, T.D. Analysis of relative gene expression data using real-time quantitative PCR and the $2^{-\Delta\Delta CT}$ method.

56. Goodstein, D.M.; Shu, S.; Howson, R.; Neupane, R.; Hayes, R.D.; Fazo, J.; Mitros, T.; Dirks, W.; Hellsten, U.; Putna, N.; et al. Phytozome: A comparative platform for green plant genomics. Nucleic Acids Res. 2012, 40, D1178–D1186. [CrossRef] [PubMed]

Chapter-11

Bacterial Microbiota Isolated from Cysts of *Globodera rostochiensis* (Nematoda: Heteroderidae)

1. Introduction

Bacteria are ubiquitous organisms, inhabiting even the most extreme environments like polar snow [1], volcanoes and acidic hot springs [2,3]. The natural soil environment, aside from other microorganisms, harbors as many as 10^6–10^8 bacterial cells and 10^6–10^7 actinomycete cells per 1 g and around 10^7 nematodes per 1 m^2 [4].

The potato cyst nematodes (PCNs) *Globodera rostochiensis* and *G. pallida* are plant parasites of potatoes and other Solanaceae plants, classified as quarantine organisms. PCN females are sedentary organisms living inside potato roots with numerous eggs within their enlarged spherical bodies called cysts. The nematodes develop within the eggs to first and second stage juveniles. The latter is the invasive stage, searching for the appropriate host plant. When they find a target host, they start to invade roots, penetrating the host tissue with their stylets and move inside it. Inside the root tissue, they develop into females and males. After mating and fertilization, new eggs and juveniles are produced within the cysts, so the parasitic cycle continues. Some juveniles do not hatch until the following season or favorable conditions, remaining in soil for a long time [5]. The potato cyst nematodes cause up to GBP 300M worth of damage to the potato crop in the EU each year [6].

Both *Globodera* species were brought to Europe with the introduction of potato from South America [7]. Because the PCNs persist in soil, the external and internal areas of cysts harbor numerous microorganisms whose presence can lead to cyst death and population decline, suggesting that they can be potential candidates for use in biocontrol. Microscopic counts using 5-(4,6-dichlorotriazine-2-yl) aminofluorescein staining and in situ hybridization (EUB 338) revealed that cysts contain 2.6×10^5 bacteria [8].

Diverse bacterial species have been reported as nematode antagonists. *Streptomyces avermitilis* and *Pseudomonas fluorescens* were found to possess anthelmintic properties [9]. Nine isolates belonging to *Pseudomonas* and *Streptomyces* species were found to control both fungal pathogens and *Meloidogyne incognita* and were considered as promising biological control agents [10]. Bacterial isolates that inhibited egg hatching of the potato cyst nematodes were mostly from the genus *Bacillus* [11]. Bacterial species of the genus *Pasteuria* were found to be parasites of *Meloidogyne, Belonolaimus, Pratylenchus, Heterodera,* and *Globodera* spp. [12]. The Gram-negative bacterium *Stenotrophomonas (Xanthomonas) maltophilia* G2 was found to have a high nematotoxic activity against the free-living nematode *Panagrellus redivivus*, and the plant parasitic nematode *Bursaphelenchus xylophilus* [13]. *Serratia, Curtobacterium, Pseudomonas, Pantoea,* and *Rhanella* species were nematotoxic toward *B. xylophilus* [14]. Treatment with *B. cereus* strain S2 had a lethal effect on *Caenorhabditis elegans* and *M. incognita* [15].

This study aims to: (i) identify bacterial species associated with two PCN populations, (ii) infer phylogenetic relationships of the bacteria based on the maximum likelihood (ML) and Bayesian inference (BI) of 16S sequences rRNA genes, (iii) evaluate the influence of some microclimate and edaphicfactorsonbacterialdiversity.

2. Results and Discussion

The results revealed that bacterial microbiota from the locations of Pozega and Krupanj (the Republic of Serbia) generally contain similar species with varying abundance. The cysts obtained from Pozega have more diverse bacterial microbiota (Figure 1) with the presence of 74.0% of members of the class Bacilli and the order Bacillales divided into the families Bacillaceae and Paenibacillaceae. Furthermore, there are 14.0% of members of Proteobacteria, whereas Actinobacteria are present in the lowest percentage (6.0%). The Alphaproteobacteria are represented by the order Rhizobiales and the family Hyphomicrobiaceae (*Devosia* sp.), while Actinobacteria are represented by the order Micrococcales and the family Brevibacteriaceae i.e., *Brevibacterium* sp. The bacterial microbiota of Krupanj (Figure 2) is less diverse, containing the majority of the class Bacilli (40.0%), represented by the families Bacillaceae and Paenibacillaceae as well. The next group is Actinobacteria (28.0%) with the family Micrococcaceae and *Arthrobacter* spp., while the lowest percentage (20.0%) pertains to Alphaproteobacteria represented by the family Hyphomicrobiaceae and *Devosia* sp.

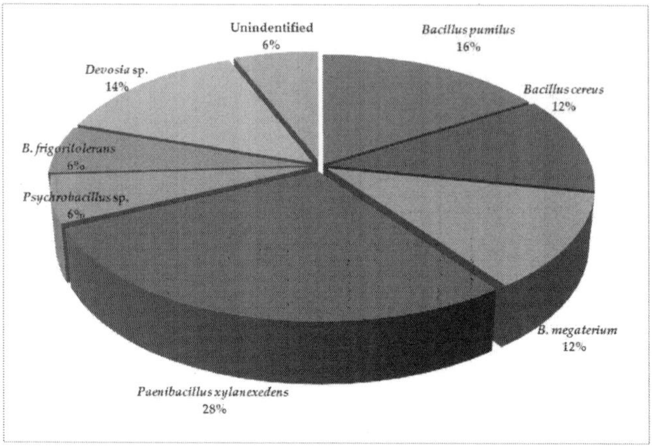

Figure 1. BacterialmicrobiotafoundincystsfromPozega.

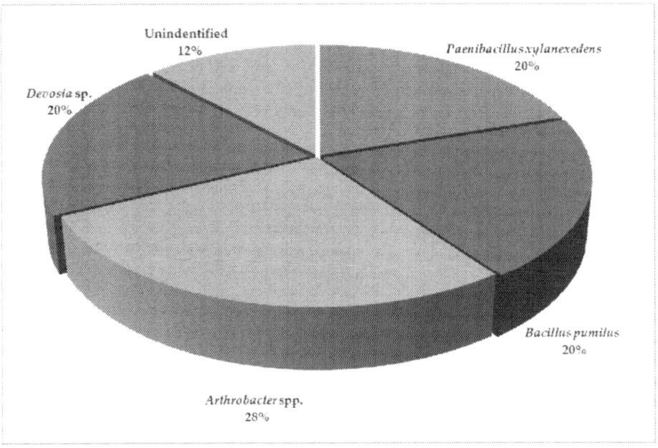

Figure 2. Bacterialmicrobiotafoundincystsfrom Krupanj.

The genus *Bacillus* was the principal genus in Pozega, which was similar to Costa et al. [16] (p. 718), who observed that *Bacillus* was present in 80% of the isolates of the bacterial microbiota of *M. exigua* egg masses in coffee plantations.

Bacillus was found not only to be prevalent in the rhizosphere, but also in the phyllosphere. Maximum colonization was shown by the genus *Bacillus* isolated from carrot, cabbage and turnip phyllosphere bacteria [17]. Members of the order Bacillales (*B. pumilus* and *P. xylanexedens*) were found in both locations. In contrast, more Actinobacteria were detected in Krupanj, suggesting that this location was probably more polluted with organic contaminants and the processes of natural bioremediation occurred. In Krupanj, *Arthrobacter* spp. corresponded to 28% of the total bacterial microbiota; likewise, the genus *Arthrobacter* comprised more than 21% of the total soil community of the burned holmoak forest [18].

In comparison with two soil samples from Spain, analyzed by the denaturing gradient gel electrophoresis of bacteria isolated from *M. incognita* and *P. penetrans*, in which the most abundant bacterial classes were Betaproteobacteria, Bacilli and Actinobacteria [19], in our study, the prevalent classes were Bacilli, Actinobacteria and Alphaproteobacteria. The dominance of the order Bacillales was evident in both locations with 80% in Pozega and twice less in Krupanj. In contrast, more Actinobacteria and Alphaproteobacteria (*Arthrobacter* spp. and *Devosia* sp., respectively) were detected in Krupanj.

The phylogenetic analyses based on 16S sequences are shown in the Figures 3 and 4. Both ML and BI trees are in agreement and generated three distinct clades. Within the first clade, there are subclades composed of *Bacillus cereus, B. megaterium, B. flexus, B. subtilis, B. pumilus* and a *Psychrobacillus* species, representing the family Bacillaceae. The other subclade with *Paenibacillus* spp. represents the family Paenibacillaceae, which, together with the family Bacillaceae, are affiliated to the order Bacillales and the phylum Firmicutes. The difference is that *Devosia* spp. are independent in the ML tree (Figure 3). The *Devosia* species clade represents the family Hyphomicrobiaceae and Alphaproteobacteria linked with the two subclades of Actinobacteria, the subclade of *Arthrobacter* spp. and the subclade of *Brevibacterium* species in the BI tree, because the

Bayesian inference considers all the species to be monophyletic (Figure 4). The sequences of *Brevibacterium frigoritolerans* were not clustered with other *Brevibacterium* species. Instead, they were grouped with *Bacillus cereus* species as the closest relatives, suggesting their affiliation to the family Bacillaceae.

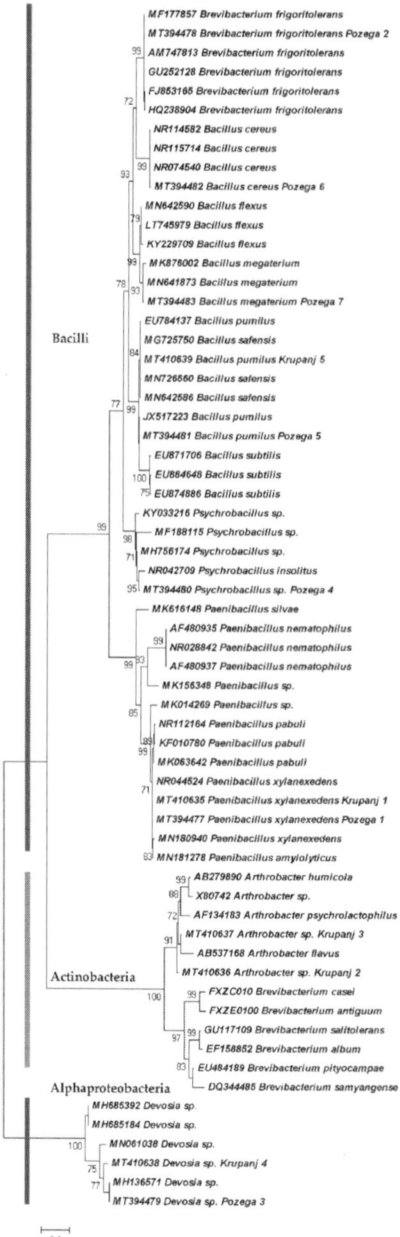

Figure 3. Maximum likelihood phylogenetic tree of bacterial microbiota isolated from *G. rostochiensis* cysts based on 16S sequence region using General Time Reversible (GTR), invariable sites and gamma distribution(GTR+ I+ G)nucleotide evolutionmodel.

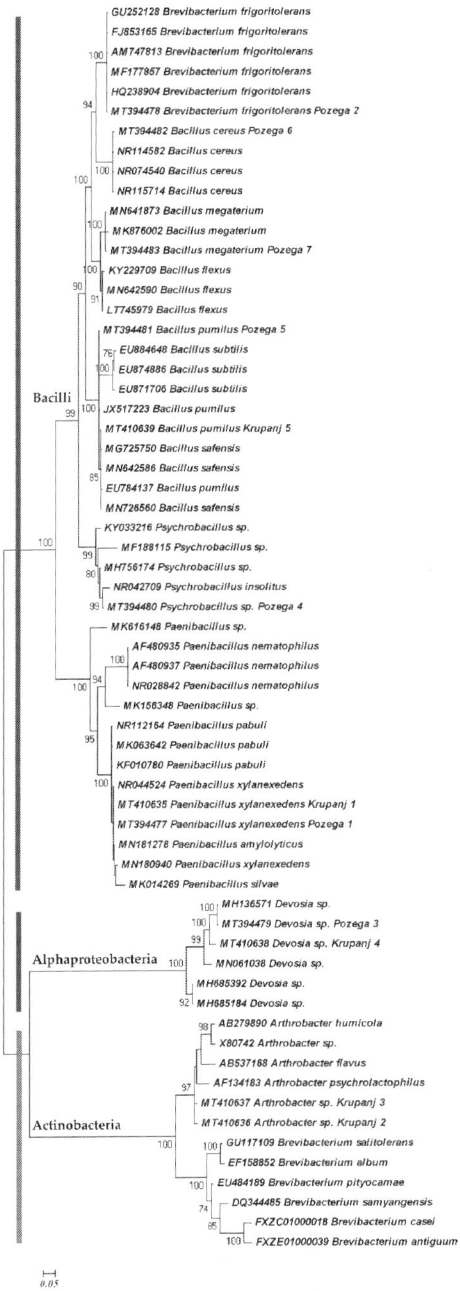

Figure 4. Bayesian phylogenetic tree of bacterial microbiota isolated from *G. rostochiensis* cysts and derived from consensus 50% majority rule based on 16S sequence region using GTR + I + G nucleotide evolution model.

Similar observations were reported by other authors. *Brevibacterium frigoritolerans* was in the same group with other *Bacillus* spp., i.e., *B. simplex, B. muralis, B.*

psychrosaccharolyticus [20–22]. This bacterium can biosynthesize silver nanoparticles and tolerate silver as some *Bacillus* species can tolerate salt [23]. In addition, *B. frigoritolerans* has the ability to sporulate, thereby providing evidence that this strain is actually a misidentified *Bacillus* sp. [20]. Recently, based on the phenotypic, chemotaxonomic, phylogenetic and genomic characteristics, it has been demonstrated that *B. frigoritolerans* DSM 8801T should belong to the genus *Bacillus*, and to be reclassified as *Bacillus frigoritolerans* [24]. Our study confirms its reclassification and genetic closeness to *B. cereus*. On the other hand, the other species of *Brevibacterium* were clustered together with *Arthrobacter* spp. within Actinobacteria. Apart from *G. rostochiensis*, this species was isolated from juveniles of *B. xylophilus* [25]. Under in vitro bioassay conditions, the isolate of *Brevibacterium frigoritolerans* exhibited bacteremia-like symptoms and induced mortality of the Coleopteran larvae of *Anomala dimidiata* and *Holotrichia longipennis* [26], suggesting its possible use in biocontrol.

Comparisons based on climate factors during the 28-year period (1990–2018) revealed differences between the two locations. Pozega shows the lower values of the air temperatures (optimum, minimum and maximum) and insolation, and the higher values of relative humidity, and cloudiness (Table 1). On the contrary, the values of temperatures, insolation hours and precipitation are higher in Krupanj, whereas the values of relative humidity are lower and there are fewer cloudy days (Table 1).

Table 1. Comparison of annual means and honest significant difference (HSD) of climate factors for two observed locations during the 28-year period (1990–2018).

Climate Factors (Units)	Locations	Means	SD	Range	HSD (p = 0.05)
Optimum Air	Pozega	10.5	0.6	9.0–10.9	a
Temperature (°C)	Krupanj	12.6	0.8	10.6–13.4	a
Maximum Air	Pozega	17.4	0.7	15.0–18.0	a
Temperature (°C)	Krupanj	18.6	1.0	15.7–19.7	a
Minimum Air	Pozega	5.1	0.7	3.4–6.6	a
Temperature (°C)	Krupanj	7.7	0.7	5.0–8.8	a
Relative Humidity (%)	Pozega	82.1	3.1	74–85	a
	Krupanj	77.8	4.0	67–86	a
Insolation *(h)	Pozega	1634.6	253.0	1110.5–2064.2	b
	Krupanj	2107.2	188.8	1701.4–2381.7	a
Cloudiness	Pozega	6.5	0.5	5.3–7.1	a
	Krupanj	5.8	0.4	4.7–6.4	a
Precipitation *(mm)	Pozega	762.3	154.4	460.6–1121.5	b
	Krupanj	902.5	160.7	529.2–1242.4	a

* *statistically significant.*

The honest significant difference (HSD) test demonstrates that there are statistically significant differences between insolation hours and precipitation values. The difference in insolation between locations is almost 500 h with more variation of this factor in Pozega. In contrast, the precipitation sum was higher in Krupanj throughout the year. The insolation itself has a direct impact on the air temperature, making the distinction of this factor between the two locations. A decrease in air temperature causes the decrease in soil temperature, which, in combination with higher relative humidity, favors the environment suitable for cold tolerant species. This fact was confirmed by the presence of *Bacillus frigoritolerans* and a *Psychrobacillus* species in Pozega. Despite the fact that there are no significant differences in air temperature at two locations, the lower annual temperatures in Pozega favored the development of psychrotolerant species. In climate studies, statistical significance does not always provide an adequate basis for decision making; for example, a rise in temperature by two degrees Celsius may not be statistically significant but it can adversely affect the vegetation growth and lead to ecological imbalances [27].

All of the physicochemical properties of the soils, except the content of potassium, were similar in both locations (Table 2). However, the content of soil organic matter in Pozega is higher than in Krupanj, which may explain the more diverse bacterial microbiota in Pozega. Soil with a higher content of organic matter is generally associated with high microbial abundance and diversity [28].

Table 2. Comparison of soil physicochemical parameters and HSD for two observed locations.

Physicochemical Parameters	Locations	Values	HSD ($p = 0.05$)
pH (H_2O)	Pozega	7.73	a
	Krupanj	7.01	a
pH (1M KCl)	Pozega	6.71	a
	Krupanj	6.26	a
Soil organic matter (%)	Pozega	5.24	a
	Krupanj	3.33	a
N (%)	Pozega	0.22	a
	Krupanj	0.23	a
P_2O_5 (mg/100 g)	Pozega	28.21	a
	Krupanj	29.80	a
K_2O * (mg/100 g)	Pozega	24.50	b
	Krupanj	61.88	a
Sand particles (>0.2 mm)%	Pozega	2.4	a
	Krupanj	1.5	a
Sand particles (0.02–0.2 mm)%	Pozega	19.9	a
	Krupanj	17.6	a
Silt (0.002–0.02 mm)%	Pozega	40.6	a
	Krupanj	35.6	a
Clay (<0.002 mm)%	Pozega	37.1	a
	Krupanj	45.3	a
Silt+Clay (<0.02 mm)%	Pozega	77.7	a
	Krupanj	80.9	a

* *statistically significant.*

The HSD test demonstrates that there is a significant difference in the amount of potassium between the two locations. Since K^+ is a major nutritional element for plants, enrichment of K^+ in the exchange sites due to fertilizer practice can be expected [29], which may indicate high potassium fertilizer inputs in Krupanj.

Regarding the granulometric content of the two examined soils, the smallest clay and silt particles (0.002–0.02 mm) are dominant: 77.7 versus 80.9%. Pozega has a higher content of silt, whereas Krupanj has a higher content of clay. With decreasing particle size, there is an increase in particle number and in the surface area per gram of soil. It is clear that the interfacial area enlarges with an increase in the proportion of the clay–size fraction and, consequently, the opportunities for sorptive interactions between microorganisms and soil particles should increase [30]. The dominance of silt and clay in both soil samples enables good interaction between bacteria and soil.

All found species of the family Bacillaceae have been reported to have high potential as biocontrol agents, which resulted in the development of commercial bionematicidal agents [12]. *Bacillus cereus* strain S2 can produce sphingosine to induce reactive oxygen accumulation, destroy the genital area in nematodes, and inhibit nematode reproduction [15].

Bacillus pumilus demonstrated its ability as a potential biocontrol agent against *M. arenaria*, causing 39.8 and 92.8% J2 mortality after three days of exposure to 2.5 and 10% concentrations of bacterial culture, respectively [31]. *Bacillus subtilis* and *B. pumilus* caused the highest reduction (82% and 81.8%, respectively) in *M. incognita* on cowpea [32]. An isolate of *Bacillus megaterium* reduced the root penetration and migration of *M. graminicola* to between 40 and 60% compared with non–treated roots of rice plants [33].

Paenibacillus nematophilus has been found to hamper more than 98% of the dispersal of the beneficial nematode *Heterorhabditis megidis* and reduce its infectivity in moth larvae [34].

Psychrobacillus species play a role in biodegradation and as antimicrobial agents. *Psychrobacillus soli* could degrade around 72% of oil components at an initial oil concentration of 1500 ppm [35]. Among ten endophytic bacteria, *Psychrobacillus insolitus* and *Curtobacterium oceanosedimentum* showed the highest anticandidal effect against *Candida albicans* and *C. glabrata* [36], while two strains of *P. insolitus* (Mam2 and Ame3) exhibited an inhibitory action against staphylococcal strains isolated from food [37].

Devosia and *Arthrobacter* species are best characterized for their bioremediation potential. *Devosia* are well known for their dominance in soil habitats contaminated with various toxins. The uptake and utilization of nutrients for growth and survival was found to be the dominant function of the genus along with the detoxification and degradation of organic pollutants [38].

Arthrobacter species were involved in biodegrading a wide variety of compounds, e.g., nicotine, organosilicon compounds, fluorene, the herbicide atrazine [39], and *m*-chlorobenzoate, the central molecule in many pesticides [40]. The majority of the selected strains exhibited a great ability to degrade organic polymers in vitro. Moreover, they possibly present a direct mechanism for plant growth promotion [18]. One of the strains of *A. nicotianae* showed 100% nematicidal activity against *C. elegans* and 91–97% nematicidal activity against *M. incognita* [41].

The higher presence of bioremediators in our samples may indicate the higher presence of pollutants in Krupanj and explain the reduced diversity of bacterial microbiota.

3. Materials and Methods

3.1 Isolation of Bacteria

The cysts of *G. rostochiensis* found in potato fields near the locations of Pozega (44°04' N 20°14' E) and Krupanj (44°18' N 19°20' E) were used as a source for screening bacterial microbiota. During the growing season, the soil samples were taken as 50 subsamples/ha in a systematic sampling pattern in order to make approximately one kilogram of composite sample [42].The cyst extraction was done with the Spears apparatus [43] and collected on a 150-µm sieve.

Fifty randomly selected cysts of different ages from each location were surface sterilized with 96% ethanol, 1.5% NaOCl and washed with sterile water according to the procedure applied for *Globodera* juveniles [44]. The cysts were placed on potato dextrose agar (PDA) and maintained for seven days at 25 °C. After the emergence of bacteria on PDA, single bacterial colonies were used to obtain pure cultures by the streak plate method [45].

3.2 Molecular Study

The extraction of DNA from bacteria was performed according to a previously described procedure [46]. The PCR reaction mixture consisted of 25 µL 2× PCR Mastermix, 0.5 µL of forward and reverse primers (10 µM), 1 µL of DNA template and PCR-grade water to a total volume of 50 µL. Amplification of the DNA region coding for 16S rRNA was performed by using P0 (5'-GAGAGTTTGATCCTGGCTCAG-3') and P6 (5'-CTACGGCTACCTTGTTACGA-3') primers.

The temperature profile for the PCR reaction was as follows: 95 °C for 90 s followed by 35 cycles consisting of 95 °C for 30 s, the annealing temperature (60 °C for the first 5 cycles, 55 °C for the next 5 cycles, and 50 °C for the last 25 cycles) for 30 s, and 72 °C for 4 min. The reaction mixture was then incubated at 72 °C for 10 min and at 60 °C for 10 min. The obtained PCR products were purified and sequenced [47]. Phylogenetic analyses were performed with sequences of the isolated bacterial species deposited under accession numbers MT394477-MT394483 (Pozega) and MT410635-MT410639 (Krupanj) and related species from the GenBank nucleotide sequence database, using maximum likelihood (ML) and Bayesian inference (BI) phylogenetic methods. The ML and BI were calculated with the help of PhyML 3.1 [48], and MrBayes 3.1.2 [49] computer programs, respectively. The sequence alignment was done with ClustalW in Mega 4 [50].

The ML tree was obtained with the General Time Reversible model (GTR), invariable sites and gamma distribution (GTR+ I+ G). The dendrogram obtained by Bayesian inference was created by 2.2×10^6 generations of Markov Chain Monte Carlo, with a sample frequency of 100, and burning function of 20%. The nucleotide evolution model was GTR + I + G as well. Branch supports higher than 70% were shown next to the node.

3.3 Statistical Data Analysis

The annual values of climate factors of Pozega and Krupanj were obtained from the official site of the Republic Hydrometeorological Institute of Serbia. The 28-year period (1990–2018) was used for calculating the means of the optimum, maximum, and minimum air temperature, the relative humidity, insolation, cloudiness and precipitation.

The units for the air temperatures were presented in degrees Celsius, the relative air humidity was expressed in percentages, while the duration of the solar radiation (insolation) was expressed in hours. Values of the cloudiness parameter lower than 2 were considered as clear days, while values higher than 6 were considered as cloudy days. The precipitation was expressed in millimeters (Table 1). Soil pH, the content of organic matter, the amount of nitrogen, phosphorus and potassium, as well as the soil granulometric composition (Table 2), were determined according to standard methods and those from the literature [51–55]. The values were compared with a post-ANOVA Tukey's honest significant difference (HSD) test using DSAASTAT computer program [56], at the 95% confidence interval. Values with the same letter were not significantly differentfromeachother.

4. Conclusions

Regarding the higher bacterial taxonomy, our results indicate that the observed locations have similar microbiota, but with a different abundance and species identity. The dominant bacterial phyla are Firmicutes, Actinobacteria and Proteobacteria. Based on 16S sequences, the maximum likelihood and the Bayesian phylogeny clustered the members of the genus *Bacillus, Psychrobacillus* and *Paenibacillus* within the family Bacillaceae. *Brevibacterium frigoritolerans* belonged to the same group with *B. cereus, B. megaterium* and *B. flexus* within the family Bacillaceae, confirming its recent reclassification. Other clades were occupied by *Devosia* and *Arthrobacter* species known for their function in environmental detoxification and the degradation of pesticides. The lower values of air temperatures, insolation, and precipitation and the higher values of relative humidity and cloudiness created conditions for the development of psychrophilic species. The location of Pozega is characterized by psychrotolerant representatives of *Bacillus frigoritolerans,* and a *Psychrobacillus* species. In contrast, Krupanj is characterized by the higher content of potassium, the lower content of organic matter and the presence of bioremediators such as *Devosia* and *Arthrobacter* species. In other words, bacterial species perform as specific indicators of microclimate properties and environmental pollution.

As efforts have been moved towards expanding the source of microorganisms involving the more complex systems in nature [57], nematodes and their related bacterial microbiota present the next biological system to explore the taxonomic diversity of soil bacteria. Nematodes, especially cyst nematodes, are a significant natural source of microorganisms due to their long persistence in soil and the specific environmental conditions inside and outside of the closed area of cysts, in which diverse bacteriaarehidden.

References

1. Schuerger, A.C.; Lee, P. Microbial ecology of a crewed rover traverse in the Arctic: Low microbial dispersal and implications for planetary protection on human Mars missions. *Astrobiology* **2015**, *15*, 478–491. [CrossRef] [PubMed]

2. Danovaro, R.; Canals, M.; Tangherlini, M.; Dell'Anno, A.; Gambi, C.; Lastras, G.; Amblas, D.; Sanchez-Vidal, A.; Frigola, J.; Calafat, A.M.; et al. A submarine volcanic eruption leads to a novel microbial habitat. *Nat. Ecol. Evol.* **2017**, *1*, 144. [CrossRef] [PubMed]

3. Gómez, F.; Cavalazzi, B.; Rodríguez, N.; Amils, R.; Ori, G.; Olsson-Francis, K.; Escudero, C.; Martínez, J.; Miruts, H. Ultra-small microorganisms in the polyextreme conditions of the Dallol volcano, Northern Afar, Ethiopia. *Sci. Rep.* **2019**, *9*, 7907. [CrossRef] [PubMed]

4. Back, M.A.; Haydock, P.P.J.; Jenkinson, P. Disease complexes involving plant parasitic nematodes and soilborne pathogens. *Plant Pathol.* **2002**, *51*, 683–697. [CrossRef]

5. Oro, V. Potato Cyst Nematodes–Morphology, Molecular Characterization and Antagonists. Ph.D. Thesis, Faculty of Biofarming, Backa Topola, Serbia, 8 September 2011.

6. Ryan, N.A.; Duffy, E.M.; Cassells, A.C.; Jones, P.W. The effect of mycorrhizal fungi on the hatch of potato cyst nematodes. *Appl. Soil Ecol.* **2000**, *15*, 233–240. [CrossRef]

7. Oro, V.; Nikolic, B.; Josic, D. The potato road and biogeographic history of potato cyst nematode populations from different continents. *Genetika* **2014**, *46*, 895–904. [CrossRef]

8. Nour, S.M.; Lawrence, J.R.; Zhu, H.; Swerhone, G.D.W.; Welsh, M.; Welacky, T.W.; Topp, E. Bacteria associated with cysts of the soybean cyst nematode (*Heterodera glycines*). *Appl. Environ. Microbiol.* **2003**, *69*, 607–615. [CrossRef]

9. Kerry, B.R. Rhizosphere interactions and the exploitation of microbial agents for the biological control of plant-parasitic nematodes. *Annu. Rev. Phytopathol.* **2000**, *38*, 423–441. [CrossRef]

10. Krechel, A.; Faupel, A.; Hallmann, J.; Ulrich, A.; Berg, G. Potato-associated bacteria and their antagonistic potential towards plant-pathogenic fungi and the plant-parasitic nematode *Meloidogyne incognita* (Kofoid& White) Chitwood. *Can. J. Microbiol.* **2002**, *48*, 772–786. [CrossRef]

11. Ryan, N.A.; Jones, P. The ability of rhizosphere bacteria isolated from nematode host and non-host plants to influence the hatch in vitro of the two potato cyst nematode species *Globodera rostochiensis* and *G. pallida*. *Nematology.* **2004**, *6*, 375–387. [CrossRef]

12. Tian, B.Y.; Yang, J.K.; Zhang, K.Q. Bacteria used in biological control of plant-parasitic nematodes: Populations, mechanisms of action, and future prospects. *FEMS Microbiol. Ecol.* **2007**, *61*, 197–213. [CrossRef] [PubMed]

13. Huang, X.; Liu, J.; Ding, J.; He, Q.R.; Xiong, R.; Zhang, K. The investigation of nematocidal activity in *Stenotrophomonas maltophilia* G2 and characterization of a novel virulence serine protease. *Can. J. Microbiol.* **2009**, *55*, 934–942. [CrossRef] [PubMed]

14. Paiva, G.; Proença, D.N.; Francisco, R.; Verissimo, P.; Santos, S.S.; Fonseca, L.; Abrantes, I.M.O.; Morais, P.V. Nematicidal bacteria associated to pinewood nematode produce extracellular proteases. *PLoS ONE* **2013**, *8*, e79705. [CrossRef] [PubMed]

15. Gao, H.; Qi, G.; Yin, R.; Zhang, H.; Li, C.; Zhao, X. *Bacillus cereus* strain S2 shows high nematicidal activity against *Meloidogyne incognita* by producing sphingosine. *Sci. Rep.* **2016**, *6*, 28756. [CrossRef] [PubMed]

16. Costa, L.S.A.S.; Campos, V.P.; Terra, W.C.; Pfenning, L.H. Microbiota from *Meloidogyne exigua* egg masses and evidence for the effect of volatiles on infective juvenile survival. *J. Nematol.* **2015**, *17*, 715–724. [CrossRef]

17. Ali, B. Functional and genetic diversity of bacteria associated with the surfaces of agronomic plants. *Plants* **2019**, *8*, 91. [CrossRef]

18. Fernández–González, A.J.; Martínez–Hidalgo, P.; Cobo–Díaz, J.F.; Villadas, P.J.; Martínez–Molina, E.; Toro, N.; Tringe, S.G.; Fernández–López, M. The rhizosphere microbiome of burned holm–oak: Potential role of the genus *Arthrobacter* in the recovery of burned soils. *Sci. Rep.* **2017**, *7*, 6008. [CrossRef]

19. Elhady, A.; Gine, A.; Topalovic, O.; Jacquiod, S.; Sørensen, S.J.; Sorribas, F.J.; Heuer, H. Microbiomes associated with infective stages of root–knot and lesion nematodes in soil. *PLoS ONE* **2017**, *12*, e0177145. [CrossRef]

20. Beesley, C.A.; Vanner, C.L.; Helsel, L.O.; Gee, J.E.; Hoffmaster, A.R. Identifcation and characterization of clinical *Bacillus* spp. isolates phenotypically similar to *Bacillus anthracis*. *FEMS Microbiol. Lett.* **2010**, *313*, 47–53. [CrossRef]

21. Tong, X.; Yuan, L.; Luo, L.; Yin, X. Characterization of a selenium–tolerant rhizosphere strain from a novel Se–hyperaccumulating plant *Cardamine hupingshanesis*. *Sci. World J.* **2014**, 108562. [CrossRef]

22. Zhang, C.; Li, X.; Yin, L.; Liu, C.; Zou, H.; Wu, Z.; Zhang, Z. Analysis of the complete genome sequence of *Brevibacterium frigoritolerans* ZB201705 isolated from drought– and salt–stressed rhizosphere soil of maize. *Ann. Microbiol.* **2019**, *69*, 1489–1496. [CrossRef]

23. Singh, P.; Kim, Y.J.; Singh, H.; Wang, C.; Hwang, K.H.; Farh, M.E.-A.; Yang, D.C. Biosynthesis, characterization, and antimicrobial applications of silver nanoparticles. *Int. J. Nanomed.* **2015**, *10*, 2567–2577. [CrossRef] [PubMed]

24. Liu, G.-H.; Liu, B.; Wang, J.-P.; Che, J.-M.; Li, P.F. Reclassification of *Brevibacterium frigoritolerans* DSM 8801T as *Bacillus frigoritolerans* comb. nov. based on genome analysis. *Curr. Microbiol.* **2020**, *77*, 1916–1923. [CrossRef] [PubMed]

25. Kwon, H.R.; Choi, G.J.; Choi, Y.H.; Jang, K.S.; Sung, N.-D.; Kang, M.S.; Moon, Y.; Lee, S.K.; Kim, J.-C. Suppression of pine wilt disease by an antibacterial agent, oxolinic acid. *Pest. Manag. Sci.* **2010**, *66*, 634–639. [CrossRef]

26. Selvakumar, G.; Sushil, S.; Stanley, J.; Mohan, M.; Deol, A.; Rai, D.; Ramkewal; Bhatt, J.C.; Gupta, S.H. *Brevibacterium frigoritolerans* a novel entomopathogen of *Anomala dimidiata* and *Holotrichia longipennis* (Scarabaeidae: Coleoptera). *Biocontrol Sci. Technol.* **2011**, *21*, 821–827. [CrossRef]

27. Mehan, S.; Guo, T.; Gitau, M.W.; Flanagan, D.C. Comparative study of different stochastic weather generators for long–term climate data simulation. *Climate* **2017**, *5*, 26. [CrossRef]

28. Li, L.; Xu, M.; Eyakub Ali, M.; Zhang, W.; Duan, Y.; Li, D. Factors affecting soil microbial biomass and functional diversity with the application of organic amendments in three contrasting cropland soils during a field experiment. *PLoS ONE* **2018**, *13*, e0203812. [CrossRef]

29. Chen, Y.; Banin, A.; Borochovitch, A. Effect of potassium on soil structure in relation to hydraulic conductivity. *Geoderma* **1983**, *30*, 135–147. [CrossRef]

30. Marshall, K.C. Clay mineralogy in relation to survival of soil bacteria. *Annu. Rev. Phytopathol.* **1975**, *13*, 357–373. [CrossRef]

31. Lee, Y.S.; Kim, K.Y. Antagonistic potential of *Bacillus pumilus* L1 against root—knot nematode, *Meloidogyne arenaria*. *J. Phytopathol.* **2016**, *164*, 29–33. [CrossRef]

32. Padgham, J.L.; Sikora, R.A. Biological control potential and modes of action of *Bacillus megaterium* against *Meloidogyne graminicola* on rice. *J. Crop Prot.* **2007**, *26*, 971–977. [CrossRef]

33. Abd–El–Khair, H.; Wafaa, M.A.; El–Nagdi, W.M.A.; Mahmoud, M.A.; Youssef, M.M.A.; Abd–Elgawad, M.M.M.; Dawood, M.G. Protective effect of *Bacillus subtilis*, *B. pumilus*, and *Pseudomonas fluorescens* isolates against root knot nematode *Meloidogyne incognita* on cowpea. *Bull. Natl. Res. Cent.* **2019**, *43*, 64. [CrossRef]

34. Enright, M.R.; Griffin, C.T. Effects of *Paenibacillus nematophilus* on the entomopathogenic nematode *Heterorhabditis megidis*. *J. Invertebr. Pathol.* **2005**, *88*, 40–48. [CrossRef] [PubMed]

35. Pham, V.H.T.; Jeong, S.-W.; Kim, J. *Psychrobacillus soli* sp. nov., capable of degrading oil, isolated from oil–contaminated soil. *Int. J. Syst. Evol. Microbiol.* **2015**, *65*, 3046–3052. [CrossRef]

36. Das, G.; Patra, J.K.; Choi, J.; Baek, K.-H. Anticandidal effect of endophytic bacteria isolated from *Equisetum arvense L.* against *Candida albicans* and *Candida glabrata*. *Braz. Arch. Biol. Technol.* **2017**, *60*, e17160433. [CrossRef]

37. Oliveira, V.F.; Abreu, Y.J.L.; Fleming, L.R.; Nascimento, J.S. Anti–staphylococcal and antifungal substances produced by endospore–forming bacilli. *J. Appl. Pharm. Sci.* **2012**, *2*, 154–157. [CrossRef]

38. Talwar, C.; Nagar, S.; Kumar, R.; Scaria, J.; Lal, R.; Negi, R.K. Defining the environmental adaptations of genus *Devosia:* Insights into its expansive short peptide transport system and positively selected genes. *Sci. Rep.* **2020**, *10*, 1151. [CrossRef]

39. Wackett, L.P. *Arthrobacter* and related genera: An annotated selection of World Wide Web sites relevant to the topics in environmental microbiology. *Microb. Biotechnol.* **2016**, *9*, 136–138. [CrossRef]

40. Jones, D.; Keddie, R.M. The Genus Arthrobacter. *Prokaryotes* **2006**, *3*, 945–960. [CrossRef]

41. Xu, Y.; Lu, H.; Wang, X.; Zhang, K.; Li, G. Effect of volatile organic compounds from bacteria on nematodes. *Chem. Biodivers.* **2015**, *12*, 1415–1421. [CrossRef]

42. Coyne, D.L.; Nicol, J.M.; Claudius-Cole, B. *Practical Plant Nematology: A Field and Laboratory Guide*, 2nd ed.; SP-IPM Secretariat International Institute of Tropical Agriculture (IITA): Cotonou, Benin, 2014; pp. 25–29.

43. Spears, J.F. *The Golden Nematode Handbook-Survey, Laboratory, Control and Quarantine Procedures, Agriculture Handbook 353*; USDA, Agricultural Research Service: Washington, DC, USA, 1968; pp. 1–82.

44. Heungens, K.; Mugniery, D.; Van Montagu, M.; Gheysen, G.; Niebel, A. A method to obtain disinfected *Globodera* infective juveniles directly from cysts. *Fundam. Appl. Nematol.* **1996**, *19*, 91–93.
45. Brown, A.; Smith, H. *Benson's Microbiological Applications: Laboratory Manual in General Microbiology*, 13th ed.; McGraw-Hill Education: New York, NY, USA, 2015; pp. 73–80.
46. Goldenberger, D.; Perschil, I.; Ritzler, M.; Altwegg, M. Simple "universal" DNA extraction procedure using SDS and proteinase K is compatible with direct PCR amplification. *Genome Res.* **1995**, *4*, 368–370. [CrossRef] [PubMed]
47. Picard, C.; Di Cello, F.; Ventura, M.; Fani, R.; Guckert, A. Frequency and biodiversity of 2,4-Diacetylphloroglucinol producing bacteria isolated from the maize rhizosphere at different stages of plant growth. *Appl. Environ. Microbiol.* **2000**, *66*, 948–955. [CrossRef] [PubMed]
48. Gunidon, S.; Gascuel, O. A simple, fast, and accurate algorithm to estimate large phylogenies by maximum likelihood. *Syst. Biol.* **2003**, *52*, 696–704. [CrossRef] [PubMed]
49. Huelsenbeck, J.P.; Ronquist, F. Bayesian analysis of molecular evolution using MrBayes. In *Statistical Methods in Molecular Evolution: Statistics for Biology and Health*; Springer: New York, NY, USA, 2005; pp. 183–226. [CrossRef]
50. Tamura, K.; Dudley, J.; Nei, M.; Kumar, S. MEGA4: Molecular Evolutionary Genetics Analysis (MEGA) software version 4.0. *Mol. Biol. Evol.* **2007**, *24*, 1596–1599. [CrossRef]
51. SRPS ISO 10390:2007. *Soil Quality-Determination of pH*; Institute for Standardisation of Republic of Serbia: Belgrade, Serbia, 2007.
52. SRPS ISO 10694:2005. *Soil Quality-Determination of Organic and Total Carbon after Dry Combustion ("Elemental Analysis")*; Institute for Standardization: Belgrade, Serbia, 2005.
53. SRPS ISO 13878:2005. *Soil Quality-Determination of Total Nitrogen Content by Dry Combustion ("Elemental Analysis")*; Institute for Standardization: Belgrade, Serbia, 2005.
54. Egnér, H.; Riehm, H.; Domingo, W.R. Untersuchungen über die chemische bodenanalyse als grundlage für die beurteilung des nährstoff-zustandes der böden II. Chemische extraktions methoden zur phosphor- und kalium bestimmung. *Kungl. Lantbrukshögsk. Ann.* **1960**, *26*, 199–215.
55. Hadzic, V.; Belic, M.; Nesic, L. Determination of the mechanical composition (texturally gravimetric) of the soil. In *Methods of Research and Determination of Physical Properties of the Soil*; Bosnjak, D.J., Ed.; JDPZ: Novi Sad, Serbia, 1997; pp. 17–32.
56. Onofri, A. Routine statistical analyses of field experiments by using an Excel extension. In Proceedings of the 6th National Conference of the Italian Biometric Society: "La Statistica nelle Scienze della Vita e Dell'ambiente", Pisa, Italy, 20–22 June 2007; pp. 93–96.
57. Shen, Y.; Fu, Y.; Yu, Y.; Zhao, J.; Li, J.; Li, Y.; Wang, X.; Zhang, J.; Xiang, W. *Psychrobacillus lasiicapitis* sp. nov., isolated from the head of an ant (*Lasius fuliginosus*). *Int. J. Syst. Evol. Microbiol.* **2017**, *67*, 4462–4467. [CrossRef]

Index